"This stunningly important volume will become essential reading for all those concerned about the state of our planet's health and that of the myriad species whose futures are intertwined with each other in their complex symbiotic, environmental, and ecological interrelations. By marshalling a remarkably broad range of disciplinary expertise—including law, social sciences, and the humanities—beyond core sciences and medicine, this volume challenges the One Health movement to take account of fundamental, new questions of inclusivity within the broader sociopolitical and colonialist arenas of practice and collective action that do, will, and must address the broadest horizons of care for the sake of our planet and all of its inhabitants."

James J. Bono, *Emeritus Professor, Department of History & School of Medicine and Biomedical Sciences, University at Buffalo, The State University of New York, USA*

"A hugely interesting book, which is making a genuinely new contribution to debates around One Health. Irus Braverman has drawn together authors from an exceptionally wide range of disciplines, including multiple humanities, multiple social sciences, key advocates of One Health, alongside critically important perspectives exploring ideas and challenges of shared health beyond the Global North—and beyond the human. The volume carefully places these contributions into productive dialogue, rather than rehearsing well-established oppositions, to develop new visions of what "One Health Otherwise" could look like. At a time when the ongoing impacts of the COVID-19 pandemic have forcibly made the case for *why* thinking about health across humans, other animals, and environments is needed, a volume which starts to explore *how* is welcome indeed."

Angela Cassidy, *Senior Lecturer in Science and Technology Studies, Department of Social and Political Sciences, Philosophy and Anthropology,University of Exeter, UK*

"As Irus Braverman and the authors she has brought together in this volume point out, softening the ontological boundaries between the categories of the human, the animal, and the environment falls far short of the fundamental transformation in health praxis that the intensifying crises all earth beings are experiencing demands. Only by bringing an ethical lens—the lens of justice—to the ways in which those boundaries have justified and policed violent, hierarchical, colonial, and extractive relationships can One Health approach its promise."

Danielle Celermajer, *Professor in the Department of Sociology and Social Policy, The University of Sydney, Australia*

"The interconnectedness between human–animal–environment health is more apparent than ever, yet, as the chapters within reveal, One Health approaches are often still too anthropocentric, still caught up amidst neo-liberal and colonial logics, still hesitant to engage with local knowledges. It is thus exciting and refreshing to see creative and critical engagement with One Health, particularly, an approach that draws on the humanities and social sciences to both expand and refine the conceptual toolkit of One Health whilst creating an agenda for transdisciplinary futures."

Richard Gorman, *Research Fellow, Brighton and Sussex Medical School, University of Sussex, UK*

"Accessibly written and brought to life through first-hand accounts from the worlds of policy and medicine, critical engagements with archival materials and multispecies ethnography, this collection takes seriously the challenge of understanding One Health in practice, in all its complex, contested, and compromised detail. Genuinely global and multidisciplinary in scale and scope, the chapters offer an important corrective to the limits of existing One Health scholarship, especially its tendencies towards anthropocentricism and its neglect of both non-Western knowledges and research in social sciences and humanities. This book is essential reading for scholars, policy makers, and anyone with an interest in the intersections between human, animal, and environmental wellbeing."

Beth Greenhough, *School of Geography and the Environment, University of Oxford, UK*

"This ground-breaking collection takes up the challenge not only of identifying the limitations of One Health, but of also of proposing how to move beyond it, and delivers urgently needed perspectives on how to approach the entanglements of human and nonhuman health."

Christos Lynteris, *Professor of Medical Anthropology, University of St Andrews, UK*

"This timely volume sketches out the potentials and limits of the One Health approach by choosing an interdisciplinary, ethical, and engaged perspective on multispecies and planetary life. In focusing on their subjects, human, animal and floral, in a relational manner, the authors also manage to combine, among other concepts, feminist theories with STS scholarship. As such, the volume provides a much-needed expansion on the interspecific thinking of health and illness."

Mieke Roscher, *Professor, Department of History and Social Sciences, University of Kassel, Germany*

MORE-THAN-ONE HEALTH

This edited volume examines the complex entanglements of human, animal, and environmental health. It assembles leading scholars from the humanities, social sciences, natural sciences, and medicine to explore existing One Health approaches and to envision a mode of health that is both more-than-human and also more sensitive to, and explicit about, colonial and neocolonial legacies—urging the decolonization of One Health.

While acknowledging the importance of One Health, the volume at the same time critically examines its roots, highlighting the structural biases and power dynamics still at play in this global health regime. The volume is distinctive in its geographic breadth. It travels from Inuit sled dogs in the Arctic to rock hyraxes in Jerusalem, from black-faced spoonbills in Taiwan to street dogs in India, from spittle-bugs on Mallorca's almond trees to jellyfish management at sea, and from rabies in sub-Saharan Africa to massive culling practices in South Korea. Together, the contributors call for One Health to move toward a more transparent, plural, and just perception of health that takes seriously the role of more-than-humans and of nonscientific knowledges, pointing to ways in which One Health can—and should—be decolonized.

This volume will appeal to researchers and practitioners in the medical humanities, posthumanities, environmental humanities, science and technology studies, animal studies, multispecies ethnography, anthrozoology, and critical public health.

Irus Braverman is Professor of Law and Adjunct Professor of Geography at the University at Buffalo, the State University of New York. Her books include *Planted Flags: Trees, Land, and Law in Israel/Palestine* (2009), *Zooland: The Institution of Captivity* (2012), *Coral Whisperers: Scientists on the Brink* (2018), and *Zoo Veterinarians: Governing Care on a Diseased Planet* (2021). Braverman's latest monograph is entitled *Settling Nature: The Conservation Regime in Palestine-Israel* (forthcoming).

Routledge Studies in Environment and Health

The study of the impact of environmental change on human health has rapidly gained momentum in recent years, and an increasing number of scholars are now turning their attention to this issue. Reflecting the development of this emerging body of work, the *Routledge Studies in Environment and Health* series is dedicated to supporting this growing area with cutting edge interdisciplinary research targeted at a global audience. The books in this series cover key issues such as climate change, urbanisation, waste management, water quality, environmental degradation and pollution, and examine the ways in which these factors impact human health from a social, economic and political perspective.

Comprising edited collections, co-authored volumes and single author monographs, this innovative series provides an invaluable resource for advanced undergraduate and postgraduate students, scholars, policy makers and practitioners with an interest in this new and important field of study.

Healthy Urban Environments
More-than-Human Theories
Cecily Maller

Climate Information for Public Health Action
Edited by Madeleine C. Thomson and Simon J. Mason

Environmental Health Risks
Ethical Aspects
Edited by Friedo Zölzer and Gaston Meskens

Climate Change and Urban Health
The Case of Hong Kong as a Subtropical City
Emily Ying Yang Chan

Environmental Health and the U.S. Federal System
Sustainably Managing Health Hazards
Michael R. Greenberg and Dona Schneider

The Politics of the Climate Change-Health Nexus
Maximilian Jungmann

Research Ethics for Environmental Health
Edited by Friedo Zölzer and Gaston Meskens

More-than-One Health
Humans, Animals, and the Environment Post-COVID
Edited by Irus Braverman

For more information about this series, please visit: https://www.routledge.com/Routledge-Studies-in-Environment-and-Health/book-series/RSEH

MORE-THAN-ONE HEALTH

Humans, Animals, and the Environment Post-COVID

Edited by Irus Braverman

Routledge
Taylor & Francis Group
LONDON AND NEW YORK

earthscan
from Routledge

First published 2023
by Routledge
4 Park Square, Milton Park, Abingdon, Oxon OX14 4RN

and by Routledge
605 Third Avenue, New York, NY 10158

Routledge is an imprint of the Taylor & Francis Group, an informa business

British Library Cataloguing-in-Publication Data
A catalogue record for this book is available from the British Library

Library of Congress Cataloging-in-Publication Data
Names: Braverman, Irus, 1970- editor.
Title: More-than-one health : humans, animals, and the environment post-COVID / edited by Irus Braverman.
Description: First edition. | Abingdon, Oxon ; New York, NY : Routledge, an imprint of the Taylor & Francis Group, an informa business, 2023. |
Series: Routledge studies in environment and health | Includes bibliographical references and index.
Identifiers: LCCN 2022030086 (print) | LCCN 2022030087 (ebook) | ISBN 9781032277868 (hbk) | ISBN 9781032277882 (pbk) | ISBN 9781003294085 (ebk)
Subjects: LCSH: One health. | Environmental health. | World health. | Globalization--Health aspects. | Communicable diseases--Prevention. | Zoonoses. | COVID-19 (Disease)
Classification: LCC RA441.3 .M67 2023 (print) | LCC RA441.3 (ebook) | DDC 362.1--dc23/eng/20221007
LC record available at https://lccn.loc.gov/2022030086
LC ebook record available at https://lccn.loc.gov/2022030087

ISBN: 978-1-032-27786-8 (hbk)
ISBN: 978-1-032-27788-2 (pbk)
ISBN: 978-1-003-29408-5 (ebk)

DOI: 10.4324/9781003294085

Typeset in Bembo
by Deanta Global Publishing Services, Chennai, India

To my parents, Aurelia and Dan

CONTENTS

CONTRIBUTORS

Warwick Anderson is the Janet Dora Hine Professor of Politics, Governance and Ethics in Health, based in the Charles Perkins Centre and in Anthropology at the University of Sydney, Australia. He is also an honorary professor in the School of Population and Global Health, University of Melbourne, Australia. In 2020–2022, he was co-chair of the steering committee on health and climate change of the Australian Academy of Health and Medical Sciences. As a historian of science and medicine, Anderson focuses on the biomedical dimensions of racial thought, especially in colonial settings, on the globalization of medicine and science, and on the development of disease ecology and Planetary Health.

Myung-Sun Chun is Associate Professor of Veterinary Humanities and Social Science at the College of Veterinary Medicine, Seoul National University, South Korea. As a member of the Human-Animal Research Network, she researches sociocultural and historical topics related to animal diseases and human–animal relations. Among other topics, she has published multiple articles on the pathogen loads and modes of pathogen transmission in feral cat populations in urban and rural habitats in South Korea.

Maneesha Deckha is Professor and Lansdowne Chair in Law at the University of Victoria, Canada, where she directs the Animal Studies Research Initiative. She publishes in the areas of feminist animal studies, animal law, law and culture, and bioethics. She is author of *Animals as Legal Beings: Contesting Anthropocentric Legal Orders* (2021).

Hannah Dickinson is a postdoctoral researcher in the Department of Geography at Durham University, UK. Her work explores the intersections of environmental governance, biodiversity conservation, and international trade. Her current research

xplores how marine biomaterials—in particular, shrimp-derived chitosan—are
eralded as a solution to a range of socioecological ills.

tephen Hinchliffe is Professor of Human Geography at the University of Exeter, JK, and a Fellow of the Academy of Social Sciences. His books include *Pathological ives* (2016) and *Humans, Animals and Biopolitics: The More-than-Human Condition* 2016). He is currently working on a number of interdisciplinary projects on disase, biosecurity, and drug-resistant infections, focusing on Europe and Asia. He a member of the Wellcome Centre for Cultures and Environments of Health t the University of Exeter and sits on the UK Government's Scientific Advisory Committee on Exotic Diseases and on the UK Department for Environment, Food nd Rural Affairs Science Advisory Group's Social Science Expert Group.

lizabeth R. Johnson is an Associate Professor of Geography at Durham University, JK. Her work explores how ties between the biosciences and technological innoation are changing how we understand life in the context of environmental prearity. She writes on biomimicry, biosensing, and biotechnology and their influence n how we inhabit our environment.

aura H. Kahn is a physician, educator, co-founder of the One Health Initiative, nd independent policy research scholar. She is the author of *Who's in Charge? eadership During Epidemics, Bioterror Attacks, and Other Public Health Crises* (2020) nd *One Health and the Politics of Antimicrobial Resistance* (2016).

rédéric Keck is Director of Research at the Laboratory of Social Anthropology CNRS) in France, where he has been researching the history of social anthropolgy and contemporary biopolitical questions raised by avian influenza. From 2014 o 2018, he was the Director of the Research Department of the Musée du quai Branly, a museum of ethnographic objects. He is the author of *Avian Reservoirs: Virus Hunters, and Birdwatchers in Chinese Sentinel Posts* (2020) and co-author of *Anthropology of Epidemics* (2019).

iheung Kim is Assistant Professor in the Division of Humanities and Social ciences, Pohang University of Science and Technology, South Korea. He is lso a member of the Disaster STS Network and the Human-Animal Research Network. His research focuses mainly on the role of infectious diseases in human nd animal relations. He is the author of *Social Construction of Disease: From Scrapie o Prion* (2016).

jørn Ralf Kristensen is a doctoral student in the Environmental Studies Program nd the Department of Philosophy, the University of Oregon, USA. His applied esearch focuses on the intersections of animal ethics, environmental justice, public ealth, sanitation, and development studies.

Susan McHugh is Professor of English at the University of New England in Main USA. She is author of three monographs, most recently *Love in a Time of Slaughter. Human–Animal Stories Against Genocide and Extinction* (2019). She has coedited sev eral academic volumes, including *The Palgrave Handbook of Animals and Literatu* (2020) and *Posthumanism in Art and Science: A Reader* (2021).

Deborah Nadal is a cultural and medical anthropologist specializing in South Asi and an affiliate researcher at the School of Biodiversity, One Health and Veterinar Medicine at the University of Glasgow, UK. She also works as a consultant for th Department of the Control of Neglected Tropical Diseases at the World Healt Organization, Geneva.

Emily Reisman is Assistant Professor in the Department of Environment an Sustainability at the University at Buffalo, USA. Her research draws on agraria political economy, agroecology, and science and technology studies to examine th politics of agricultural knowledge, most recently in almond production and emerg ing agri-food technologies.

Susan Merrill Squier is Brill Professor Emeritus of Women's, Gender and Sexualit Studies and English at The Pennsylvania State University, USA. Her books an coedited volumes include the *Graphic Medicine Manifesto* (2015), *Epigenet Landscapes: Drawings as Metaphor* (2017), and *PathoGraphics: Narrative, Aesthetic Contention, Community* (2020). Squier is co-editor of the *Graphic Medicine* boo series at the Pennsylvania State University Press and past President of the Graphi Medicine International Collective.

Abigail Woods is ProVice Chancellor and Head of College of Arts at the Universit of Lincoln, UK. She has received multiple grants from the Wellcome Trust an other funding bodies to pursue research on livestock health, welfare, and produc tion in Britain. She is co-author of *One Health and its Histories: Animals and th Shaping of Modern Medicine* (2017).

ACKNOWLEDGEMENTS

'he idea for this volume was born at the tail end of my research on wildlife eterinarians in January 2019. As it happens, this was also when the COVID andemic started percolating in Asia. The wildlife vets I was interviewing at the ime were not at all surprised. In fact, they had been talking about this for quite ome time. Their insistence that health is relational and ties together not only uman and nonhuman lives but also the ecosystem appealed to my transdiscipli-ary sensibilities. It was refreshing to realize that the connections I was intuiting etween climate change, mass extinctions, and our health were being realized by thers who were developing ways to express and address these connections in the vorld. The vets referred to this approach as One Health—and I was immediately ıtrigued. At that time, One Health was not the popular public health approach : has become since COVID hit the human world. I decided to try to learn more.

I was fortunate to obtain a grant to convene a diverse group of scholars in a vorkshop setting. At that point, we were not sure if we would be able to meet in erson. But it was the beginning of the first lockdown and so we hoped that we vould be out within a few weeks. When the few weeks extended into months, ıe workshop took place virtually in April 2021, and many of the contributors f this volume participated in it. I am grateful to the members on the organizing ommittee from the University at Buffalo who envisioned the workshop and ıade it possible: James J. Bono (history), Jennifer Surtees (biochemistry), and aul Vanouse (art); as well as to Lucinda Cole from the University of Illinois. I m also thankful to the Baldy Center for Law and Social Policy for funding the vorkshop and especially to Caroline Funk for her warm support.

The workshop was recorded and made public online. It received immedi-te responses from a wide range of scholars and laypersons who were thrilled bout it and asked for more. At that point, I was also approached by Routledge ɔ consider publishing the contributions as a book collection. Grace Harrison,

Routledge's acquisitions editor, made me feel that this book would have a health home. Thank you, Grace, for your support along the initially bumpy path—an thanks to Matthew Shobbrook from Routledge for taking this project forwar from there. My fantastic research assistant Margaret Drzewiecki put in dozens c hours into researching, transcribing, and formatting this book.

As always, I am deeply grateful to my two children, River and Tamar, fc their inspiration and support. I would like to dedicate this book to my parents Aurelia and Dan Braverman. As health care providers in large hospitals, the have each dedicated more than forty years of their lives to care for the health c others. May they continue to live a long, healthy, and joyful life.

FOREWORD

The Lure of One Health

Stephen Hinchliffe

A lure is something that entices, tempts, or appeals. It's also a trap, a decoy.

Introduction

In order to survey wild bird populations for avian influenza, conservation managers and volunteers use lures to round up waterfowl on various wetlands in the UK. The aim is to estimate the environmental load of avian influenza, sending samples back to the government's veterinary laboratories. If all goes well in terms of sample quality and laboratory analyses, officials within the relevant ministry will be able to understand the threat posed to poultry farming, and make some sense of the risk (by no means improbable) that the avian influenza virus would evolve or adapt into a pathogen of pandemic potential.[1]

The lure of One Health is of course more than placing food on a platform at a wildfowl reserve at the break of dawn. It is the easily appreciated sense that people, plants, animals, and their environments, share health outcomes. It is the positive-sum game, wherein gains in environmental and nonhuman animal health benefit humans, and vice versa. It's a collective approach to inextricably shared fortunes: *one for all and all for one.* It is an ecological extension to the ancient Roman dictum, "*Salus populi suprema lex esto*"—the health of the people should be the supreme law—only translating people or population as a more-than-human matter.[2]

FIGURE F.1 Women practicing One Health in a peri-urban village near Dhaka, Bangladesh. The women are drawing maps and timelines of disease events and related practices in the village, a method used by the Food and Agriculture Organization of the United Nations to encourage ownership of One Health by local villagers. Photo by author, February 19, 2019.

The trap of One Health is also more substantial than the temporary discom- fort experienced by wild birds prior to being released back onto the wetland Indeed, One Health might obscure some important questions, or even imply that something can be delivered smoothly and in an uncontentious manner, when in fact there are bound to be uncomfortable trade-offs and compromises. The cal- culus may be more complex than the positive-sum version implies. By bundling everything and everyone together, are we missing something, skating over ques- tions that still need to be asked regarding how to approach questions of more- than-human life and health?

I will start this Foreword by focusing on an emblematic case for One Health demonstrating its attraction and utility. In the process and in what follows, I wil open up some less clear-cut issues. I will outline some of the conditions for One Health and some of the matters that may not quite fit or that remain part of a more uncomfortable calculus and politics of health. In doing this, I hope to set up some resources for reading the chapters in this book. The question to take to this and the pieces that follow might be: what is gained and what might be missec when we adopt the One Health signature?

Definitively One Health: Rabies

Rabies is a viral disease that can be transmitted to people by a range of mam- mals, notably canids (dogs, foxes) but also bats and rodents.[3] The infection, fol- lowing a bite from an infected animal, almost always a dog in those countrie without rabies controls, can cause inflammation of the brain and, if untreated will be fatal. Treatment relies on rapid diagnosis (which can be difficult as early symptoms are often unspecific) and timely post-exposure prophylaxis (PEP) PEP involves multiple doses of immunoglobulin and vaccine, is expensive, and often impractical in low-resource settings and in remote locations.[4] As a result infection prevention remains the most effective form of disease management Measures include managing dog populations through culls and neutering pro- grams, though this is rarely as effective as it may sound.[5] The basic reproduction number (R_0) for rabies is largely independent of canine population density, mean- ing that population reduction may have little effect on disease risk. Culling may also impact negatively on vaccination programs as it tends to remove the most easily accessible members of a population. Around 70% of dogs within a given population need to be vaccinated in order to disrupt transmission cycles and eliminate the disease in dogs and other mammalian hosts.[6] Anti-rabies vaccines have been available for well over a century, and in many countries with rabies transmission risks, vaccination of companion animals is mandatory. Outside of these areas of successful disease prevention, and particularly in sub-Saharan Africa and parts of Asia, alarming numbers of human cases and associated suffer- ing are almost all associated with transmission from dog bites. In these settings vaccination programs can achieve huge benefits to human and animal health.

Improving canine health through vaccination reduces the health burden on people, relieves the pressure on under-resourced healthcare systems, and removes a key reservoir of infection from the environment. It is clearly a win-win situation, a paradigmatic One Health venture. Just as characteristically for One Health cases, there are also potential gains that relate to the roll-out of such a program. These include, first, the need for and benefits from working across established disciplinary and health provider boundaries. Second, there is the utility in developing community engagement and participation in the delivery of collective health gains. Third, there is the need to secure and develop innovative forms of financial support necessary to initiate and sustain a One Health program. I will address each of these in turn.

Cooperation between human, animal, and environmental health professions is a key element of any One Health venture. Sharing expertise on a disease, its etiology and transmission mechanisms, pooling experience concerning biomedical and other health-related interventions, and devising innovative ways of working across institutional landscapes are key areas for collaboration. Nevertheless, these forms of co-working are often more difficult than they might be seen. As others have noted, barriers exist in part because of the evolution of biomedical sciences, and, in particular, the divergence of medical, veterinary, and environmental sciences in the nineteenth and twentieth centuries.[7] Just as significantly, medical and veterinary services have tended to develop quite different institutional practices, with unique mixes of state-based and or private forms of service delivery and cost allocation. As a result, there tend to be distinct resource allocations, budgeting practices, sector-specific norms in terms of service delivery, variations in terms of the designation of public and private goods, and distinctions in terms of payment vehicles and forms of cost recovery. As One Health programs are cross-departmental or inter-ministerial in complexion, the result is that numerous services, products, and health care practices need to be aligned. This is especially difficult when the costs of intervention fall on one department or sector while the benefits accrue elsewhere. As Cleaveland et al. explain in relation to rabies control in sub-Saharan Africa:

> It is the human health sector that derives the public health and economic benefits from canine rabies control—the reduction or elimination of human rabies deaths, and [the] reduced need for costly PEP for people bitten by suspected rabid animals. However, it is the veterinary services that generally incur the costs for canine rabies control, but derive few economic benefits, as domestic dogs are not an economically valuable species. Without a sharing of costs and benefits across sectors, there may be little incentive for veterinary services in low-income countries to prioritise investments for rabies control.[8]

The distribution of costs and benefits clearly depends on the specific disease system, host species, transmission pathways, and so on. For example, vaccination of livestock for a zoonotic disease may result in improved productivity and lower mortality, benefiting farmers, reducing pressure on veterinary services, providing gains for the food production sector, or the relevant ministry for livestock or food. In such cases, it may be reasonable and practical for the veterinary service, farmers, and the relevant sector or ministry to contribute to an intervention which provides direct benefits to those sectors. For companion animals, wildlife, and other environmental health interventions, and, indeed, when the agricultural sector is characterized by large numbers of smallholders, cost recovery is likely to be more complicated, especially in lower-income settings. Where animal ownership is uncertain (where there are high proportions of street or free-roaming dogs) and/or where keepers do not have the resources to pay for animal medicine, there are real barriers to reaching the required levels of compliance to make a program effective. In these circumstances, alternative means of delivering cross-sectoral health gains need to be developed.

This cross-sectoral issue and the need for a joined-up approach to health service delivery extends to a transdisciplinary, participatory, or engaged approach to public health. One Health initiatives, in other words, may be more successful when a broad coalition takes shared ownership and can recognize a collective benefit. Some of this is a matter of good communication, while in other cases there are more active attempts at co-production of health initiatives and outcomes.[9] For the rabies case, in resource-scarce environments with stretched veterinary services, engagement may help the vaccination process—training local auxiliaries or para-vets to vaccinate dogs as well as generating community ownership of a program can help to increase vaccination rates. Similarly, once the program has started, health service workers and communities need to be confident that there is indeed a diminished risk of infection. Community engagement with the program and its results can help to alter health-seeking practices and relieve pressure on health systems. People need to be confident that the risk of infection with rabies has decreased substantially in order to refrain from administering or demanding expensive PEP. In turn, this requires ongoing investment in veterinary surveillance and diagnostic capacity to maintain robust disease transmission risk estimates and develop appropriate evidence for vaccination effectiveness. Sharing scarce laboratory facilities in ways that benefit both human and animal health care sectors will aid this joined-up approach to One Health.

There are good reasons to suppose that some of this community engagement and ownership of a One Health issue can increase in the future—better access to stable vaccines, rapid or point-of-care diagnostic tests, widely available mobile telephony, and so on are all potential contributors to the vision of a One Health and community-based solution to a zoonotic disease. As Cleaveland and colleagues note: "community-directed interventions may be feasible, and deliver more cost-effective and sustainable approaches to rabies control in Africa than centrally coordinated strategies implemented by the veterinary services alone."[10]

Although community engagement may help to improve efficacy and reduce the burden on poorly resourced and over-extended veterinary and health services, financial support remains crucial to One Health initiatives. One-off or catalytic funding may be useful in terms of purchasing vaccines and training dog handlers, but subsequent vaccinations, antibody surveillance, and maintenance of disease control programs, require ongoing funding. As with other, solution-based, and cross-domain forms of development, One Health increasingly tends to be linked to new kinds of pharmaceutical, health care, and development financing.[11] In the rabies case, Cleaveland et al. suggest that some of the issues with supporting a programme of rabies control might be solved by development impact bonds (DIBs). Given the intrinsic sectoral barriers to One Health initiatives, the uneven distribution of costs and benefits, and the delay to realizing downstream benefits, impact bonds seem custom made for One Health initiatives.

These packages are part of the globalization and financialization of health and development[12] and part of a more general shift in the relationships between states, non-bilateral agencies and aid organizations.[13] Like catastrophe and vaccine bonds, they allow for the generation and release of larger amounts of finance than would normally be made available from state lenders or official development assistance.[14] As Mawdsley notes,[15] conventional catalytic funds are effectively used to "leverage private sector investment," transforming "aid oriented" into "growth oriented" financing. Whereas catastrophe and vaccine bonds allow investors to hedge their investment by covering a wide geographical area, DIBs are designed to generate returns on investments once a pre-determined and externally verified set of objectives have been achieved. The host state effectively sells an option on its future health or development status, with capital investments earning returns for the investors based on the delivery of those health and development gains. The resulting bonds are fungible and tradable assets, making them economically efficient in the terms set by international financial markets. In effect, they attract the levels of investment that investors regard as appropriate to the risks and time periods of returns. They also benefit by being under-written by state or philanthropic aid donations, state lenders, or national and international banks, and allow investors to securitize a bundle of risks and investments.

Despite the obvious attraction of novel financing arrangements to One Health type issues, questions remain. First, there is little or no transparency around some of these packages, with finance "leveraged behind closed doors."[16] Second, financialization involves turning health outcomes into market opportunities, arguably altering the conventional humanitarian- and security-based regimes of global health.[17] As Mawdsley captures it, this is a "re-configuration of parts of the developing world as the risky frontiers of profitable investment [...] Converting the 'mundane' into investable objects and tradable commodities."[18] It is also a reconfiguration of the state, which rescinds its role as service provider and remains only as a deliberate absence in what Geissler calls the para-state (an architecture of non-governmental, calculative and market based institutions that rely on the notion of the state but that act beyond its jurisdiction).[19] Third, despite the win-win rhetoric of both One Health and financial leverage, marketization of health investments will inevitably distort resource allocations. Investment decisions based on risk and returns may mean that disadvantaged regions remain unsupported. Riskier investments exist in those very areas where there is social instability and where the need for health investments is greatest. Likewise, financialization is bound to gravitate toward those problems that are amenable to technological and or pharmaceutical interventions, to relatively simple solutions and to returns that are easily measured. Many One Health programs are likely to fall short of these criteria and, despite their value and importance, will become difficult to operationalize.

For Cleaveland and colleagues, the rabies case is definitive: "One Health principles characterize all successful rabies efforts: effective intersectoral partnerships and communication; high levels of community ownership and participation; and strong political support at local and national levels."[20] To this list of principles, we might add "the international" and a financialized "global" market for development. The requirement to secure new forms of financial support for One Health initiatives in lower-income countries, an endeavor that is difficult to operationalize on account of cross-sectoral cost and benefit streams, occurs within a competitive funding environment (with investments from China and India as well as old-world financial centers). The point of note is that this form of activity is increasingly operating beyond conventional state-based coordination and oversight, involving the release as well as servicing of private finance and capital.

At this juncture, we can list some of the characteristics that mark the One Health paradigm:

- Clear benefits to both human and nonhuman animals, as well as environments;
- An identified need to work around existing service delivery and funding models (especially when costs and benefits fall unequally between existing sectors);
- Increased public and community engagement as an opportunity for improved health practices, skills development and as a means to overcome health care resource scarcity;
- An exploration of new forms of finance and calculus to generate necessary funds and overcome previously circumscribed budgets and inflexible cost recovery programs.

This is the lure of One Health, and, although I have my reservations (particularly around the issue of financialization), it is an attractive and, in many senses, self-evident case. In what follows, I'd like to briefly touch on some other disease situations.[21] The point is not to undermine the importance of One Health; it is to augment this list with some potential traps that we may, if we are not careful, fall into.

Indefinitely One Health: Guinea Worm Disease (GWD) in Chad

Guinea worm disease (GWD) or dracunculiasis is a parasitic disease caused by the nematode *Dracunculus medinensis*, with a water-based disease cycle involving larvae passing from open wounds in infected hosts into watercourses. The released larvae are transmitted to new hosts orally via intermediate water fleas, which are copepods of the Cyclopidae family, through consumption of water or via fish and amphibians acting as transport hosts—as depicted in

Figure F.2. There are no vaccines or treatments and the results of the disease can be debilitating and may permanently affect the limbs of those infected. Disease management involves water treatment and community engagement, both of which are important measures in interrupting and preventing transmission.

In the mid-1980s, there were around 3.5 million cases of GWD each year worldwide. In recent years, and largely as the result of a coordinated global campaign, the number of cases has been around 20–30 per annum. This phenomenal success was a major achievement of the Carter Center (founded by US President Jimmy Carter), who lead the Guinea Worm Eradication Program (GWEP), building local, national, and international partnerships and coordinating numerous donors. GWD was all set to become the first human disease to be globally eradicated since smallpox in 1980.[22] The effort to do so was based on well-informed assumptions that eradication was biologically feasible and clinically verifiable and that the program would produce numerous additional benefits in terms of improved sanitation and health care training opportunities. Eradication within a state or region required verification and certification, and was closely managed by international health institutions. The World Health Assembly definition of successful interruption of transmission and elimination of GWD stipulated that "there have been no reports of GWD for three or more years."[23]

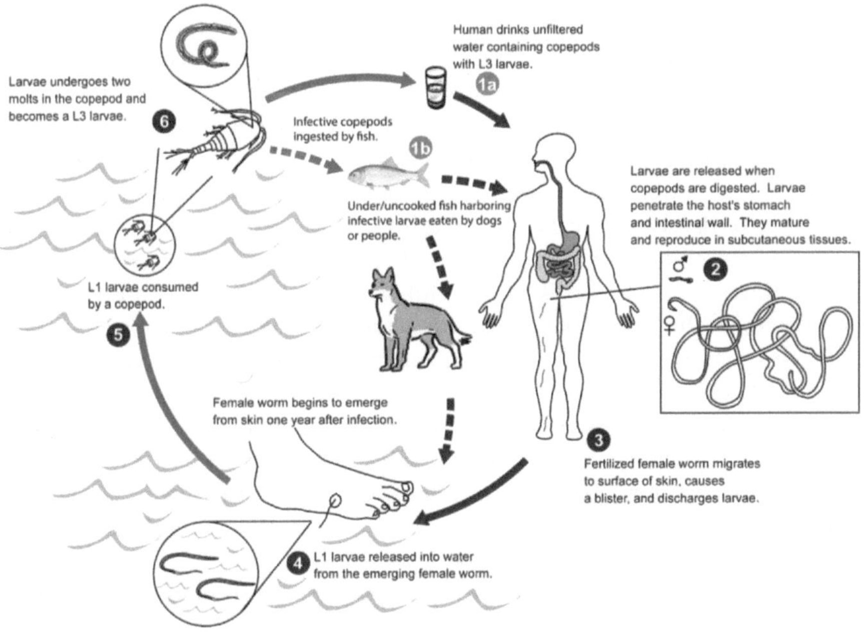

FIGURE F.2 Life cycle and transmission pathways of *Dracunculus medinensis*. Source: CDC, public domain. This use does not constitute an endorsement by the US Government, Department of Health and Human Services, or Centers for Disease Control and Prevention.

Dracunculiasis was, in several respects, a straightforward public health issue, its eradication conforming to a so-called first wave of public health interventions: improving sanitation and interrupting transmission. But the issue has recently taken a turn to become a One Health matter. In the mid-2010s, infections were reported in dogs in Chad and in several other states in Africa. The canine parasites were genetically identical to those in humans, indicating that the disease had become or should now be recognized as a zoonosis (an infectious disease that conventionally speaking has jumped from nonhuman animals to humans, though, in this and many other cases, the direction of travel has been reversed).[24] Canine hosts, it was supposed, were now playing a part in maintaining a reservoir of this parasite.[25]

In terms of resulting actions, there are parallels to the rabies case. Targeting water treatment programs in areas close to dog populations and encouraging community engagement with disease prevention were advised. The latter included education programs on transmission, discouraging the feeding of dogs with fish catch remains, rewarding those who reported canine cases, and preventing infected dogs from visiting water courses—a controversial practice involving tethering of dogs.[26] But a new issue started to arise in terms of verification and certification of eradication. Once the disease became zoonotic, the logic of disease certification started to look less secure.

In human disease systems, verification and certification may be relatively straightforward, especially in easily diagnosed diseases like GWD, with clinical observation, case reporting, and centralized records generating a disease incidence picture.[27] However, once a pathogen is in nonhuman hosts with wide spatial ranges[28] and the potential exists for further spillback into wildlife, then the calculative environment may alter. If human health, and the logic of eradication, deals in absolute numbers (based on case reports), animal health tends to follow a probabilistic logic. As animals do not take themselves to a clinic, and as dog ownership may be unclear, there is likely to be underreporting of disease. In this case, establishing case numbers or environmental pressure is based on developing a robust diagnostic test, sampling a population, and estimating incidence. Robust pond-side tests exist,[29] but the Popperian point here is that surveillance for eradication becomes embroiled in a new logic, as we move from clinical cases and numbers to statistical inference and likelihoods or risks.

This is robust knowledge, but it is important to note that the logic has shifted from definitive absence of a condition to diminishing risks of infection. The issue is partly epistemological, marking differences between clinical medicine, population medicine, and ecological sciences. But it is also more sweeping than a simple alteration of survey methods. The shift to disease ecology—or the relational science of hosts, environments, and pathogens[30]—is marked by an alteration in logic that unsettles the rationales of eradication or disease-free schema. There is no longer a finite end or vanishing point. Time, in this probabilistic landscape, is now infinite. The implication is that One Health can start to open up some important questions concerning the "ends" of health interventions, their rationale, and their measurement.

Definitively More-than-One Health: Ponds, Prawns, and Antimicrobial Resistance

This logical conundrum opens up health as a conditional process, rather than an absolute endpoint. And it may contain a crucial lesson if we are to make good on the promise or the lure of One Health. I can illustrate this by taking the issue of disease management in relation to the threats posed by, and the solutions offered to, the risks of drug-resistant microbes, and specifically antimicrobial resistance (AMR). AMR is becoming one of the leading causes of mortality globally, particularly in parts of Africa and Asia,[31] and has been described as a quintessential One Health issue.[32] Human, animal, and environmental health are all implicated in the use of antimicrobial compounds, in the emergence of resistant organisms, their persistence, and transmission. Like other One Health issues, there is a clear need to develop inter- and transdisciplinary working, as well as cross-sectoral collaboration, in order to meet the complex challenges of managing disease risks, altering health-seeking practices, and reducing transmission of resistant microbes. Yet, as an esteemed clinical and scientific team of authors suggest (notably without social science representation), a major gap is the appreciation of the socioeconomic drivers of both disease- and health-seeking practices. "One Health science seems to fall short when it comes to understanding the economic forces behind many emerging infectious diseases; there is a further gap to be bridged between the biomedical, environmental, and animal sciences, and the social sciences."[33]

The point can be underlined with case studies of disease management in lower-income countries. For example, in recent decades, a rapid expansion of inland aquaculture has produced large gains in access to dietary proteins and revenue, especially in Asia.[34] Most production is undertaken by a "missing or squeezed middle" of commercial producers who "enjoy none of the benefits of investments in biosecurity or pathogen control characteristic of intensive systems nor the low input/low risk/low output typical of extensive systems."[35] These producers are adopting practices such as commercial feed use, water, and livestock treatments, including antimicrobial uses, but are loosely tied to value chains, subject to little or no veterinary oversight and are weakly regulated by buyers and/or state-based organizations. Disease is a persistent threat—representing an estimated US$6 billion in losses per annum to the global industry.[36] One means of reducing disease burden is to improve the quality and sources of seed stock to farmers. This may involve funding hatcheries to produce specified pathogen-free and/or certified or tested stock. In Bangladesh, for example, farmers used to pay collectors of wild fry and larvae to stock their ponds—a practice that not only led to environmental damage within fragile mangrove wetlands but also recycled disease between the riverine environment and the ponds. The disease burden was particularly high, with farmers reporting frequent disease problems in their ponds.[37] Farm supply shops in Bangladesh were increasingly operating concessions and were pressured by wholesalers and commercial salespeople to increase sales of treatments to farmers (who have little access to other expert advice). Given the disease burden and the availability of over-the-counter antimicrobials, even small amounts of which are thought to be potentially significant in aqueous environments,[38] there would seem to be a strong case for a One Health intervention.

In this situation, it may seem a straightforward statement to suggest that certified disease-free hatchery seeds would reduce disease burden, interrupt transmission from farms to rivers and its recycling to livestock via wild seed, and lower the impact on mangrove habitats. The reduction in disease would in turn reduce the likelihood that farmers would have use for medicines and treatments that are potentially damaging to environmental and human health. From a One Health perspective, this looks like a win-win situation. Finance for the hatcheries and relevant NGOs was available in the form of international loans and grants, and hatchery owners secured tax concessions that allowed the technical innovations to proceed.

Yet, at the pond side, farmers initially saw little change, in terms of performance or disease, with the new seed. They were concerned that the hatchery-bought larvae weren't as well acclimatized to the water and temperature conditions in their ponds.[39] To realize (and for farmers to see) the benefits of disease-free seed, NGOs encouraged farmers to change their farming practices, to farm in a more biosecure or disease-free fashion. These "missing middle" farmers had largely survived and farmed in a system characterized by frequent health and disease problems by developing a form of multi-cropping, farming a variety of species, and frequently re-stocking their ponds. Where possible, ponds were multi-use—the same *gher* or flooded paddy field produced rice crops in winter, as well as finfish, mud frogs, shrimp, and prawn. Farmers bought their seed and fry at regular intervals, topping up stock throughout the production period. After the summer monsoon, for example, farmers could take advantage of changing water conditions (the rains would alter water depth, as well as the salinity, oxygen content, and temperature of the ponds) by re-stocking or adding new species. Farmers would, as a result, spread the cost of seed. An added benefit was that they could harvest at various points in the season in order to take advantage of variable market prices and secure regular income to service the weekly collections of payments on their microfinance loans.[40] All this tinkering, of course, ran counter to the ideals of disease-free farming, as each introduction to the pond risked adding disease and undermining the integrity of the system (and the point of supplying disease-free seed). So, farmers were encouraged to buy all their seed in one transaction, and to follow a batch-like production system. Doing so would reduce disease and, in turn, lessen the need for disease treatments.

Although lowering stocking frequency seemed to reduce mortality in the ponds (though this was probably a result of reduced predation of larvae by other pond inhabitants), the effect on disease incidence was less clear. Perhaps more importantly, qualitative research suggested that the farmers were more rather than less exposed to livelihood risks once they followed the stocking guidance. By disrupting the vernacular system of managing disease and financial risks, using frequent stocking and multi-cropping, farmers were more rather than less likely to experience an economically ruinous disease event and so turn to disease treatments as a means to rescue their livelihoods. Whereas farmers were exasperated that these treatments were unreliable and expensive, they also said that, if they didn't use them, they "would be left with nothing." In other words, by concentrating their disease-free farming on a single species with limited numbers of harvests, they exposed themselves to more rather than less livelihood risk. In so doing, they would be more, rather than less, likely to turn to antimicrobial and other disease treatments.

In terms of One Health, the lesson is that the health of the system is more than simply incidence of disease. It is how that disease is managed that matters—and whether it is seen as something to adjust to or as something that can potentially lead to economic ruin. Although reducing disease transmission is a good thing, the assumption that farmers needed to adopt a "disease-free" farming system in that case produced a risk gradient that encouraged more rather than less treatment with antimicrobials. In an economically precarious situation, there is a clear need not only to understand disease incidence, but also to appreciate the socioeconomic risks faced by the farmers. More broadly, the point of this case may be that One Health is delivered in ways that may *not* be optimal for human, animal, and environmental health, at least if the optima are measured in absolute terms. The best One Health outcome in this case might be a tolerable and manageable background noise of disease. It is the lessons learnt from the farmers—multi-cropping, frequent stocking, and agro-ecological approaches to production—that suggest that living with disease and adapting farming practices to those diseases, and to other environmental challenges, offer the more sustainable pathways compared to imposed norms of biosecure farming and disease-free solutions.[41] The distinction is both clear but also subtle. One Health is not an absolute state (disease freedom)—it is a process, an approximation, where the optimum may involve several suboptima within the components.

Rethinking Optimal Outcomes

The American Veterinary Medical Association defines One Health as "the collaborative effort of multiple disciplines—working locally, nationally, and globally—to attain optimal health for people, animals, and our environment."[42] Optimization of health, it should now be clear, cannot be reduced to a simple metric, or to a state of being. The presence or absence of illness, or even more reductively, the presence or absence of specific pathogens, is an insufficient guide to the health of an environment, organism, or society. Moreover, although health outcomes are undoubtedly shared across human and nonhuman communities, it may not be possible to ensure that there will be optimization of every sector or subsector of the One Health circle.

The 1948 World Health Organization Constitution declared that health is the absence of illness as well as "a state of complete physical, mental, and social well-being."[43] There is,of course, a lot more that could be said here about health—for example, how contemporary senses of immunity draw attention not only to the constitutive role played by illness and pathogens in the continuous production of an immuno-competent body,[44] but also to the need to understand health as an ability to recover, or to adapt to disease.[45] As Porter put it so succinctly, health is not a matter of approximations to the average, or a norm, but a matter of appropriate adaptation to environment,[46] and so, we could add, a matter of geographical specificity. We can extend this definition of human health to one that befits environments (the adaptation of ponds in Bangladesh) and of course nonhuman species. In this case optimization is not a matter of *being* disease-free, but a relational process in which the *becoming* or dynamics of microbiomes and interspecies relations are key. One Health becomes a matter of working with rather than against environments. Disease is, in this sense, about understanding not only the mechanisms associated with causative agents, but also how an environment and a host play their parts in producing pathogenicity. Health in this sense is a matter of the mixtures and patterns of multispecies assemblages. It is the health and regulatory work of the microbiome in guts, on skin, in the soil, and in ponds that matters. For example, research on what have been called pathobiomes—"the set of host-associated organisms (crucially encompassing prokaryotes, eukaryotes, and viruses) associated with reduced (or potentially reduced) health status, as a result of interactions between members of that set and the host"[47]—and the development of metagenomics and metabolomics starts to open up new ways of assessing health across human–nonhuman–environment interfaces. In the ponds in Bangladesh, for example, the health of the system may be less a matter of the presence or absence of a livestock pathogen, and more a matter of the relative balances between various populations of microbes, macrobes, and other, physiochemical parameters. While the language of the normal and the pathological persists, the step change is clear in the sense that a healthy biome involves a range of assemblages, which may tip[48] into pathogenic arrangements through processes that involve a reduction in self-regulating behaviors.

Similar examples are found in studies of human, plant, and environmental health. The research of those in a newly re-energized field of pollution and discard studies, as well as plant health, is exemplary.[49] Notions of acceptable levels and assimilative capacities are being displaced by work inflected with readings of the colonial logics that stem from and fuel the errors of assuming stable chemistries and relatively fixed thresholds between contaminated and polluted media. Likewise, a "fumigatory" approach to plant and soil life in the name of crop health and market return has been fuelled by this license to pollute as well as a reductive approach to living and soil processes. Toxicology has long since moved on to understanding the synergies of materials and pollutants, while subtle shifts in holobiont relations are increasingly shown to reconfigure health and wellbeing. A holobiont is a unit of biological organization composed of a host and its associated bacteria, archaea, viruses, protists, and fungi—it is a useful term to displace the sense of there being discrete organisms of "monogenomic differentiated cell lines"—"we," like other organisms, are instead "far-from-equilibrium assemblies of highly heterogeneous cells."[50]

Exemplary holobiont work includes historical research on the under-reported illnesses of downwinders at twentieth-century plutonium producing districts in Russia and the US, and, more recently, the challenge to official statistics associated with the 1986 Chernobyl accident.[51] In all cases, radioactive materials remain in the soil, water, and environment, but at concentrations and producing emissions that are conventionally reported as within acceptable levels. In these locations, health complaints including chronic syndromes, digestion problems, immune disorders, and so on, were historically dismissed as individual illnesses rather than environmentally or industrially produced conditions. For the science-studies scholar and historian Katrina Brown, working with microbiologists, the patterns of downwinders' ill-health could be linked to mutations not in the patients' bodies or monogenomic differentiated cell lines, but to damage within their intestinal bacteria. Ill-health, as a result, was related not so much to an acute dose of pathological radiation, but to chronic sub-lethal exposures that affected the relations between the patient and their microbiological companions. As Tsing and colleagues summarize,[52] ill-health thus becomes a matter of suffering the ills of another.

On this reading, health is a matter of entanglements—being entangled with others of various kinds is key to being a healthy holobiont.[53] This social science refrain of the importance of entanglement is, of course, important ontologically. But the difficult question remains, how to optimize entanglements, and which of these numerous entangled relationships matter? In turn, how can a more capacious, relational, and inclusive One Health be made to work? Pathobiomes and other microbiome imaginaries can, despite the scientific enthusiasms, become bogged down in data. The promise of machine learning and high-throughput data processing may suggest that signals can be discerned in future.[54] But the multifactorial nature of health means that the evidence base for establishing proximate and distal causes that lead to ill-health may remain illusory. They may also be unlikely to follow the same norms and criteria of the states and standards associated with legal and regulatory processes that have been utilized in the past. Rather like the dogs in Chad, the calculus may be changing from clear states or levels that are easily translated into rules and legislature, to more complex statements around healthy assemblages. The implications include a need to expand the tool kit for the governance of One Health: considering different forms and formats of evidence; adopting, where appropriate, precautionary approaches to regulation; adopting open forms of monitoring and data generation that can trigger reviews of processes or products that were previously considered safe or inconsequential.

So, in terms of optimization, there may be no straightforward answer to what is effectively a set of relations and a situation where there will be relative gains *and* losses. In other words, we may be moving away from the win-win rhetoric of One Health, and edging closer to something that is more familiar to politicians. One Health might be better characterized as something that involves winners and losers, trade-offs, and compromises, and the aim will then be to find solutions that are not so much optimal but geographically appropriate, politically just, and biologically, socially, and ecologically legitimate.

A possible opening to this version of One Health would be to look toward other areas where similar issues of conditional optimization have been discussed. For example, in the related field of animal welfare studies,[55] a heuristic model can help to shape discussions of the quantity and quality of welfare. Although there are reasonably robust definitions of animal welfare (The World Organisation for Animal Health, formerly the Office International des Epizooties, defines this as applying when an animal is healthy, comfortable, well-nourished, and able to express innate behavior and not suffering from pain, fear, or distress),[56] actual or delivered welfare will depend on a trade-off between a variety of factors including markets for produce, desires of a society to sponsor or tolerate certain levels of welfare, and so on. John McInerney's[57] simple model makes this trade-off function clear—as livestock productivity increases, animal welfare may initially increase in the form of improved diet, husbandry, and removal of environmental stress, but will soon start to fall as the biological stresses of production, housing, and so on start to take their toll. At some point, the animals are pushed to their biological limits and there is a collapse—see point E on the graph in Figure F.3. For McInerney, the actual welfare that is deemed appropriate is not necessarily the optimal for the animal in question (although, of course, this position may be supported by many). It is rather the level, close to C on the curve, that offers the acceptable compromise between human and animal benefit. This is the point where human health and welfare may gain from food security and accessibility, while animals experience what are agreed to be acceptable welfare standards. The graph in Figure F.3 is, of course, too simple, and there will be many more dimensions to this issue than the two represented on the axes. The intention of the model is not necessarily to argue over the details of the metrics, or indeed to deny that many societies and livestock systems get this wrong. It is to underline that any system of "optimization" will, in all likelihood, involve complex trade-offs and compromises. These may not be acceptable to all parties and will be subject to claims and counter-claims. For our purposes, we could say that when welfare or health become more than a state of being, but involve some measure of quantity (how much health?) then there are bound to be public disputes over priorities and emphases.

When the metaphysics change from things to processes, and organisms are understood as holobionts, when the evidence and knowledge base shifts and when we re-insert health into a social and economic context, then One Health becomes a more variable set of outcomes. Rather than a technical matter of

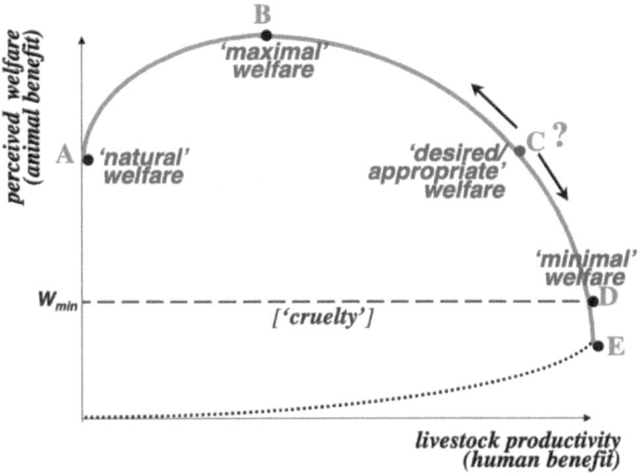

FIGURE F.3 Conflicts between animal welfare and productivity. Source: McInerney, "Animal Welfare, Economics and Policy," The National Archives, February 2004, p. 30.

assuming a positive-sum game and a win-win situation, we enter a world of competing interests, alternative interventions, scarce public resources, different investment opportunities, and uncertain outcomes. It is, in other words, a public matter—One Health is part and parcel of what might be understood as an ongoing contest about what counts as health. It is a matter for, and of, public interest.[58]

Some Conclusions

I started this Foreword by referring to the lures that are used to draw in wildfowl to a wild bird survey. The rationale for the research was to detect avian influenza in migrating birds, as they moved south and west from their summer feeding sites in Scandinavia, Siberia, and the Arctic. The reason for doing so was largely to gauge the threat to UK poultry farming, providing early warning of transmission risk to domestic birds. In this sense, the lures configured wild birds as a threat to the economy, as reservoirs of infection and onward transmission.

The lures may also have been something of a diversion. Perhaps they, or the survey apparatus more generally, partially obscured the more worrying sense that wild birds were not so much the source *of* danger, but *in* danger. As the timing and routes of avian migrations started to alter in response to changing climates and diminished food availability; as previously discrete populations mixed in increasingly isolated and pressured feeding sites, and as global poultry expands and alters viral opportunities, we were (and are) witnessing a shift in planetary biology. Framing migrant wild birds as threats while failing to divert resources to changing livestock production and habitat depletion is akin to blaming a river for flooding. It is to mistake symptoms with causes. One Health should be an opportunity to question how human–animal–environment relations are being formatted. What, in other words, is being missed or silenced?

This widening of scope is evident in the chapters of this book, and, as several authors make clear, it is in some ways an outcome of the experience of a global pandemic. Whereas there is, and will undoubtedly continue to be, investments in searches for the origins and some quite legitimate concern with the spill-over events that enabled a relatively innocuous bat coronavirus to jump species and become highly pathogenic and transmissible in people, the pandemic has also raised other concerns about shared health outcomes. Perhaps most notable of these has been the role of health inequalities in the transmission, infectivity, and health outcomes that relate to infection with SARS-CoV-2. The high prevalence, in many badly affected states, of metabolic diseases (obesity, diabetes, hypertension) and their uneven distribution in terms of racial and socioeconomic inequalities,[59] has opened up pressing questions for One Health. The micro- and macro-impoverishment of holobionts and the effects on metabolic pathways and processes[60] has produced a slow violence,[61] gradually and unevenly eroding the capabilities of groups of people, environments, nonhumans, and health care systems, to adapt to disease and illness.

The questions for One Health, then, might be more than the economic, social, and environmental drivers of spill-overs, important though these are. Critical questions include the conditions that produce infectibility and increasingly uneven experiences of vulnerability. This applies to food systems and metabolic justice, dietary patterns, economic and healthcare systems that are all changing what it is to be a human being (a being that sleeps fewer hours, eats more carbohydrates and refined sugars, maintains a lower body temperature, and is increasingly suffering the ills of another). In other words, as COVID-19 has troubled any hard distinction between communicable and non-communicable diseases, with morbidity and mortality rates strongly correlated with those already living with chronic illnesses, it's time to loosen the obsession with pathogenic microbes, and consider what is driving pathological lives.[62]

The trap of One Health is perhaps to seduce, in its language of wholeness and integration, and to lose sight or sense of the core issues of both planetary change and of health. The veterinary science-based promoters of One Health are in some cases too wedded to an ethos of making and mending the worlds of human extraction and exploitation of nonhuman animals and environments. We need, urgently, a One Health that is not only able to take health more seriously—as an inevitably patchy process of piecing together compromises and adaptations to the conditions of life.[63] We also need a One Health that can, and is not afraid to, open up radical questions concerning the acceleration of poor health opportunities for many, if not most, of the earth's inhabitants and environments.

Notes

1 Stephen Hinchliffe and Stephanie Lavau, "Differentiated Circuits: The Ecologies of Knowing and Securing Life," *Environment and Planning D: Society and Space* 31, no. 2 (2013): 259–274; Stephen Hinchliffe, "Sensory Biopolitics: Knowing Birds and a

Politics of Life," in *Humans, Animals and Biopolitics: The More-than-Human Condition*, eds. Kristin Asdal, Tone Druglitro, and Stephen Hinchliffe (Abingdon: Routledge, 2017), 152–170. See also Keck, this volume.
2 Cassidy rightly identifies the anthropocentrism of much of the early One Health literature in comparative medicine and epidemiology. Much of that literature treats nonhuman health as being in the service of human health. As should become clear, this instrumentalism negates to ask the critical questions that mark a more radical sense of the One Health project. Angela Cassidy, "Humans, Other Animals and 'One Health' in the Early Twenty-First Century," in *Animals and the Shaping of Modern Medicine: One Health and Its Histories*, eds. Abigail Woods et al. (London: Palgrave Macmillan, 2017), 193–236.
3 Melanie J. Rock, Dawn Rault, and Chris Degeling, "Dog-Bites, Rabies and One Health: Towards Improved Coordination in Research, Policy and Practice," *Social Science and Medicine* 187 (2017): 126–133. See also Nadal, this volume.
4 Sarah Cleaveland et al., "Rabies Control and Elimination: A Test Case for One Health," *Veterinary Record* 175, no. 8 (2014): 188–193.
5 Krithika Srinivasan et al., "Reorienting Rabies Research and Practice: Lessons from India," *Palgrave Communications* 5, no. 1 (2019): 152.
6 Katie Hampson et al., "Transmission Dynamics and Prospects for the Elimination of Canine Rabies," *PLOS Biology* 7, no. 3 (2009): e1000053.
7 Jakob Zinsstag et al., "From 'One Medicine' to 'One Health' and Systemic Approaches to Health and Well-Being," *Preventative Veterinary Medicine* 101, nos. 3–4 (2011): 148–156.
8 Cleaveland et al., "Rabies Control," 189.
9 Alicia Davis and Jo Sharp, "Rethinking One Health: Emergent Human, Animal and Environmental Assemblages," *Social Science & Medicine* 258 (2020): 1–8.
10 Cleaveland et al., "Rabies Control," 192.
11 Susan Craddock, "Precarious Connections: Making Therapeutic Production Happen for Malaria and Tuberculosis," *Social Science and Medicine* 129 (2015): 36–43; Linsey McGoey, *No Such Thing as a Free Gift: The Gates Foundation and the Price of Philanthropy* (London: Verso, 2015).
12 Emma Mawdsley, "Development Geography II: Financialization," *Progress in Human Geography* 42, no. 2 (2016): 264–274.
13 Vincanne Adams, Dominique Behague, Carlo Caduff, Ilana Löwy & Francisco Ortega, "Re-imagining Global Health through Social Medicine," *Global Public Health* 14, no. 10 (2019): 1383–1400.
14 Susan Erikson, "Global Health Futures?," *Medicine Anthropology Theory* 6, no. 3 (2019): 77–108.
15 Mawdsley, "Development Geography II," 268.
16 Erikson, "Global Health Futures?"
17 Andrew Lakoff, "Two Regimes of Global Health," *Humanity: An International Journal of Human Rights, Humanitarianism, and Development* 1, no. 1 (2010): 59–79.
18 Mawdsley, "Development Geography II," 271.
19 Paul Wenzel Geissler, "Introduction: A life Science in its African Para-State," in P.W. Geissler (Ed.) *Para-States and Medical Science: Making African global health* (Durham, Duke University Press, 2015): 1–44.
20 Cleaveland et al., "Rabies Control," 192.
21 The term is introduced in Stephen Hinchliffe et al., *Pathological Lives: Disease, Space and Biopolitics* (London: Wiley Blackwell, 2016).
22 Nancy Leys Stepan, *Eradication: Ridding the World of Diseases Forever?* (Ithaca: Cornell University Press, 2011).
23 Donald R. Hopkins et al., "Dracunculiasis Eradication: Neglected No Longer," *American Journal of Tropical Medicine and Hygiene* 79, no. 4 (2008): 474–479.
24 See the different treatments in the chapters by Squier and Kristensen, this volume.
25 Ewen Callaway, "Dogs Thwart Effort to Eradicate Guinea Worm," *Nature* 529, no. 7584 (2016): 10–11.

26 Robbie A. McDonald et al., "Ecology of Domestic Dogs *Canis familiaris* as an Emerging Reservoir of Guinea Worm *Dracunculus medinensis* Infection," *PLOS Neglected Tropical Diseases* 14, no. 4 (2020): 1–12; Donald R. Hopkins et al., "Dracunculiasis Eradication: Are We There Yet?," *American Journal of Tropical Medicine and Hygiene* 99, no. 2 (2018): 388–395, 388.

27 Even in human health cases, verification of the ends of a disease may be less straightforward. See Vargha's account of eradication of Polio in Hungary. Dóra Vargha, *Polio Across the Iron Curtain: Hungary's Cold War with an Epidemic* (Cambridge: Cambridge University Press, 2018).

28 Jared K. Wilson-Aggarwal et al., "Spatial and Temporal Dynamics of Space Use by Free-Ranging Domestic Dogs *Canis familiaris* in Rural Africa," *Ecological Applications* 31, no. 5 (2021): 1–12.

29 Neil Boonham et al., "A Pond-Side Test for Guinea Worm: Development of a Loop-Mediated Isothermal Amplification (LAMP) Assay for Detection of *Dracunculus medinensis*," *Experimental Parasitology* 217 (2020): 1–4.

30 Warwick Anderson, "Natural Histories of Infectious Disease: Ecological Vision in Twentieth-Century Biomedical Science," *Osiris* 19 (2004): 39–61.

31 Christopher J. L. Murray et al., "Global Burden of Bacterial Antimicrobial Resistance in 2019: A Systematic Analysis," *The Lancet* 399, no. 10325 (2022): 629–655.

32 T. P. Robinson et al., "Antibiotic Resistance is the Quintessential One Health Issue," *Transactions of the Royal Society of Tropical Medicine and Hygiene* 110, no. 7 (2016): 377–380.

33 Ibid., 379.

34 G. D. Stentiford et al., "Sustainable Aquaculture through the One Health Lens," *Nature Food* 1, no. 8 (2020): 468–474.

35 David C. Little et al., "Sustainable Intensification of Aquaculture Value Chains Between Asia and Europe: A Framework for Understanding Impacts and Challenges," *Aquaculture* 493 (2018): 338–354.

36 Grant D. Stentiford et al., "New Paradigms to Help Solve the Global Aquaculture Disease Crisis," *PLoS Pathogens* 13, no. 2 (2017): 1–6.

37 Stephen Hinchliffe, Andrea Butcher, and Muhammad Meezanur Rahman, "The AMR Problem: Demanding Economies, Biological Margins, and Co-Producing Alternative Strategies," *Palgrave Communications* 4 (2018): 1–12.

38 Felipe C. Cabello et al., "Aquaculture as Yet Another Environmental Gateway to the Development and Globalisation of Antimicrobial Resistance," *The Lancet Infectious Diseases* 16, no. 7 (2016): e127–e133; Nick G. H. Taylor, David W. Verner-Jeffreys, and Craig Baker-Austin, "Aquatic Systems: Maintaining, Mixing and Mobilising Antimicrobial Resistance?," *Trends in Ecology and Evolution* 26, no. 6 (2011): 278–284.

39 Stephen Hinchliffe et al., "Production Without Medicalisation: Risk Practices and Disease in Bangladesh Aquaculture," *Geographical Journal* 187, no. 1 (2021): 39–50.

40 Ibid.

41 Hinchliffe et al., "The AMR Problem"; Stephen Hinchliffe, "Postcolonial Global Health, Post-Colony Microbes and Antimicrobial Resistance," *Theory, Culture and Society* (2021): 145–168.

42 One Health Initiative Task Force, "One Health: A New Professional Imperative," American Veterinary Medical Association, 2008, https://www.avma.org/sites/default/files/resources/onehealth_final.pdf.

43 "Constitution of the World Health Organization," World Health Organization, 2022, https://www.who.int/about/governance/constitution.

44 Warwick Anderson and Ian R. MacKay, *Intolerant Bodies: A Short History of Autoimmunity* (Baltimore: Johns Hopkins University Press, 2014).

45 Georges Canguilhem, *The Normal and the Pathological*, trans. Carolyn R. Fawcett (New York: Zone Books, 1991).

46 Roy Porter, "Before Foucault," *London Review of Books* 12, no. 2 (1990), https://www.lrb.co.uk/the-paper/v12/n02/roy-porter/before-foucault.

47 David Bass et al., "The Pathobiome in Animal and Plant Diseases," *Trends in Ecology and Evolution* 34, no. 11 (2019): 996–1008.
48 The notion of tipping points was used in relation to disease and livestock in Stephen Hinchliffe et al., "Biosecurity and the Topologies of Infected Life: From Borderlines to Borderlands," *Transactions of the Institute of British Geographers* 38, no. 4 (2013): 531–543.
49 Max Liboiron, *Pollution is Colonialism* (Durham: Duke University Press, 2021). Julie Guthman, *Wilted: Pathogens, Chemicals and the Fragile Future of the Strawberry Industry* (Oakland: University of California Press, 2019).
50 John Dupré, *The Metaphysics of Biology* (Cambridge: Cambridge University Press, 2021).
51 Kate Brown, *Manual for Survival: A Chernobyl Guide to the Future* (New York: W.W. Norton, 2019).
52 Anna Lowenhaupt Tsing et al., eds., *The Arts of Living on a Damaged Planet* (Minneapolis: University of Minnesota Press, 2017).
53 Alex M. Nading, "Humans, Animals and Health: From Ecology to Entanglement," *Environment and Society: Advances in Research* 4, no. 1 (2013): 60–78; Alex M. Nading, *Mosquito Trails: Ecology, Health, and the Politics of Entanglement* (Oakland: University of California Press, 2014).
54 Kristen D. Curry, Michael G. Nute, and Todd J. Treangen, "It Takes Guts to Learn: Machine Learning Techniques for Disease Detection from the Gut Microbiome," *Emerging Topics in Life Sciences* 5, no. 6 (2021): 815–827.
55 Henry Buller and Emma Roe, *Food and Animal Welfare* (London: Bloomsbury Publishing, 2019).
56 "Introduction to the Recommendations for Animal Welfare," OIE, 2018, https://www.oie.int/fileadmin/Home/eng/Health_standards/tahc/2018/en_chapitre_aw_introduction.htm.
57 John McInerney, "Animal Welfare, Economics and Policy," The National Archives, February 2004, https://webarchive.nationalarchives.gov.uk/ukgwa/20110318142209/http://www.defra.gov.uk/evidence/economics/foodfarm/reports/documents/animalwelfare.pdf.
58 Stephen Hinchliffe, Lenore Manderson, and Martin Moore, "Planetary Healthy Publics After COVID-19," *The Lancet Planetary Health* 5, no. 4 (2021): e230–e236.
59 Anthony Ryan Hatch, *Blood Sugar: Racial Pharmacology and Food Justice in Black America* (Minneapolis: University of Minnesota Press, 2016).
60 Hannah Landecker, "Postindustrial Metabolism: Fat Knowledge," *Public Culture* 25, no. 3 (2013): 495–522.
61 Lauren Berlant, "Slow Death (Sovereignty, Obesity, Lateral Agency)," *Critical Inquiry* 33, no. 4 (2007): 754–780; Rob Nixon, *Slow Violence and the Environmentalism of the Poor* (Cambridge, MA: Harvard University Press, 2013).
62 Hinchliffe et al., *Pathological Lives.*
63 Stephen Hinchliffe, "More than One World, More than One Health: Re-Configuring Interspecies Health," *Social Science and Medicine* 129 (2015): 28–35.

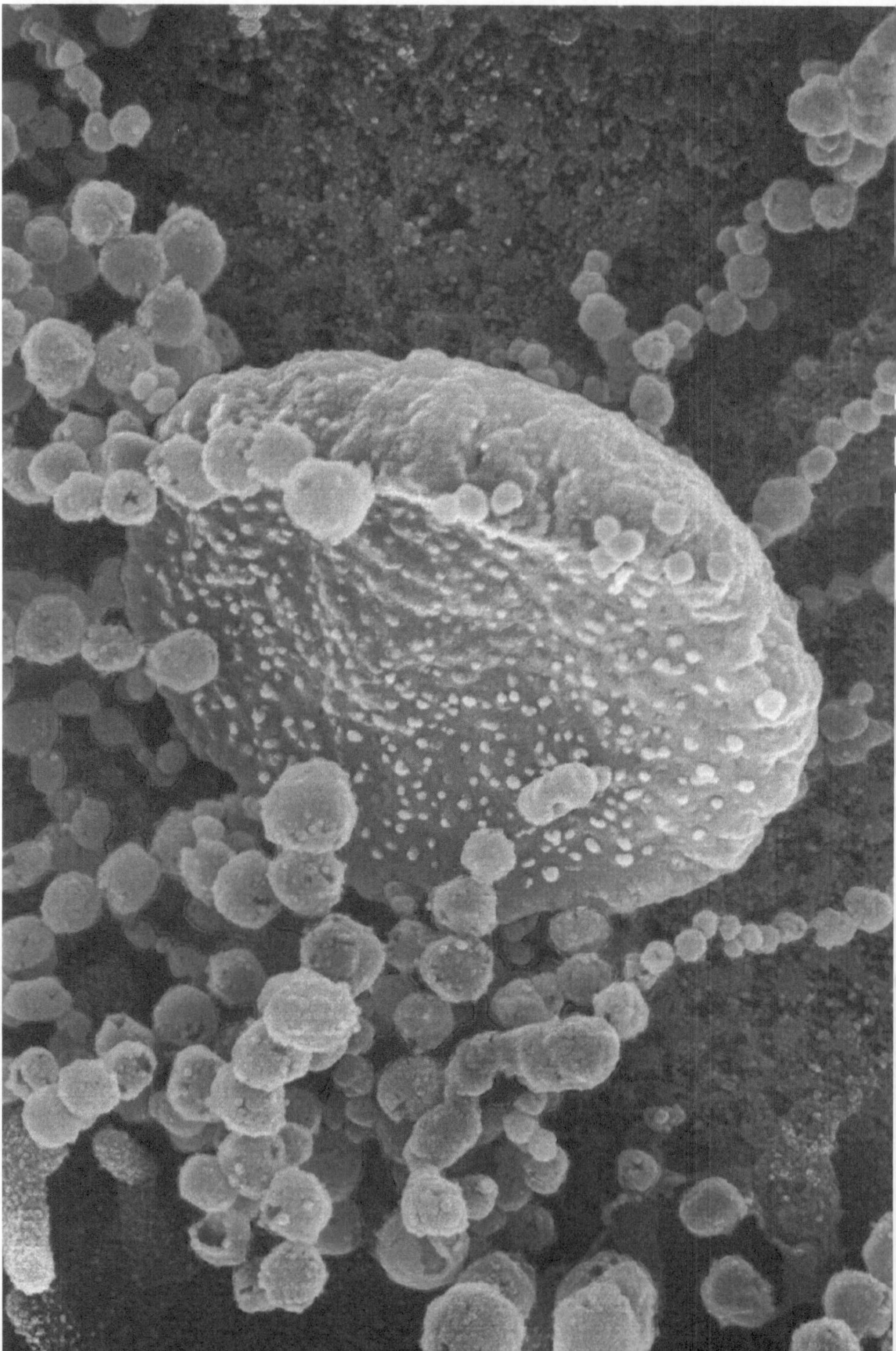

INTRODUCTION

More-than-One Health, More-than-One Governance

Irus Braverman

The concept of One Health rose to prominence in the past decade, particularly since the eruption of the COVID-19 pandemic toward the end of 2019. Its premise is that humans, animals, plants, and the environment form an interdependent system that should be considered in a coordinated, interdisciplinary, and intersectional fashion. Although there is more than one definition of One Health, they all share the basic idea of a triadic human–animal–environment connection. The World Health Organization, a global leader in public human health, recently defined One Health as "an integrated, unifying approach that aims to sustainably balance and optimize the health of people, animals, and ecosystems."[1] There is, indeed, a growing awareness that the three are not only interconnected, but also interdependent—that human, nonhuman animal, and ecological health are relational and "coproduced" in multifold ways.[2] Consider this brief, released by leading global health institutions in 2021:

> The COVID-19 pandemic and other emerging infectious diseases (EIDs), as well as well-established endemic zoonoses, and the continuing threat of antimicrobial resistance (AMR), demonstrate the importance of the connections between the health of animals and humans, as they interact with and within their shared environment, and of the urgent need to address human, animal, and environmental health challenges holistically.[3]

FIGURE I.1 COVID-19 brought One Health to the forefront of the global public health agenda. Here, an electron micrograph scan shows SARS-CoV-2 emerging from cells cultured in the lab. National Institute of Allergy and Infectious Diseases, National Institutes of Health (CC BY 2.0).

DOI: 10.4324/9781003294085-1

One Health proponents thus insist that the global governance of diseases cannot be concerned with human health alone, emphasizing that the health of this planet and all living organisms upon it are also at stake.

This volume provides an insider perspective into One Health, while also challenging One Health from multiple disciplinary orientations. Specifically, the volume assembles prominent scholars from the humanities, the social sciences, and the natural sciences and medicine to explore existing One Health approaches to health. Building on these explorations, we envision modes of health that are both more-than-human and also more sensitive to, and explicit about, One Health's colonial and neocolonial legacies and contemporary practices. The volume is distinctive in its geographic breadth. It travels from Inuit sled dogs in the Arctic to rock hyraxes in Jerusalem, from black-faced spoonbills in Taiwan to street dogs in India, from spittle-bugs on Mallorca's almond trees to jellyfish management at sea, and from rabies in sub-Saharan Africa to massive culling practices in South Korea.

Together, the volume's contributions underscore known, and reveal lesser-known, aspects of One Health, rendering the current challenges of health to a planet in crisis more accessible to wider audiences. The volume's contributions at the same time expose the anthropocentric, Western, and neoliberal properties of global health governance, illuminating the structural biases and power dynamics that are still at play in this context and pointing to ways in which we may seek to decolonize One Health. While endorsing the basic principle of interconnectivity that underlies this approach to health, at the same time we aim here to push the current thinking about One Health so that it is more transparent, relational, transdisciplinary, plural, processual, and just. We refer to this critical orientation toward One Health as "more-than-One Health."[4]

After a brief discussion of the recent history and legacies of One Health, this Introduction will explore three ways to strive for "more-than-One" Health: first, by endorsing a "more-than-human" approach that would insist that One Health depart more fully from its anthropocentric biases and transition toward multibeing forms of justice; second, by employing a "more-than-zoonotic" approach that would urge One Health to shift its focus from emerging infectious diseases—such as West Nile virus, pandemic influenza, and extensively drug-resistant tuberculosis—toward neglected tropical diseases and non-communicable diseases, and even beyond disease into the realm of optimal wellbeing; and third, by embracing a "more-than-science" critique of One Health that would encourage other ways of knowing the world, incorporating insights and methods from the humanities and the social sciences and highlighting Indigenous and local worldviews.

Underlying these various challenges to, and criticisms of, One Health is an overarching concern with its "one world-ism,"[5] which reduces differences and flattens multiplicity in ways that present themselves as holistic but are often reifications of toxic power dynamics. I will end this Introduction with a discussion of the recent call by One Health advocates to chart a Pandemic Treaty that would purportedly promote a more effective regulation, surveillance, and monitoring of health on a global level. The "one-treaty" mindset closely resembles One Health's "one world-ism," only this time the one-ness manifests itself in the governance arena. Whereas the goals of such planetary governance are obviously valid and important, we must also recognize the problems of this approach and introduce safeguards against its misuse. Finally, in its last part, the Introduction will review the volume's structure.

One Health: A Brief Overview

In late 2019, a novel coronavirus—the severe acute respiratory syndrome coronavirus 2, or SARS-CoV-2—began to spread globally, leaving the disease known as COVID-19 in its wake. At the time of writing this Introduction, the SARS-CoV-2 pandemic had resulted in more than 6.2 million confirmed human deaths and 527 million confirmed cases in humans across the globe.[6] Because it was defined as a zoonotic disease that most likely originated in bats, a major reservoir of coronaviruses, SARS-Cov2 has prompted an unprecedented recognition of One Health by political leaders at the highest levels, to the point that it has recently become a buzzword for anticipatory approaches to post-COVID pandemics. Accordingly, on June 13, 2021, One Health was referred to by the heads of G7 countries as "critical to everyone's health." This endorsement was preceded by the formal Carbis Bay G7 Communiqué, which stated, in Article 16, the need to strengthen the One Health approach "across all aspects of pandemic prevention and preparedness, recognizing the critical links between human and animal health and the environment."[7]

For many observers outside and even inside the public health arena, this was a first encounter with One Health. But One Health has been around for a while, and some even trace its central ideas to Greek philosophy.[8] In modern times, multiple terms have been used to refer to the notion of One Health, each capturing a different variant of this idea.[9] The term One Medicine was coined in the early 1960s to highlight the connections between human and animal medicine.[10] Since then, novel initiatives have taken One Medicine into the realm of conservation, resulting in what is often referred to as "conservation medicine." Among the many definitions of conservation medicine, some have offered that it is "a transdisciplinary approach to study the relationships among the health states of humans, animals, and ecosystems to ensure the conservation of all."[11]

The book *Conservation Medicine: Ecological Health in Practice*[12] further defined this new discipline, arguing that "scientists and practitioners in the health, natural, and social sciences [must] think about new, collaborative, transdisciplinary ways to address ecological health concerns in a world affected by complex, large-scale environmental threats."[13] Ten years later, ecologist Thomas Lovejoy wrote in the foreword to *New Directions in Conservation Medicine*:

> We have come to understand that the health of humans, animals (and plants!), and ecosystems are all inextricably intertwined. Indeed, given the

> major disturbance to the biophysical system of the planet itself through climate change and ecosystem destruction and degradation, we also must include the "health" of the biosphere.[14]

Parallel to the rise of conservation medicine, in 2006 the American Veterinary Medical Association (AVMA), together with the American Medical Association, the Centers for Disease Control, and myriad other health-related institutions, sponsored the One Health Initiative—an effort to promote work on health across disciplinary boundaries and species divisions.[15] An early definition of the One Health Initiative stated that it seeks to merge human and nonhuman animal health sciences to benefit both.[16] Along these lines, the AVMA's 2008 executive summary called for a collaborative, holistic, and interdependent approach to health that has a strong academic presence,[17] emphasizing that we are facing "demanding, profound, and unprecedented challenges."[18]

As a concept, One Health has reached particular prominence in the past decade, in part as a response to the perceived rise in emerging and re-emerging infectious diseases, the majority of which are zoonotic, and many either food- or vector-borne or involving animal and environmental health.[19] According to the AVMA, of the 1,461 diseases recognized in humans, approximately 60 percent are due to multi-host pathogens characterized by their "spillover" across species divides. Over the past three decades, approximately 75 percent of emerging human infectious diseases were defined as zoonotic. According to the AVMA: "Our increasing interdependence with animals and their products may well be the single most critical risk factor to our health and wellbeing with regard to infectious diseases."[20]

Whereas past definitions have focused on the animal–human nexus, more recent definitions of One Health include ecosystems and perceive their health as equally, if not more, important than that of human and nonhuman animals.[21] To highlight this difference, disease ecologists working in the field of biodiversity conservation coined yet another term—EcoHealth—not to be confused with Planetary Health, which maintains a focus on human health, only from a global perspective.[22] Australian historians of medicine James Dunk and Warwick Anderson argued in this context that "Planetary Health represents our current response to what might be called the dark side of development, progress, and the 'civilizing process,' a means of contending with the consequences of our species' incessant assault on the planet's life support systems."[23] Anderson therefore suggested that "ecology" is a more adequate method of engaging with the biological configurations and complexities of emerging diseases. He explains:

> Ecology—as a mode of configuring—is a more intricate means of collecting epidemic intelligence; so much so that ecological explanations have tended to displace older sociological insight into economic and political

> factors, which also influence the distribution and abundance of germs, and, hence, patterns of infectious disease.[24]

An important event in the evolution of the One Health approach was the 2004 meeting hosted by the Wildlife Conservation Society in New York, which adopted the Twelve Manhattan Principles.[25] Branded as One World—One Health, the Principles defined cross-sectoral and integrated approaches to health, immediately gaining traction in national and international human and animal health-related arenas.[26] On October 25, 2019, just one month before the reported emergence of COVID-19, the Wildlife Conservation Society, together with other global health leaders, issued the Berlin Principles on One Health, cautioning that this is an "Urgent Call for a United Effort to Stop Diseases Threatening All Life on Earth." The following is an excerpt from the call:

> To address the myriad of health challenges of the twenty-first century, while ensuring the biological integrity of the planet for current and future generations, we need to strengthen existing interdisciplinary and cross-sectoral approaches that address not only disease prevention, surveillance, monitoring, control, and mitigation, but also biodiversity conservation. The quality of current, and future, human and animal health, and well-being, depends on the success (or failure) of humanity's environmental stewardship.[27]

Proponents of the One World—One Health approach have explained that "effectively tackling health at the nexus of climate change, biodiversity loss, and a global pandemic requires public action, shared commitment, and positive identification with others, especially those disproportionately affected by these three global emergencies."[28] As Thomas Lovejoy and others have noted, climate change and ecological degradation contribute to the ill-health of nonhumans and humans alike.[29]

The same period also saw an increase in the use of the "One Welfare" concept,[30] with many calling publicly for its adoption alongside One Health so as "to promote the direct and indirect links of animal welfare to human welfare and environmentally friendly animal-keeping systems under the banner of passion and compassion."[31] Signed by 170 nongovernmental organizations in 2020, the "Animals' Manifesto: Preventing COVID-X" sought to incorporate this approach into policy.[32]

Over the past decade, and especially since the COVID-19 outbreak, One Health has influenced decision-making processes in the World Health Organization, the World Bank, the Food and Agricultural Organization of the United Nations, the World Organisation for Animal Health, and the US Centers for Disease Control and Prevention.[33] By linking human, animal, and environmental health, One Health strategies have been central to discussions about mitigating the next zoonotic outbreak. Its proactive, interdisciplinary, and collaborative approach promises earlier detection of risks and threats, as well as more effective responses, in part by engaging community-level stakeholders in all stages of the process. The next section will revisit and problematize the promise of a "one" One Health in various contexts, after which I will circle back to the promise of a one governance of health.

"More-than" One Health

Stephen Hinchliffe, a British geographer and the author of this volume's Foreword, pointed out elsewhere that "[n]o one can be against a set of discourses and practices that emphasize the shared health of people, animals, and [the] environment."[34] "There's a matter of fact-ness or common sense [about it] that is immediately appealing," he explained.[35] Yet there are also important qualifiers to the "truism of One Health," as he calls it. Alongside the opportunity for transdisciplinary and intersectional cooperation and the welcome recognition of the complexity of health, global health also flattens the world and undermines differences. The argument that there is indeed more than one health is tied to the geographical argument that borders between species are borderlands where many things happen beyond contamination and infection.[36]

Hinchliffe, in other words, criticizes One Health for being a "one-world-ist ontology."[37] He draws, in particular, on the work of science and technology studies (STS) scholar John Law, who argues that "one-world metaphysics are catastrophic."[38] From Law's perspective, the political implications of such an approach are immediate:

> If we live in a single Northern container-world, within a universe, then we might imagine a liberal way of handling the power-saturated encounters between different kinds of people. But if we live, instead, in a multiple world of different enactments, if we participate in a fractiverse, then there will be, there can be, no overarching logic or liberal institutions to mediate between the different realities. There is no "overarching." Instead, there are contingent, local, and practical engagements.[39]

Whereas the "one" in One Health has been depicted by critics as being too narrow to incorporate plural notions of health, others have pointed out that this concept is so broad in its scope that it is hard to imagine anything that it would not include, potentially rendering it meaningless. Wildlife veterinarian Chris Walzer articulated this latter sentiment in our interview for this volume when he told me that: "I often joke that [even] an aspirin is now labeled as a One Health product."[40]

Despite the expansiveness of the term, however, scholars have pointed out that much of the research on One Health is still too anthropocentric—and, relatedly, also too vertebrate-centric and eukaryo-centric—and that it has typically focused, accordingly, on zoonotic and vector-borne diseases and on health services that integrate only human and animal health, without considering their environment. Indeed, although One Health prides itself on aspiring to decenter the human,[41] its practices have not always followed suit. This is possibly most apparent in the context of industrial animal agriculture, which is responsible for the annual killing of tens of billions of terrestrial animals and a trillion aquatic animals and is the focus of animal rights scholar Maneesha Deckha's contribution to this volume.[42] Then there are the mass killings of nonhuman animals as means of disease prevention, discussed in this volume in the context of bird management in Taiwan and Hong Kong,[43] domesticated animals in South Korea,[44] and Inuit sled dogs in Arctic Canada.[45]

This combined discussion of the two first and interrelated "more-than" One Health critiques—namely, the "more-than-human" and the "more-than-zoonotic"—brings me to the third criticism of One Health: this time, for its less-than-ideal inclusion of "nonscientific" discourses. Contemporary stories of rabies control in particular highlight the challenges of decolonizing planetary-scale initiatives like One Health, especially because rabies elimination strategies seem to conflict with traditional ecological knowledges about human–animal relationships. As literary theorist Susan McHugh explains in this volume, the recent inquiry by the Qikiqtani Truth Commission—an Indigenous-led social justice project that began in 2005—into why Inuit sled dogs disappeared from their homeland "provides a poignant window into how cultural and historical factors complicate … the Eurowestern conventional separate-and-exterminate approach to zoonotic disease control."[46] Several contributions to this volume thus invite One Health to carve out more just paths that consider marginalized human and nonhuman lives and habitats through the lens of care[47] and camaraderie.[48]

From a broader perspective, One Health policies deployed thus far have rarely incorporated Indigenous vocabularies and insights about colonial structures and ethnic, racial, or gender power dynamics.[49] For their part, Indigenous scholars have emphasized that "scientists and policy-makers often overlook evidence that the One Health paradigm is already embedded in Indigenous values, worldviews, and laws."[50] These scholars have thus called for a more explicit integration of Indigenous principles and inputs into One Health academic programs and applications.[51] Physician Rupa Marya and political ecologist Raj Patel asserted, along these lines, that "most doctors—most humans, really—have unwittingly inherited a colonial worldview that emphasises individual health, disconnecting illness from its social and historical contexts."[52] They argued, further, that as a result of historical enslavement and appropriation of Indigenous lands and natural food sources, "rates of diabetes, cardiovascular disease and depression skyrocketed in the indigenous communities of Salmon Nation," a region extending from northern California to the north slope of Alaska.[53]

Alongside its failure to adequately include Indigenous and localized forms of knowledge, the said expansiveness of One Health typically stops abruptly at the disciplinary divides, with the social sciences and the humanities strangely absent from most of the One Health research.[54] Instead, and much too often, the content of One Health has been written through the lens of policy documents or with a strong orientation toward the natural sciences. And so, despite its call for interdisciplinarity, One Health has not engaged too closely or too readily with other-than-scientific disciplinary orientations, and especially not with critical studies in the humanities such as the environmental humanities, postcolonial studies, posthumanism, and science and technology studies (STS).

One Health has also not been too successful in integrating the social sciences. Social scientists generally, and anthropologists and scholars in allied subdisciplines (e.g., anthropology of nature, medical anthropology, and human–animal studies) in particular, are well-positioned to respond to the need for in-depth and immersive research that is more attuned to local knowledges.[55] In its sensitivity to places, landscapes, materialities, and the time–space nexus more generally, the discipline of geography, too, could provide a powerful mode of intervention into One Health. The multiple episodes of Ebola crises across West Africa have highlighted the importance of ethnographic work that is sensitive to such local spaces and knowledges and that integrates them into the management of health.[56] Working through multispecies ethnography and more-than-human geographies in particular, some social scientists have stressed the importance of including not only local human knowledges but also other-than-human knowledges of health, which brings us back to the more-than-human challenge to One Health.[57] Giving voice to these knowledges, which have often been framed as obstacles to the rational management of zoonotic outbreaks, would pluralize the ontologies of One Health.

Both methodologically and conceptually, transdisciplinarity is at the heart of this volume, and is indeed this volume's direct response to the recognized tendency of the established One Health discourses to focus on a limited range of expert knowledges. At an immediate level, then, our collection demonstrates how interesting and exciting One Health can be when using the vocabularies, styles, and methods of the humanities and the social sciences. Specifically, colonialism, capitalism, and neoliberalism are key areas of study within the social sciences and the humanities. Viewing interdependencies as historical processes, this volume is especially committed to engaging anticolonial scholarship and other-than-Western perspectives and experiences that problematize capitalist and neoliberal logics. Such transdisciplinary exchanges would arguably encourage multifaceted, nuanced, and multibeing accounts, both from within and from outside of One Health. The problematic governance of the COVID-19 pandemic has indeed revealed the urgent need for recognizing multibeing environmental injustice and its highlighting of the unfair distribution of environmental harms and the burdens placed on marginalized populations, both human and nonhuman.

Along these lines, Women's Studies and STS scholar Susan Merrill Squier suggests in her contribution to this volume that we approach One Health "not as a field but as a process: an interdisciplinary or multidisciplinary mode of drawing together insights relevant to health and welfare from proximate and even distant scientific, social scientific, and humanistic fields."[58] In his Foreword to this volume, Hinchliffe puts it this way: "optimization is not a matter of *being* disease-free, but a relational process in which the *becoming* or dynamics of microbiomes and interspecies relations are key. One Health becomes a matter of working with rather than against environments." For Hinchliffe, then, disease is not only a mechanical issue with causative agents, "but also how an environment and a host play their parts in producing pathogenicity."[59] And if disease is relational, health, too, is a matter of multibeing relations and assemblages.

Whereas not much of One Health has engaged with the humanities and with the social sciences—the reverse is also true: while One Health has made a big splash in the advocacy and policy world, and even more so in the wake of COVID-19, surprisingly little academic literature has been published on this concept in the "nonscientific" world, especially in the humanities and social sciences.[60] Drawing on their insights into scientific logics and modes of classification, posthumanist and STS explorations of One Health would be especially helpful for problematizing the foundational concepts and classifications of One Health.

Specifically, certain studies in the posthumanities have questioned the idea that the body—both that of humans and that of other organisms—is a singular materiality, instead embracing notions of symbiosis, multitude, and the holobiont.[61] Such perspectives emphasize the body as a historically constructed system that is entangled in myriad ways with other bodily and non-bodily systems—medical, political, ecological, and technological.[62] The COVID-19 pandemic has indeed exposed the social, biological, and ecological vulnerabilities that exist on a global scale, inviting and even forcing us to resist rigid notions of material separateness between humans, animals, and the environment.

The problematization of human individuality in particular would bring about a welcome questioning of Western scientific narratives that assume the power, authority, and moral prerogative of humans to control both human and nonhuman lives and environments.[63] Given that the human individual has long served as the subject of liberal societies and the systems of law to which they have given rise, such a critique of human individuality would radically alter the governance of health generally, and definitions of personhood and legal rights alongside matters of copyright and intellectual property in particular.[64]

Applying a posthumanist perspective would also question other central concepts and distinctions utilized by One Health proponents—and here the terms "zoonotic," "pathogen," "emerging infectious diseases,"[65] "spillover,"[66] and "preparedness" come to mind.[67] Consider the term "zoonosis," for example. This term has often promoted a politics of blame, casting certain nonhumans as the culprits and the causes of diseases and, as a result, setting them up as "killable."[68] A contributor to this volume, environmental philosopher Bjørn Ralf Kristensen wrote elsewhere that:

> The term zoonotic, which refers to a disease transmitted from animals to humans, masks the relational elements at the core of this pandemic. Covid

> and zoonotic diseases generally are not truly diseases caused by animals, but rather are rooted in the deeply fraught relationship that humans have with the more-than-human world, and with those at the periphery of our own species.[69]

In this volume, Kristensen makes the case "for a more sympathetic and reflective approach to zoonotic disease mitigation policy that is mindful—not just with respect to the impact that animal health and wellbeing have on humans but also to more-than-human dependencies developed through anthropogenic conditions."[70]

Similarly, in her reflections on One Health through the lens of Graphic Medicine for this volume, Squier offers to reverse the typical zoonotic directionality from animals to humans and to instead assume a nonhuman perspective—in her case, the perspective of Elmer the rooster, among other comics figures. When Elmer and his conspecifics become aware of the horrors of humans' industrial agriculture, they unite and rebel, creating a multispecies world government in which chickens live as equals with humans. Taking such a reverse perspective, referred to by Squier as the "necropastoral," an engagement with art highlights the metaphoric potential of zoonosis, "exposing the negative effects that have already spread from humans to the other life forms around us."[71]

Likewise, in her contribution to this volume, political ecologist Emily Reisman urges us to reconsider conventional views on pathogens used widely by scientists and policymakers, which position pathogens as an outside threat. For her, "One Health must go beyond pathogens and parasites to address the socio-economic and political factors impacting the wellbeing of all."[72] Using her case study of Spain's *Xylella* plant disease epidemic, Reisman argues, specifically, that this epidemic was produced "not only by a bacterium, but also by the conditions of possibility created by tourism, unstable land tenure, histories of marginalization, and retreat of government from farm advising."[73] Viewing plant epidemics from the vantage of care, Reisman demonstrates that the "pathogen" is only part of the picture and advocates avoiding "a good–bad dichotomy, asking difficult questions about how to care."[74] Care is not a shorthand for universal values, Reisman clarifies, and in this sense, too, she departs from One Health's globalism. Rather, care is "a call to give relational maintenance work the political weight it is due."[75] Similarly, in their contribution on oceanic health, human–environment geographers Elizabeth R. Johnson and Hannah Dickinson eschew assigning the blame of pathogenicity to particular organisms, suggesting that "rather than 'good' or 'bad' organisms, the entanglements that make up more-than-human relations can become healthful or pathological, depending on composition."[76] Hinchliffe put it succinctly in his Foreword to this volume: "it's time to loosen the obsession with pathogenic microbes, and consider what is driving pathological lives."[77]

A final example for how transdisciplinary engagements can shed light on and bring pause to One Health's conventional use of concepts and categories regards the basic distinction between multicellular eukaryotic and single-cell prokaryotic organisms. In his Afterword to this volume, Warwick Anderson argues that One Health relies on typologies of living beings that are problematic from both biological and historical perspectives. In his words: "It may be a useful exercise to try to imagine the 'animal' otherwise, as borderlands not boundary, as co-immune not immune, as interconnected and networked, hybrid and cyborg."[78] Whereas some propose the phrase "multispecies health" as an alternative to One Health's monist and animalistic approach, Anderson argues that species typologies are problematic too, because they occlude the full variety of ecological interrelationships. As he puts it: "this persistence of typological thinking in One Health, 'black-boxing' animals and species as boundary *objects*, distinct from evolutionary *processes*, surprises and disturbs me."[79]

And if the categories of the "human" and the "animal" cannot stand on their own, so too is the case with the category "environment" for many of the same reasons. Such posthumanities-inspired perspectives, then, would question the triadic approach at the heart of One Health—namely, the practice of situating care at the interface of human, animal, and the environment. Employing the insights of posthumanism as well as of critical animal studies and multispecies approaches in the social sciences, this volume aspires to expand the conceptual spaces of One Health, with immediate implications for our understandings of what is human, what is animal, and what is the environment as well as what would be the most suitable regulatory structures and institutions to govern these categories.

The Pandemic Treaty: Ushering "More-than" One Governance

The spread of COVID-19 around the world saw a flurry of monographs—most notably, Mark Honigsbaum's *The Pandemic Century* (2019),[80] Lyle Fearnley's *Virulent Zones* (2020),[81] Frédéric Keck's *Avian Reservoirs* (2020),[82] and Gavin Andrews et al., *Covid 19 and Similar Futures* (2021).[83] Like our collection, these studies are timely in that they help us understand the global response to, and the mobilization around, the pandemic. These monographs also indicate the heightened contemporary interest in pandemic-related literature and in critical studies of governance. It is toward such a critique of governance that the Introduction will now turn.

Today, a wide scope of fields—including, but not limited to, comparative medicine, public health, environmental sciences, biochemistry, and plant pathology—define themselves under the umbrella of One Health, and its integrative approaches are institutionalized in a variety of international organizations. One Health's overwhelming focus on emergency governance has manifested in a hyper focus on preparedness by these organizations. Investigating the history of preparedness as a set of governmental techniques for approaching uncertain threats, sociologist Andrew Lakoff distinguishes between a "potential threat," which he defines as a regularly occurring event with a probability that "can be calculated based on known patterns of historical incidence," and an "unprecedented event" that can only be managed by "imaginative enactment."[84]

For example, whereas the swine flu outbreak of 1976 was managed by government officials through the existing public health framework, three decades later a new regime of public health preparedness has emerged to address the avian influenza threat of 2006 that harnessed techniques of imaginative enactment such as scenario-based exercises and the stockpiling of vaccines. Not even one decade later, the failure of global health governance to manage the Ebola crisis of 2014 led to widespread criticisms of the World Health Organization and other public health authorities. "The demand was for more and better preparedness, in anticipation of the next emergency."[85] Within a span of a few decades, then, preparedness has become the prevalent form of global health governance for future pandemics. Such a focus on preparedness is part of a wider globalization and securitization of health, a focus that redefines disease outbreaks not only as a threat to local human health but also as a global health concern that poses a threat to world security.[86]

In December 2021, four major international health organizations—the Food and Agriculture Organization of the United Nations (FAO), the World Organisation for Animal Health (formerly OIE, now WOAH), the United Nations Environment Programme (UNEP), and the World Health Organization (WHO)—embarked on negotiations toward a Pandemic Treaty. The 2021 brief released to the public articulates the urgent need for such a treaty, explaining that there is not enough integrative governance, not enough sharing of information, and not enough surveillance and preparedness, effectively aiming to amend the separation between the myriad health apparatuses. According to the brief: "despite growing attention to [One Health] at all levels of governance, there is emerging consensus that [One Health] principles remain insufficiently embedded in existing treaties and institutions."[87] The brief lists, in particular, three limitations to the current One Health mode of governance. First, it states that while there is significant focus on health emergency response, there has been too little focus "on proactive prevention efforts at the human–animal–environment interface."[88] Second, it laments that "existing governance mechanisms are siloed, with inadequate attention to the key [One Health] principle of coordinated multisectoral action to safeguard human, animal, and environmental health."[89] Third and finally, it points out that "there is insufficient global solidarity, including an unmet need for redistributive mechanisms to support One Health implementation in [low- and middle-income countries] before outbreaks or other crises occur."[90]

The proposed One Health Pandemic Treaty would include massively integrated surveillance and monitoring systems to gather and merge "data that identify risk factors for disease emergence in wildlife, companion animals, livestock, the environment (e.g., soil and water), and humans."[91] Global entities like the World Health Organization Hub for Pandemic and Epidemic Intelligence, formed in September 2021 to gather data from government, public, and private sector sources, would act as central clearinghouses for data and would perform the integrated analyses essential for effective early warning and active pandemic response systems.[92]

Ghanaian public health official John H. Amuasi is a leading advocate for the One Health Pandemic Treaty. Drawing on his leadership role in global pandemic governance, he explains in this volume the importance of comprehensive surveillance and monitoring. In his words: "surveillance entails a lot of looking. I think people get tired of looking—why are you looking when there's nothing? But that's exactly the point: we must be integrating these surveillance systems into life as it is."[93] At the same time, Amuasi is also concerned about the dangers inherent in this mode of power: "I worry … about the safety of the data and who has access to it and how it will be used. I don't have the answer to that."[94] Although preparedness and even ecology may be part of this move toward new forms of governance, the increasing cooptation of these discourses by Big Pharma and neocolonial projects must be considered when moving forward.

Furthermore, while we recognize that surveillance can be suitable for certain situations, several of the contributors to this volume suggest undertaking a radically different perspective to pandemic governance: rather than a shallow "collation" of what makes "us" all ill, replete with quantitative calculi that are easily translated into rules, we propose here a deep dive into the pluralities and processes that render health a fluid project involving complex and fluctuating assemblages. The governance of health, in other words, would need to find new methods and modalities and an expanded regulatory toolkit. Such a toolkit would include, for example, "different forms and formats of evidence; adopting, where appropriate, precautionary approaches to regulation; [and] adopting open forms of monitoring and data generation that can trigger reviews of processes or products that were previously considered safe or inconsequential."[95]

The current focus on one treaty, for one world, with one health, brings me back to the fundamental critique expressed at the outset: that One Health too often pursues a "one-ism" agenda, both on the ontological front and in the policy and governmental arenas, thus ignoring bottom-up ways of relating to multibeings. Doing more work to connect One Health with more-than-human frameworks is thus a timely and valuable undertaking, as are the efforts to make visible some of the capitalist, neoliberal, and neocolonial logics at play within the current articulations, methods, and practices of One Health. Recognizing One Health's historical and contemporary rootedness in systems of imperial exploitation, this volume begins to identify the work that is required to decolonize One Health and actively pluralize it.[96] This work is particularly prescient when global health institutions tend to look to One Health as a silver-bullet solution to global problems.

This Volume's Structure: An Outline

More-than-One Health includes voices that are firmly situated within the One Health community: Laura H. Kahn is a medical physician by training and a co-founder of the One Health Initiative; Chris Walzer is a veterinarian and Executive Director of Health in the Wildlife Conservation Society, an environmental organization that has played an important role in ushering One Health beyond the human–animal nexus into the domain of ecological health; and John H. Amuasi is a prominent One Health leader, the director of the African Research Network, and the co-editor of *The Lancet*'s One Health Commission, and has advocated for broadening the One Health focus to include neglected tropical diseases as well as non-communicable diseases. Joining these One Health experts is a diverse set of scholars from multiple disciplines and geographies, which can be roughly divided into the environmental humanities (Bjørn Ralf Kristensen, Elizabeth R. Johnson, Hannah Dickinson), the posthumanities (Susan Merrill Squier, Emily Reisman, Stephen Hinchliffe), critical animal studies (Maneesha Deckha), science and technology studies and medical humanities (Abigail Woods, Kiheung Kim, Myung-Sun Chun, Warwick Anderson), as well as cultural anthropology and multispecies ethnography (Frédéric Keck, Susan McHugh, Deborah Nadal, Irus Braverman).

The volume's thirteen contributions are grouped into four parts: situating one health, expanding one health, othering one health, and decolonizing one health. Before we embark on this four-part journey, however, a few words about Stephen Hinchliffe's Foreword to this volume. Entitled "The Lure of One Health," Hinchliffe's contribution argues that the trap (that other sense of a lure) of One Health is its tendency to conceal some important questions and insist on all-or-nothing calculations of health. Starting with the emblematic poster child for the success of One Health—rabies control in sub-Saharan Africa—and then advancing to less clear-cut examples, Hinchliffe outlines the case for, as well as the limitations of, One Health. Setting the stage for the other contributions, he asks: what is gained and what might be lost when one adopts the One Health signature? At times, living with diseases and adapting ourselves to those diseases, and to other environmental challenges, offer more sustainable pathways compared to disease-free solutions. For him, then, "One Health is not an absolute state (disease freedom)—it is a process, an approximation, where the optimum may involve several suboptima within the components."[97]

If Hinchliffe's contribution is forward looking, then Abigail Woods kicks off Part I of the collection—"Situating One Health: Histories and Practice"—with a medical history of One Health that goes back to the eighteenth and nineteenth centuries. In her chapter, entitled "One Health: A 'More-than-Human' History," Woods points to the basic assumption of contemporary One Health: that human and veterinary medicine are discrete domains devoted to different species, whose separation must be overcome to achieve health benefits for all. Her contribution problematizes this assumption by demonstrating that, until recently, their boundaries were in fact extremely fluid. Referring to specific examples from the period between 1790 and 1900, Woods shows that human medicine was once deeply zoological and encompassed a host of species, practices, and social relations that overlapped with those of veterinary medicine. In time, the boundaries between medical physicians and veterinarians hardened, resulting in the contemporary efforts to transcend these boundaries through the banner of One Health.

On the heels of Woods' contribution, the following three chapters move from history to contemporary practice, presenting three different insider takes on One Health that illustrate the divergent approaches even within this arena. In Chapter 2, "The Case for a One Health Approach from a Physician's Perspective," Laura H. Kahn examines the topic of fecal waste through the nexus of food safety and security, antimicrobial resistance, and climate change. Humans and their domesticated livestock produce at least four trillion kilograms of fecal waste annually, she notes. This fecal waste then not only contaminates food, water, crops, soil, and the atmosphere, but also worsens antimicrobial resistance and enhances climate change. Kahn's study of fecal waste supports her argument that understanding the root causes of health concerns is essential for developing effective and equitable public policies under One Health.

Chapters 3 and 4 are presented in the form of interviews that I carried out with two prominent leaders of One Health. As a wildlife veterinarian who has worked for most of his career in central and southeast Asia, Chris Walzer reflects in Chapter 3 on "Spillover Interfaces from Wuhan to Wall Street." From Walzer's ecological perspective, the increasing incidence of zoonotic viral "spillover" events such as COVID-19 is a symptom of ailing Planetary Health. As human activities and encroachment increasingly undermine the integrity of naturally balanced ecosystems, he argues, environmental health and resilience are compromised. For Walzer, the current three entwined emergencies of public health, biodiversity loss, and climate change clearly illustrate the impossibility of protecting human health in isolation from the health of other animals and the environment. At the same time, these three emergencies also highlight the need to incorporate solidarity and environmental justice into the One Health agenda.

In Chapter 4, "One Health, Surveillance, and the Pandemic Treaty," John H. Amuasi explores the formation of the quadripartite One Health partnership in the context of COVID-19. Explaining the notion of global interdependence and its importance for governing climate change and neglected tropical diseases, Amuasi also considers the human barriers to progress in the regulation of these areas. Finally, he discusses the importance of a Pandemic Treaty and of surveillance, prevention, and preparedness for the future governance of pandemics.

After outlining the diverse ideas and agendas behind the current practices of One Health, Part II of the volume, "Expanding One Health: Beyond the Human–Animal–Environment Triad," considers spaces and materialities that have typically not been included under the One Health umbrella, specifically calling to expand One Health by considering oceans, plants, and art. Accordingly, in Chapter 5, "Between Healthy and Degraded Oceans: Promising Human Health through Marine Biomedicine," Elizabeth R. Johnson and Hannah Dickinson turn our attention toward ocean health. Arguing that oceans have been relatively marginal in One Health discourses, their contribution explores how the uses of jellyfish and shrimp have evolved under a One Health mode of thinking, producing novel ways of consuming these organisms through extractivist neoliberal logics. Specifically, they show how, in response to a rising frequency of jellyfish blooms, their bodies are reimagined as the food and pharmaceuticals of the future and revalorized as health-promoting rather than damaging. Similarly, the shrimping industry is redirecting its indigestible by-product—shrimp shells—into an array of sustainable innovations including chitosan: a biomaterial which could purportedly help to ameliorate the "obesity" crisis in the West.

Chapter 6, entitled "More-than-Almonds: Plant Disease and the Politics of Care," is similarly critical toward the neoliberal discourses underlying many One Health approaches, which manifest in this case in the aggressive tourism industry in Mallorca, the largest of Spain's Balearic Islands. In this contribution, Emily Reisman details the case of *Xylella fastidiosa*, an introduced bacterium enabled by a spittle-bug vector and emboldened by climate change, depicting how this disease has infected almond trees throughout the island to the point that they are expected to perish within five years. Although plant diseases are well known for their world-shaping impacts on human, animal, and ecological health, they have received little attention in One Health. Building on feminist theories of care, Reisman calls to examine the relational conditions of possibility for disease and to make visible the more-than-human maintenance work which is required so that our interconnected lives may persist.

Next, in Chapter 7, "What Can Graphic Medicine Contribute to One Health?," Susan Merrill Squier argues that works of Graphic Medicine—comics about illness, medical treatment, and disability—are powerful vehicles for spreading the One Health message. Furthermore, she argues that Graphic Medicine could provide a template for moving beyond a unitary understanding of One Health, thus decolonizing health care. Using images as well as words, comics can present a multiscaled understanding of health, bridging the disciplinary divides between the life sciences and human and veterinary medicine. Through her explorations of comics, Squier identifies three ways in which microbes affect health: zoonoses—illnesses caused by microbes that pass between humans and animals; zooeyia—feelings of joy or wellbeing arising from the connection between humans and animals; and zooambivalence—the mingled positive and negative impacts of the microbes connecting humans and animals.

From an expansion of One Health toward alternative spaces and domains, Part III of this volume—"Othering One Health: Toward Multibeing Justice"—moves to consider alternative normative trajectories of care, solidarity, and justice and to envision their practical implementation in One Health. This third part kicks off with Chapter 8, entitled "The One Health Initiative and a Deeper Engagement with Animal Health and Wellbeing: Moving Away From Animal Agriculture." Maneesha Deckha asks in this contribution: what does a commitment to One Health mean in the context of the unprecedented levels of the breeding of animals and their deliberate exposure to disease, disability, and death in land-based agriculture? Such a commitment to One Health, Deckha argues, would acknowledge that commercially-intensive animal agriculture is incompatible with the One Health vision, planning in turn for an accelerated transition away from intensive farming practices. She thus calls on the One Health community to commit itself to an ethical vegan outlook or, at the very least, to the goal of eliminating the leading land-based industry that maims and kills animals.

Chapter 9 is entitled "Can Camaraderie Help Us Do Better than Compassion and Love for Nonhuman Health? Some Musings on One Health Inspired by the Case of Rabies in India." In this contribution, cultural and medical anthropologist Deborah Nadal draws on the heated debate in India about whether and how to coexist with free-roaming dogs to question the usefulness of compassion and love as guiding policy principles. She then advocates for "interspecies camaraderie"—namely, a spirit of empathy, respect, and support among beings with a shared experience of life. In addition to the conceptual discussion, Nadal's aim in this chapter is practical: to propose the addition of interspecies camaraderie to the list of One Health values guiding our actions toward the achievement of optimal health outcomes for all.

Closely aligned with Deckha's vegan ecofeminist approach and with Nadal's vision of camaraderie, in Chapter 10, "Anthrodependency, Zoonoses, and Relational Spillover," Bjørn Ralf Kristensen draws on John Dewey's philosophy of moral deliberation to bring about an adaptive flexibility to complex situations that defy habitual understandings of wild animals. Documenting two case studies—rock hyraxes carrying the zoonotic disease leishmaniasis in Jerusalem and the decline of seabird populations on a Swedish island in the Baltic Sea during the COVID-19 pandemic—this contribution makes the case for a more sympathetic and reflective approach to zoonotic disease mitigation policies. For Kristensen, this approach must be mindful not just of the impact that animal health and wellbeing has on humans but also of more-than-human dependencies developed through anthropogenic conditions—what he refers to as "anthrodependency."

If Part III of the volume is mainly concerned with how to do justice toward other-than-humans, then its final part, Part IV—"Decolonizing One Health: Toward Postcolonial and Indigenous Knowledges"—brings humans back into focus to consider the ways in which multibeing vulnerabilities and violence are intrinsically linked. Opening this part is Chapter 11, entitled "Birds as Sentinels of the Environment in Hong Kong and Taiwan." In this contribution, anthropologist Frédéric Keck discusses the enrollment of birdwatchers in the monitoring of avian influenza in Hong Kong and Taiwan. Using the concept of "sentinel" as the imagination of vulnerabilities shared by humans and animals, Keck explores how birdwatching societies in Hong Kong and Taiwan have shifted from a military model of biosecurity to a democratic model of biodiversity. For him, the democratic collaboration between birdwatchers across political borders provides an example for how One Health can integrate transnational modes of animal surveillance as an alternative to animal killing as a biosecurity intervention.

Next, in Chapter 12, "The Spatialization of Diseases: Transferring Risk onto Vulnerable Beings," science and technology studies scholars Kiheung Kim and

Myung-Sun Chun discuss how preventive culling has become the preferred way of controlling infectious zoonotic diseases in South Korea. In the last decade alone, over 74 million domestic animals were culled by the South Korean government for the purpose of preventing foot and mouth disease, avian influenza, and African swine fever. While this preventive measure has been controversial, it was justified by the South Korean government as the scientific and economic way of ensuring biosecurity for both humans and animals. In their intervention, Kim and Chun argue that such culling practices are not only violent toward nonhuman animals but also impute the risk to vulnerable humans such as foreign workers and untrained temporary workers. Examining the historical origins of Korea's authoritarian and centralized sanitary policy in the Japanese colonial era of the late nineteenth and early twentieth centuries, they propose that the colonial strategies of controlling cholera, Manchurian plague, and rinderpest played an important role in shaping South Korea's current centralized and space-based modes of control. The authors thus call on One Health to break from such longstanding legacies of colonial and neoliberal governance that are embedded within public health systems the world over.

In the volume's last chapter, entitled "Rabies on Ice: Learning from Interspecies Suffering in Arctic Canada," Susan McHugh picks up the discussion about colonial legacies in health, this time in the context of settler colonial policies of rabies control in Arctic Canada. She argues that for cultures where the arrival of rabies coincided with settler colonialism, One Health practices were never as simple as extending care to animals. Building on the Qikiqtani Truth Commission's inquiry into the colonial massacre of qimmit, or Inuit sled dogs, her contribution explores how cultural and historical factors have complicated both settler colonial "separate-and-exterminate" logics as well as the traditional ecological knowledges of disease control. Successful new strategies of managing rabies in the region are causing a shift from One Health to EcoHealth, she argues, and thus to more site-specific and multispecies understandings of health.

Warwick Anderson's Afterword, "Among Animals, and More: One Health Otherwise," brings this multivocal volume to a conclusion. In this final contribution, Anderson asks us to consider what, or who, counts as an "animal" in One Health. The field, he argues, remains "oriented around a kind of vertebrate charisma, or even eukaryotic glamor, that limits its organismal reach and ecological range."[98] Specifically, Anderson laments that One Health still regards microbes as transmissible items of no particular value. From his perspective as well as from that of many of the other contributors to this volume, the imperial and settler colonial foundations of contemporary One Health have resulted in a greater sensitivity to human entanglements with a limited range of charismatic animals and celebrity species. At the same time, they have suppressed other ways of appreciating the relatedness of organisms and environments. This volume seeks to highlight such failings in order to encourage modes of governing health that celebrate the diversity of beings and modes of knowledge, thereby moving beyond One Health into the realm of many more-than-one healths.

Notes

1 "Tripartite and UNEP Support OHHLEP's Definition of 'One Health,'" World Health Organization, December 1, 2021.
2 Sheila Jasanoff, *States of Knowledge: The Co-Production of Science and Social Order* (London: Routledge, 2004).
3 Global Health Centre Policy Brief I, "One Health as a Pillar for a Transformative Pandemic Treaty" (Geneva: Global Health Centre, 2021), 5.
4 In using this concept, I draw on Stephen Hinchliffe, "More than One World, More than One Health: Re-Configuring Interspecies Health," *Social Science and Medicine* 129 (2015): 28–35.
5 John Law, "What's Wrong with a One-World World?," *Heterogeneities.net*, September 25, 2011, 1–14.
6 "COVID-19 Dashboard (Center for Systems Science and Engineering (CSSE) at Johns Hopkins University (JHU)," accessed May 26, 2022, https://www.arcgis.com/apps/dashboards/bda7594740fd40299423467b48e9ecf6. See also Elizabeth Lunstrum et al., "More-than-Human and Deeply Human Perspectives on Covid-19," *Antipode* 53, no. 5 (2021): 1503–1525.
7 "Carbis Bay G7 Summit Communiqué," The White House, June 13, 2021, https://www.whitehouse.gov/briefing-room/statements-releases/2021/06/13/carbis-bay-g7-summit-communique/.
8 Christos F. Kleisiaris, Chrisanthos Sfakianakis, and Ioanna V. Papathanasiou, "Health Care Practices in Ancient Greece: The Hippocratic Ideal," *Journal of Medical Ethics and History of Medicine* 7 (2014): 6.
9 Jakob Zinsstag et al., "From 'One Medicine' to 'One Health' and Systemic Approaches to Health and Well-Being," *Preventive Veterinary Medicine* 101, nos. 3–4 (2011): 148–156; Jakob Zinsstag et al., "Mainstreaming One Health," *EcoHealth* 9, no. 2 (2012): 107–110.
10 Zinsstag et al., "From 'One Medicine.'" See also Sharon L. Deem, "Conservation Medicine to One Health: The Role of Zoologic Veterinarians," in *Fowler's Zoo and Wild Animal Medicine Current Therapy, Volume 8*, eds. Eric R. Miller and Murray E. Fowler (Saint Louis: Elsevier, 2015), 698–703, 699.
11 Ibid., 699.
12 A. Alonso Aguirre et al., *Conservation Medicine: Ecological Health in Practice* (Oxford: Oxford University Press, 2002).
13 A. Alonso Aguirre, Gary M. Tabor, and Richard S. Ostfeld, "Conservation Medicine: Ontogeny of an Emerging Discipline," in *New Directions in Conservation Medicine: Applied Cases of Ecological Health*, eds. A. Alonso Aguirre, Richard S. Ostfeld, and Peter Daszak (New York: Oxford University Press, 2012), 3–16, 3.
14 Thomas Lovejoy, Foreword to *New Directions in Conservation Medicine: Applied Cases of Ecological Health*, eds. A. Alonso Aguirre, Richard S. Ostfeld, and Peter Daszak (New York: Oxford University Press, 2012), xi–xii, xii.
15 Laura H. Kahn et al., "History of the One Health Initiative Team and Website (April 2006 through September 2015)," One Health Initiative, February 4, 2021, https://onehealthinitiative.com/history-of-the-one-health-initiative-team-and-website-april-2006-through-september-2015-and-the-one-health-initiative-website-since-october-1-2008-revised-to-june-2020-and-again-to-date-februar/.
16 Deem, "Conservation Medicine," 699.
17 Lonnie J. King et al., "Executive Summary of the AVMA One Health Initiative Task Force Report," *Journal of the American Veterinary Medical Association* 233, no. 2 (2008): 259–261, 260. See also "A Perspective … Some Significant Historic Inspirations for Advancing the One Medicine/One Health Concept & Movement—19th, 20th and 21st Centuries," One Health Initiative, June 8, 2020, https://onehealthinitiative.com/a-perspective-some-significant-historic-inspirations-for-advancing-the-one-medicine-one-health-concept-movement-19th-20th-and-21st-centuries/; Bruce Kaplan,

"'One Medicine-One Health': An Historic Perspective," One Health Initiative, accessed May 30, 2022, https://onehealthinitiative.com/wp-content/uploads/2021/02/One-Medicine-One-Health-An-Historic-Perspective-FEB1-2021-1.pdf; Kahn et al., "History of the One Health Initiative Team."

18 King et al., "Executive Summary," 260.

19 Sarah B. Paige et al., "Uncovering Zoonoses Awareness in an Emerging Disease 'Hotspot'," *Social Science and Medicine* 129 (2015): 78–86, 78; Louise H. Taylor, Sophia M. Latham, and Mark E. J. Woolhouse, "Risk Factors for Human Disease Emergence," *Philosophical Transactions of the Royal Society of London* 356, no. 1411 (2001): 983–989; Hinchliffe, "More than One World, More than One Health."

20 King et al., "Executive Summary," 261.

21 Paige et al., "Uncovering Zoonoses"; Taylor, Latham, and Woolhouse, "Risk Factors"; Hinchliffe, "More than One World."

22 Henrik Lerner and Charlotte Berg, "A Comparison of Three Holistic Approaches to Health: One Health, EcoHealth, and Planetary Health," *Frontiers in Veterinary Science* 4, no. 163 (2017): 1–7.

23 James Dunk and Warwick Anderson, "Assembling Planetary Health: Histories of the Future," in *Planetary Health: Protecting Nature to Protect Ourselves*, eds. Samuel Myers and Howard Frumkin (Washington, DC: Island Press, 2020), 17–35.

24 Warwick Anderson, "Think Like a Virus," Public Books, January 7, 2021, https://www.publicbooks.org/think-like-a-virus/.

25 "About Us: The Manhattan Principles," One World—One Health, accessed May 1, 2022, https://oneworldonehealth.wcs.org/About-Us/Mission/The-Manhattan-Principles.aspx.

26 "Contributing to One World, One Health: Strategic Framework for Reducing Risks of Infectious Diseases at the Animal-Human-Ecosystems Interface," FAO et al., 2008, https://www.preventionweb.net/files/8627_OWOH14Oct08.pdf; Diane Frank, "One World, One Health, One Medicine," *Canadian Veterinary Journal* 49, no. 11 (2008): 1063–1065; Wildlife Conservation Society, "One Health in Action," November 2020, https://c532f75abb9c1c021b8c-e46e473f8aadb72cf2a8ea564b4e6a76.ssl.cf5.rackcdn.com/2020/12/07/3y67zbyeuj_One_Health_in_Action_final.pdf, 4.

27 Wildlife Conservation Society, "One Health in Action."

28 Kim Gruetzmacher et al., "The Berlin Principles on One Health—Bridging Global Health and Conservation," *Science of the Total Environment* 764, no. 10 (2021): 1–4, 3.

29 Thomas E. Lovejoy, Lee Hannah, and Edward O. Wilson, eds., *Biodiversity and Climate Change: Transforming the Biosphere* (New Haven: Yale University Press, 2019); Thomas E. Lovejoy, "To Fight for a Living Planet, Restore its Biology," *New Security Beat* (blog), February 25, 2020, https://www.newsecuritybeat.org/2020/02/fight-living-planet-restore-biology/.

30 One Welfare Phoenix Advisory Board, "One Welfare Phoenix—Supporting the Dairy Industry to Recognise the Interconnections Between Animal and Human Abuse and Neglect," *IDF Animal Health Report* no. 14 (2020): 34; Rebeca García Pinillos, "One Welfare Impacts of COVID-19—A Summary of Key Highlights within the One Welfare Framework," *Applied Animal Behavior Science* 236 (2021): 105262.

31 Quoted from "One Welfare," 2022, https://www.onewelfareworld.org/nye2020.html. See also Joann M. Lindenmayer and Gretchen E. Kaufman, "One Health and One Welfare," in *One Welfare in Practice: The Role of the Veterinarian*, ed. Tanya Stephens (Boca Raton: CRC Press, 2022), 1–30.

32 See "Animals' Manifesto," World Federation for Animals, 2020, https://wfa.org/animals-manifesto/.

33 Maria Knight Lapinski, Julie A. Funk, and Lauren T. Moccia, "Recommendations for the Role of Social Science Research in One Health," *Social Science and Medicine* 129 (2015): 51–60, 51–52.

34 Hinchliffe, "More than One World," 28.
35 Ibid.
36 See also Warwick's Afterword, this volume.
37 Hinchliffe, "More than One World," 34.
38 Law, "What's Wrong," 9.
39 Ibid., 2; see also Annemarie Mol and John Law, "Regions, Networks and Fluids: Anaemia and Social Topology," *Social Studies of Science* 24, no. 4 (1994): 641–671.
40 See my interview with Walzer in Braverman, this volume, 58.
41 On the problems of such a decentering of the human, see Lunstrum et al., "More-than-Human and Deeply Human."
42 Deckha, this volume.
43 Keck, this volume.
44 Kim and Chun, this volume.
45 McHugh, this volume.
46 McHugh, this volume, 252.
47 Reisman, this volume.
48 Nadal, this volume.
49 Joe Copper Jack et al., "Traditional Knowledge Underlies One Health," *Science* 369, no. 6511 (2020): 1576; Lunstrum et al., "More-than-Human and Deeply Human."
50 Copper Jack et al., "Traditional Knowledge Underlies One Health."
51 Ibid.
52 Quoted in a review of Rupa Marya and Raj Patel, *Inflamed: Deep Medicine and the Anatomy of Injustice* (New York: Farrar, Straus and Giroux, 2021), in *The Guardian*, August 17, 2021, https://www.theguardian.com/books/2021/aug/17/inflamed-by-rupa-marya-and-raj-patel-review-covid-race-colonialism.
53 Ibid.
54 Karsten Hueffer et al., "One Health in the Circumpolar North," *International Journal of Circumpolar Health* 78, no. 1 (2019): 1–10, 1; Lapinski, Funk, and Moccia, "Recommendations for the Role of Social Science," 53; Meike Wolf, "Is There Really Such a Thing as 'One Health'? Thinking about a More-than-Human World from the Perspective of Cultural Anthropology," *Social Science and Medicine* 129 (2015): 5–11, 6; Hinchliffe, this volume.
55 Nicolas Lainé and Serge Morand, "Linking Humans, Their Animals, and the Environment Again: A Decolonized and More-than-Human Approach to 'One Health,'" *Parasite* 27 (2020): 55. https://doi.org/10.1051/parasite/2020055.
56 Emilie Venables and Umberto Pellecchia, "Engaging Anthropology in an Ebola Outbreak: Case Studies from West Africa," *Anthropology in Action* 24 (2017): 1–8.
57 Ibid.
58 Squier, this volume, 138.
59 Hinchliffe, this volume, xxx.
60 That said, two scholarly projects in the social science and humanities stand out for their engagement with One Health. First, Abigail Woods et al., *Animals and the Shaping of Modern Medicine: One Health and Its Histories* (London: Palgrave Macmillan, 2018); and second, Susan Craddock and Steve Hinchliffe, "One World, One Health? Social Science Engagements with the One Health Agenda," *Social Science and Medicine* 129 (2015): 1–4. While the first takes a medical humanities approach that is mostly historical, the second focuses mainly on biosecurity. See also Jakob Zinsstag et al., eds., *One Health: The Theory and Practice of Integrated Health Approaches* (Wallingford: CABI, 2021), which contains a few interdisciplinary contributions although it is primarily aimed toward the natural sciences.
61 See, e.g., Scott F. Gilbert, Jan Sapp, and Alfred I. Tauber, "A Symbiotic View of Life: We Have Never Been Individuals," *The Quarterly Review of Biology* 87, no. 4 (2012): 325–341; see also Hinchliffe, this volume.
62 As William Viney, Felicity Callard, and Angela Woods argue in their editorial introduction to the issue, the medical humanities have been dominated by problematic

humanist "models of the self, of the ill and suffering body, and of modes of intervention and care." William Viney, Felicity Callard, and Angela Woods, "Critical Medical Humanities: Embracing Entanglement, Taking Risks," *Medical Humanities* 41, no. 1 (2015): 2–7, 2. See also Lucinda Cole, "Zoonotic Shakespeare: Animals, Plagues, and the Medical Posthumanities," in *The Routledge Handbook of Shakespeare and Animals*, eds. Karen Raber and Holly Dugan (New York: Routledge, 2021), in Chapter 8; Luna Dolezal, "Morphological Freedom and Medicine: Constructing the Posthuman Body," in *The Edinburgh Companion to the Critical Medical Humanities*, eds. Anne Whitehead et al. (Edinburgh: Edinburgh University Press, 2016), in Chapter 19.

63 See, e.g., Kristensen's discussion of Dewey's moral philosophy, in Kristensen, this volume.

64 See, e.g., Irus Braverman, "Law's Underdog: A Call for Nonhuman Legalities," *Annual Review of Law and Social Science* 14 (2018): 127–144.

65 "If the category of emerging disease seemed self-evident by early 2016, it is important to underline its relatively recent invention." Andrew Lakoff, *Unprepared: Global Health in a Time of Emergency* (Stanford: University of California Press), 5.

66 See, e.g., the discussion of "spillovers" in my interview with Chris Walzer. Braverman, this volume.

67 See, e.g., Wolf, "Is There Really Such a Thing," 6.

68 See Donna Haraway's concept of "making beings killable." Donna Haraway, *When Species Meet* (Minneapolis: University of Minnesota Press, 2008), 80. Another deployment of Haraway's concept in the context of animal ethics can be found in Kathryn A. Gillespie and Patricia J. Lopez, "Introducing Economies of Death," in *Economies of Death: Economic Logics of Killable Life and Grievable Death*, eds. Patricia J. Lopez and Kathryn A. Gillespie (New York: Routledge, 2015), 1–13. See also Irus Braverman, "Is the Puerto Rican Parrot Worth Saving? The Biopolitics of Endangerment and Grievability," in *Economies of Death: Economic Logics of Killable Life and Grievable Death*, eds. Kathryn A. Gillespie and Patricia J. Lopez (New York: Routledge, 2015), 73–94.

69 Bjørn Ralf Kristensen, "Welcome to the Viralocene: Transcorporeality and Peripheral Justice in an Age of Pandemics," *Medium* (blog), May 22, 2020, https://medium.com/@bjornkristensen/viralocene-66a954260487; see also Najmul Haider et al., "Covid-19—Zoonosis or Emerging Infectious Disease?," *Frontiers in Public Health* 8 (2020): 596944.

70 Kristensen, this volume, 194.

71 Squier, this volume, 143, drawing on Joyelle McSweeney, *The Necropastoral* (Ann Arbor: University of Michigan Press, 2014).

72 Reisman, this volume, 119.

73 Ibid., 121.

74 Ibid.

75 Ibid.

76 Johnson and Dickinson, this volume, 100.

77 Hinchliffe, this volume, xxxiv.

78 Anderson, this volume, 269. See also Haraway, *When Species Meet*.

79 Anderson, this volume, 269.

80 Mark Honigsbaum, *The Pandemic Century: One Hundred Years of Panic, Hysteria, and Hubris* (New York: W.W. Norton & Company, 2020).

81 Lyle Fearnley, *Virulent Zones: Animal Disease and Global Health at China's Pandemic Center* (Durham: Duke University Press, 2020).

82 Frédéric Keck, *Avian Reservoirs: Virus Hunters and Birdwatchers in Chinese Sentinel Posts* (Durham: Duke University Press, 2020).

83 Gavin J. Andrews et al., eds., *Covid-19 and Similar Futures: Pandemic Geographies* (Cham, Switzerland: Springer, 2021).

84 Lakoff, *Unprepared*, 8.

85 Ibid., 12.

86 Adam Kamradt-Scott and Colin McInnes, "The Securitisation of Pandemic Influenza: Framing, Security and Public Policy," *Global Public Health* 7, supp. 2 (2012): 95–110; Tim Brown, Susan Craddock, and Alan Ingram, "Critical Interventions in Global Health: Governmentality, Risk, and Assemblage," *Annals of the Association of American Geographers* 102, no. 5 (2012): 1182–1189.
87 Global Health Centre Policy Brief I, 2021, 7.
88 Ibid.
89 Ibid.
90 Ibid.
91 Ibid., 15.
92 World Health Organization, "WHO Hub for Pandemic and Epidemic Intelligence," 2022, https://www.who.int/initiatives/who-hub-for-pandemic-and-epidemic-intelligence.
93 See Braverman's interview with Amuasi. Braverman, this volume, 92.
94 Ibid., 94.
95 Hinchliffe, this volume, xxxi.
96 See, similarly, in Venables and Pellecchia, "Engaging Anthropology."
97 Hinchliffe, this volume, xxix.
98 Anderson, this volume, 268.

PART I

Situating One Health: Histories and Practice

T.

. BUT NOT HAPPY.

THE MONKEY DENTIST;
OR,
TEETH EXTRACTED GRATIS.

1

ONE HEALTH

A "More-than-Human" History

Abigail Woods

Introduction

The term "One Health" was first adopted in 2003 to describe an agenda for collaborative, cross-disciplinary working that "promotes health through interdisciplinary study and action, across all animal species."[1] The precipitating factor was the resurgent threat of zoonotic disease—initially SARS, and subsequently highly pathogenic avian influenza. Outbreaks of these diseases focused scientific and policy attention on the transmission of infections from animal to human populations. They also highlighted constraints to the sharing of knowledge and coordination of policy across international health organizations, which were caused by the disciplinary and institutional silos that separated human and animal health. As a consequence, calls were made for more effective, integrated working across health domains. These included a 2008 joint paper by the World Health Organization, World Organisation for Animal Health (formerly the Office International des Epizooties), and the Food and Agriculture Organization of the United Nations, and a follow-up statement in 2010, which reaffirmed the commitment of these organizations to One Health. New research and advocacy groups were formed to advance this agenda, and postgraduate training courses in One Health were established.[2]

Since then, the focus of One Health has expanded beyond zoonotic diseases to include numerous other threats to human and animal health, such as climate change, food insecurity, mental health, chronic diseases, and antimicrobial resistance.[3] Nevertheless, the analysis by Cassidy in 2017 of scientific citation databases revealed that zoonoses continue to dominate, and that, of the publications

FIGURE 1.1 Interior of a dentist's surgery with animal participants. Reproduction of a colored wood engraving. Credit: Wellcome Collection, public domain.

DOI: 10.4324/9781003294085-3

discussing One Health, over 60 percent appeared in veterinary science journals.[2] The topic has made limited inroads into infectious diseases and public health publications and features only infrequently in biomedical or environmental science journals. These findings suggest that One Health is struggling to achieve its interdisciplinary aspirations, and that, in spite of its intention to advance health across all species, it is primarily humans that are benefiting.[4]

Historically, however, there are many examples of health experts working in precisely the ways that are aspired to by today's One Health advocates, and with less anthropocentric objectives. Their activities are not represented adequately in the historical timelines that typically feature in One Health publications. These refer to the work of just a few, unrepresentative historical figures, such as Edward Jenner, Rudolph Virchow, Louis Pasteur, Robert Koch, William Osler, and Calvin Schwabe, whose selection is driven not by empirically grounded research into the history of One Health, but by the desire to raise its present-day profile.[5] Although medical historians have produced some accounts of zoonotic diseases and animal experimentation,[6] they have not gone nearly far enough in correcting the historical misrepresentation of One Health. This is probably because they share the mis-founded assumption, held also by One Health advocates, that medicine in the past was a human-centered endeavor, which took an interest in animals only when they threatened or had the potential to advance human health.[7] Here, I will challenge this assumption through a series of examples of how British practitioners of human medicine engaged with diseased animals over the period circa 1790–1900. These will reveal that in the relatively recent past, medical professionals paid far more attention to animal diseases and for a greater variety of reasons than are acknowledged today by historians and One Health advocates.

Various sources have been used in this analysis, including medical journals (notably *The Lancet* and *The British Medical Journal*), the records of medical and zoological societies, medical texts, public health reports, and the records of government commissions of enquiry into particular diseases. My approach to these sources aligns with recent aspirations to create "more-than-human" approaches to One Health.[8] It is informed by recent developments in human–animal studies, particularly animal history. Drawing on elaborations of actor–network theory, scholars in these fields have sought to challenge established anthropocentric worldviews by establishing animals as historical subjects and demonstrating their capacity to shape human society. Although animals are nonverbal creatures, they have left myriad traces on the medical historical record. Digital search tools are particularly useful in revealing these traces, as the insertion of species names produces numerous hits that would not be retrieved if the names of humans or diseases were used as the starting point of enquiry. Examination of these animal traces brings to light the zoological nature of human medicine, by revealing the extensive roles that animals played in shaping medical knowledge, practice, politics, and relationships. While this chapter is based on British sources from the period 1790–1900, its approach could easily be extended to reveal the intersections of human and animal health that existed in different times and places.[9]

I will open by sketching out the contexts in which medically trained individuals engaged with diseased animals. This will reveal how the social worlds of doctors intersected with those of various nonhuman animal species. In Section Two, I direct attention to the scientific rationales and practices that characterized these engagements. While acknowledging the importance of zoonotic diseases, I also draw attention to the practice of comparing nonhuman and human diseases and to the ways in which medical activities aligned and intersected with those of the veterinary profession. Section Three will examine the inter-professional relationships that developed through this work. It will reveal examples of collaboration and—toward the end of the nineteenth century—growing conflict, as the veterinary profession attempted to extend its influence by claiming diseased animals for themselves. This conflict led to the hardening of professional boundaries, and subsequently to efforts to transcend them through precursors to today's One Health. In conclusion, I reflect on how this history might inform present-day efforts to develop post-anthropocentric approaches to health.

Section One

Analysis of surviving historical sources reveals that in Britain during the long nineteenth century, humans were not the only species to attract medical attention. Medical investigations and practices incorporated a host of nonhuman animals and were far more zoologically grounded than they are today. Doctors' interest in and approaches to diseased animals were powerfully influenced by their lived relations and experiences, as well as their intellectual interests and desire for social and scientific status both within and outside the profession. Heavy societal dependence on animals—as food, transport, companionship, entertainment, and cultural capital—made their diseases problematic, and meant there was much to be gained by addressing them. It also meant that animal bodies and habits were familiar to and readily accessible by doctors.[10]

Throughout this period, all medical men rode, and often had their own stables and carriages. Consequently, injured and diseased horses had a very real impact on their life and work. The late eighteenth century witnessed growing societal interest in large-scale horse racing, selective horse breeding, hunting on horseback, and the performance of cavalry horses. In this context, some surgeons began to specialize in farriery, to create infirmaries for horses, and establishments for training learned farriers. These establishments predated and formed the model for some of the first veterinary schools, in which doctors also played an influential shaping role.[11] Comparative anatomy was another cutting-edge field of medical enquiry at the turn of the nineteenth century.[12] One of its most avid students was the surgeon John Hunter, who amassed a famous collection of 13,682 specimens, representing 500 different species.[13] Interest in the subject drove medical men to participate in establishing zoological gardens as sites of scientific enquiry, symbols of imperial conquest, and places for public education and entertainment. Zoos also offered opportunities for dissecting exotic animals that died there, and for overseeing their management in health and disease.[14]

Until the middle of the nineteenth century, more British people lived in rural than in urban areas. This granted many doctors, most famously Edward Jenner, plentiful opportunities for studying the natural history of animals, and making incidental discoveries about their diseases in the course of capturing, killing, dissecting, and displaying them.[15] As the century drew on, hunting, shooting, and fishing became increasingly popular leisure pursuits. Some medical participants were inspired to study the diseases of salmon, grouse, and other species which impacted on these sports by decimating populations of their animal subjects.[16] Doctors were also exposed frequently to agricultural animals, not simply in the countryside but also in towns, where horses and livestock were encountered regularly on their journeys to and from markets, and where milk was supplied largely by cows housed in urban dairies. Consequently, doctors took an interest in the mid-nineteenth century appearance of new and apparently contagious animal diseases like foot and mouth disease, contagious bovine pleuro-pneumonia, and in the devastating 1865–1867 epidemic of cattle plague (rinderpest).[17]

The impacts of urban slaughterhouses, cowsheds, and pigsties on human, animal, and environmental health attracted the attention of the emerging public health profession, as did growing evidence that the meat, milk, and wool of diseased livestock could spread diseases like tuberculosis and anthrax to humans.[18] Toward the end of the nineteenth century, additional paid opportunities for studying animal diseases opened up in universities and research institutions like the Brown Institute of Comparative Pathology.[19] However, throughout the period under investigation, the vast majority of doctors' enquiries into animal disease were pursued privately, in their own time and at their own expense. They shared their findings by creating specimens of animal diseases for inclusion in medical museums,[20] reporting their findings to meetings of medical societies,[21] and publishing books and articles in medical journals. They also received invitations from government to sit on, supply evidence to, and conduct scientific investigations on behalf of official committees of enquiry into animal diseases like cattle plague and bovine tuberculosis.[22]

Records of the Pathological Society of London, which was established in 1846 for the "cultivation and promotion of pathology," offer an illuminating insight into the diverse species that attracted medical attention. Membership of this popular organization peaked at over 700 in the 1880s and included members ranging from elite consultants to grassroots general practitioners, based in London, the provinces, and the colonies. Regular meetings were held to exhibit and discuss morbid specimens, which were then reported upon in the society's annual published *Transactions*.[23] Over the period 1846–1881, 230 separate reports appeared under the sub-heading "specimens of the lower animals." These ranged from individual case reports to lengthy expositions covering multiple species. Although they made up only 5 percent of the total specimen reports, 70 different authors were involved. This suggests that animals must have featured much more widely in members' specimen-gathering activities, because the society only invited presentations of interesting or unusual cases.[24]

Many of those who presented animal specimens were not well known in their profession. However, those who exhibited most frequently did have a substantial profile. They included men like Charles Murchison, Jonathan Hutchinson, Richard Quain, Thomas Cobbold, and John Burdon Sanderson, who held positions in London hospitals, the Royal Colleges of Surgeons, the Royal Colleges of Physicians, and research institutes. This spread of participation suggests that engaging with animal diseases was not a marginal but a mainstream pathological activity for nineteenth century doctors. Farm animals featured most frequently in their reports, followed by zoo animals, pets, horses, wildlife (usually subjects of hunting, shooting, and fishing) and, most infrequently, experimental subjects. The accompanying text reveals the role of serendipity in bringing animal bodies to the attention of doctors. Some were doctors' own pets and livestock, or belonged to friends, family, or human patients. Some were literally stumbled upon in the street. There is also evidence of doctors actively seeking out diseased animals by means of familial, social, and professional contacts, which enabled them to access bodies on farms, zoos, grouse moors, and other settings.[25]

Section Two

Of the various reasons advanced by doctors to justify their scientific interest in diseased animal bodies, the one which overlaps most directly with present-day One Health was to understand and prevent the spread of disease from animals to humans. At the time, the most problematic zoonotic diseases were recognized as anthrax, glanders, tuberculosis, and rabies. Anthrax or "splenic fever" was a sporadic but potentially devastating disease of horses, sheep, and cattle, that was originally thought to be associated with particular soils. During the 1870s and 1880s, medical scientists discovered that it had the same bacterial cause as two diseases associated with the expanding textile industry: "woolsorters disease" (a fatal pneumonia) and "malignant pustule" (a skin disease). It transpired that the growth of the global wool trade was exposing western wool workers to anthrax spores contained in the fleeces of Asian and South African sheep. This discovery generated a range of responses: disinfection of fleeces offered direct protection to humans, while the development of serum and vaccines benefited both animal and human health.[26]

Glanders was a fatal respiratory disease spread by horses to humans who worked closely with them. Initial symptoms in horses were not obvious—they had runny noses and were slightly off-color—but in humans the disease was fatal. It was particularly a problem in cities like London, where horse numbers expanded in the nineteenth century alongside the development of railways and steamships. The 1892 discovery of mallein, a diagnostic product that could identify infected but asymptomatic horses facilitated its control. Produced by government laboratories and applied by civilian and military officials under compulsory test-and-slaughter policies, it benefited human and horse health. By WWII, glanders had been eradicated from most of Europe and North America.[27]

Suspicions that tuberculosis could spread from cows to humans via meat predated Robert Koch's 1882 claim that the same bacterium was responsible for disease in both species. Late in the nineteenth century, its spread via milk attracted medical attention, as did the possible role of this substance—and the cows that produced it—in the spread of human typhoid, scarlet fever, and diphtheria. These health scares coincided with a growth in milk consumption that was driven by growing affluence, the development of the railway milk trade, the popularity of dairy farming, and use of poorly ventilated urban dairies that became known "hot beds" of the disease. Tuberculosis also had implications for meat consumption, which was highly valued, even by the lower classes, who were prepared to purchase meat of dubious provenance, provided it was cheap. The control of bovine tuberculosis was problematic because its symptoms in cows were not obvious until the disease was well advanced. Butchers, vets, and doctors laid rival claims to expertise in the identification and handling of diseased carcasses. Koch's controversial announcement in 1901, that tuberculosis in humans and cows were not, after all, identical diseases, created further confusion and controversy over the management of tuberculosis.[28]

Rabies aroused disproportionate fear and attention in nineteenth century Britain, owing to the horrific manner of death and its potential conveyancing by "man's best friend." Rabies scares coincided with the evolution of pet keeping and the Victorian "pedigree dog fancy." By transforming dogs into bestial killers, rabies challenged human efforts to reshape and domesticate them. In blaming urban street dogs for rabies spread, commentators drew on wider fears of their "human equivalents," the undisciplined, threatening lower and criminal classes. Efforts to control rabies through the enforced muzzling of dogs revealed marked contrasts in how public health doctors and dog owners perceived dogs. For doctors, dogs were potential conduits of disease, so they had to be disciplined, but, for many owners, dogs were family members whose compulsory muzzling by government diktat amounted to unjustifiable state intervention in the private sphere.[29]

A quite different reason for studying diseased animals was to compare and contrast their diseases by means of observation and dissection, and to use the findings to work out the relationships between species. Interest in interspecies relationships long predated the work of Charles Darwin. For example, for the eighteenth century surgeon John Hunter—whose extensive multispecies museum was donated after his death to the Royal College of Surgeons in London—differences in bodily structure and function translated into variations in disease. Through the cross-species comparison of normal and pathological bodies, the principles of life, disease, and death were revealed.[30] In the 1830s, the emergence of a philosophical form of comparative anatomy suggested that humans and animals were formed on the same general plan. In their efforts to comprehend it, medically trained comparative anatomists compared the anatomy and pathology of the bodies and embryos of species ranging from simple, single-celled creatures up to humans.[31] This activity was facilitated by colonial conquest, because the desire to learn about and exploit the resources of colonial territories led to the importation of many exotic animals. These symbols of European sovereignty over alien and potentially dangerous environments were housed in menageries and zoological gardens, where doctors oversaw their health, and transformed them by dissection into museum specimens after death.[32]

The practice of interspecies comparisons was further boosted by the publication of Darwin's *On the Origin of Species*.[33] Its claim that all living organisms descended by evolution from a common ancestor encouraged doctors to study particular cases of animal pathology as exemplars of general pathological processes like inflammation or degeneration. They asked what bodily structures and functions were fundamental to all life, and how and why they went wrong in disease. What factors accounted for the differential expression of diseases in humans and other animals, and what did this mean for the classification of diseases? The comparative pathological project reached its zenith in the work of late nineteenth century London surgeon, John Bland Sutton, who set about tracing the evolutionary history of disease. Taking advantage of the high death rate and rich species coverage of the London Zoological Gardens, he dissected some 12,000 human and animal subjects between 1878 and 1886, and wrote lengthy articles on their organ systems that compared the manifestations of disease in different mammalian and some non-mammalian species.[34] He saw disease as both a driver to and a product of evolution, for "The laws of evolution apply to pathology as well as to the ordinary events of animal life."[35] This work led him to deduce the causes of rickets in humans from studying its spontaneous occurrence in monkeys living in the London Zoological Gardens.[36]

Comparing across species generated not only insights into disease pathology. Prior to the germ theory, some doctors engaged in a kind of comparative epidemiology. The fact that animal epidemics often seemed to coincide with—or precede—human epidemics suggested that there was a common atmospheric influence on them. Studying animal epidemics could therefore provide an early warning of human epidemics, and also generate general knowledge about the factors that were responsible for the rise and fall of epidemics in all species.[37]

Another mode of medical engagement with diseased animals, which is today regarded as the prerogative of the veterinary surgeon, was to diagnose, treat, and prevent their ailments. Prior to the creation of a British veterinary profession, eighteenth century surgeons-turned-farriers proclaimed that "physic" (conventional medicine) was the same whether practiced on humans or horses. They argued that farriery formed part of comparative anatomy, and was therefore a polite practice, well suited to a gentlemanly surgeon or physician. Britain's first veterinary school (The London Veterinary College) which was established in 1791, was informed by this thinking and by the medical practice of comparative anatomy. For the first 50 years of its existence, its activities and personnel were largely continuous with human medicine. Edward Coleman, a surgeon who directed the school for over 40 years, modeled the structure and organization of veterinary education on that of human medicine. Prominent London doctors sat on the college's "medical examining committee," which examined students for their veterinary diplomas. They also invited veterinary students to the lectures they gave to London medical students. Some London medical students also took Coleman's optional courses in equine medicine and surgery. Many surgeons went further and enrolled at the London Veterinary College (later the Royal Veterinary College). By 1830, around 130 of them had qualified. Similar overlaps occurred in Edinburgh, where William Dick founded a veterinary school (later The Royal (Dick) School of Veterinary Studies) in 1823.[38]

According to one prominent commentator, Delabere Blaine, veterinary medicine was "a branch that has sprung from, and must grow with medicine as its parent stock," and in which advances were made "usually by the exertions of some enlightened physician or surgeon."[39] Surgeons' knowledge was highly relevant to the veterinary "art" because it enabled them to advance the "study of disease by analogy," to care for the sick bodies of their own horses, to treat other people's horses when no skilled farrier was available, and to enter veterinary practice. Another compelling reason arose in the context of the French revolutionary wars when the need for skilled equine care led to the commissioning of veterinary surgeons as officers to each regiment. This was a period in which medicine really was "One."[40]

During the 1820s and 30s, a number of qualified vets began to develop a shared identity separate from that of human medicine, and to agitate for formal recognition as a distinct profession. This was granted in 1844 with the creation by Royal Charter of a veterinary regulatory body, the Royal College of Veterinary Surgeons. However, doctors continued to bring the expertise they had honed on sick human bodies to bear on those of animals—particularly pets and zoo animals, of which vets had little knowledge or experience.[41] When the epidemic of cattle plague broke out in the London dairies in 1865, public health doctors, who were already supervising these institutions as part of their efforts to promote human and environmental health, were among the first to raise the alarm. They went on to study the cause and course of the disease, and to query its implications for humans who consumed meat and milk from infected animals. They brought specimens of diseased cows before the Pathological Society of London. Some drew parallels between the causes of cattle plague and a recent cholera outbreak in humans. They also reflected on the relationship between cattle plague in cows and typhoid and smallpox in humans, and carried out trials of therapies and preventives, including the use of smallpox vaccination.[42]

Section Three

Positioned at the nexus of human medicine, veterinary medicine, and comparative biology in the nineteenth century, diseased animals often inspired collaboration between members of the medical and veterinary professions. The prime reason for coming together was to advance knowledge of the relationships between human and animal diseases. This included diseases that were suspected of transmission between animals and humans. For example, in his 1830 book on rabies, the veterinary surgeon, William Youatt, invited medical practitioners to "kindly send to my dissecting room quadrupeds labouring under rabies, or destroyed by it, that we may experiment on, or examine them together … I should feel exceedingly grateful, and both human and veterinary practice might probably be benefitted."[43] He subsequently superintended the dissection of a dog thought to have infected a child with rabies, in the presence of the doctors who attended the child.[44] Three years later, the roles were reversed when, at the invitation of the surgeon John Elliotson, Youatt attended an autopsy on a human patient thought to have died of glanders.[45] In each case, the pathological appearances of the dissected body were scrutinized to determine whether the disease was identical to that witnessed in the affected human or animal.

On other occasions, doctors and vets collaborated on studies of particular animal diseases that were not thought to transmit to humans, but whose pathology appeared to resemble that of a human disease, suggesting relationships between their causes or consequences. In 1848 and 1864, they worked together to investigate sheep suffering from a condition that strongly resembled smallpox in humans. Experiments highlighted key differences between the diseases, as well as points of similarity.[46] During the 1860s, medical investigators into "roaring" in horses and rickets in dogs sought out veterinary perspectives on these conditions, which enabled them to draw general conclusions about their pathology, causation, and expression in all affected species. Veterinary insights into cattle plague and swine fever also informed medical understandings of the relationships between these diseases and typhoid in humans.[47] Typically, these sorts of enquiries were motivated by personal interest and were conducted in private. They were characterized by harmonious working relationships that were grounded in mutual respect. Each party recognized the complementary nature of the other's skills and insights: veterinary surgeons tended to possess specific knowledge of a particular animal disease, which doctors incorporated into a wider comparative framework. Vets were often flattered to be consulted by doctors, whose profession had a much higher status than their own. This mode of collaborative working did not have a specific label attached to it—it simply represented a problem-driven response to a particular set of disease circumstances.

On other occasions, however, medical interest in animal diseases could provoke competition and conflict with members of the veterinary profession. This was more likely to happen when doctors investigated animal diseases as problems of animal health, rather than as points of comparison with human diseases. It also occurred in discussions over zoonotic diseases, when vets challenged the right of public health doctors to make recommendations relating to the management of animal as well as human bodies. Conflict was fueled by efforts to win legal recognition for veterinary medicine as a profession, and to improve its scientific and social standing. During the late nineteenth century, as the veterinary profession grew more self-confident and ambitious, its members began to criticize medical pronouncements on animal diseases, to claim these diseases as exclusively veterinary subjects, and to seek recognition for themselves as experts capable of protecting humans from zoonotic diseases.[48]

Notable flash points included the 1865–1867 cattle plague epidemic, the management of bovine tuberculosis, and the prevention of milk-borne epidemics. As outlined above, cattle plague stimulated a raft of medical interventions that were aimed not simply at protecting the public from the effects of consuming meat and milk from infected animals, but also at preserving bovine life through preventive inoculations and therapeutic remedies. These latter activities stimulated widespread criticism from veterinary leaders, who argued that the disease was incurable and could only be managed by legislation that should be implemented by newly appointed veterinary inspectors.[49] In the case of bovine tuberculosis, inter-professional conflicts developed over the management of meat and milk from infected cows, as vets and doctors held different perceptions of the risks they posed to humans, and how they should be managed.[50] During the 1880s, leading public health doctors claimed to have identified a new condition in cows that they labeled "Hendon disease," and which appeared to be implicated in the transmission of scarlet fever, diphtheria, and possibly typhoid to humans via milk. Their veterinary opponents rubbished this diagnosis, claiming that the disease was simply cowpox, and that doctors had no clinical or epidemiological evidence for their claims.[51]

The early twentieth century development of new government funding streams for agricultural scientific research encouraged further veterinary efforts to solidify and police the profession's boundaries with human medicine. Leading vets sought to restrict funds for animal disease research to institutions created and run by vets, thereby excluding doctors. These efforts were not always successful. Nevertheless, when considered alongside the growing exclusion of animals from cities, and the professionalization of research activities, they did contribute to the lessening of private, spontaneous, medical engagements with diseased animals.[52] At the same time, medical attention was diverted away from animals as disease subjects, or points of comparison with humans, by the growth of experimental medicine. This narrowed the range of species studied, and refashioned animal subjects into "model" humans whose purpose was to illuminate and advance human health.[53]

Medical engagements with diseased animals did not disappear entirely. Throughout the twentieth century, as disease knowledge and research practices evolved, doctors pursued new animal disease subjects and modes of investigation. However, compared to its nineteenth century heyday, medicine became distinctly less zoological, and its boundaries with veterinary medicine were more distinct and difficult to traverse. Significantly, it was the hardening of these boundaries in the late nineteenth century that precipitated the first self-conscious efforts to encourage shared approaches to problems that crossed professional and disciplinary boundaries. Various labels were appended to this activity: "comparative pathology," "comparative medicine," "veterinary public health," or "One Medicine." Its main protagonists and participants were veterinary surgeons, who were motivated by the desire to raise their professional profile through greater participation in human health agendas. Notably, although today's One Health is not a direct descendent of these movements, and claims a wider remit, it has retained some of their features. It is still a veterinary-driven activity that focuses particularly on zoonotic diseases and the advancement of human health.[54]

Conclusion

This brief survey has revealed that, in Britain during the long nineteenth century, the diseases of diverse animal species were of enduring interest to certain members of the medical profession. This interest was undoubtedly fueled by doctors' familiarity with, dependence on, and ease of access to animals, and by the historical continuities between human and veterinary medicine, which encouraged collaborative working across emerging professional boundaries. It enabled them to work unself-consciously across species boundaries, without feeling the need to justify or label their activities. Studying infectious diseases that were transmitted from animals to humans formed only one aspect of their search for cross-species insights and interventions, which aimed to advance both animal and human health, and achieve understandings of their similarities and differences.

It would be ahistorical to label these ways of working as "One Health." The concept did not yet exist, and the disciplinary boundaries which One Health aspires to overcome were much more porous than they are today. Nevertheless, for contemporary actors who are seeking to develop post-human, post-anthropocentric forms of health knowledge and practice, looking backwards to the history of British nineteenth century medicine may offer a more convincing model than the human-focused and veterinary-dominated activities that dominate current One Health agendas. Such agendas still carry the legacy of veterinarian attempts to enhance their professional standing by first erecting, and then spearheading attempts to transcend barriers between human and animal health, particularly through the medium of zoonosis control.

Through reframing what One Health is, was, and could become, this historical perspective has highlighted its unrealized, contemporary "more-than-human" possibilities. It has made visible the politics of present-day One Health, and the narrowness of its approach to interspecies health when compared with past ideas and practices. In demonstrating alternative ways of conceptualizing and practicing One Health, it has shown what this self-consciously interdisciplinary project has to gain by incorporating previously neglected perspectives from the humanities. Its historical examples also prompt questions about what other activities might already be underway, which transcend the boundaries of species but without feeling the need to brand themselves as "One Health." In the search for an anti-anthropocentric mode of health research and practice, it may prove more fruitful to look at how experts are working rather than what agendas they claim to be pursuing.

Acknowledgements

This research was funded by the Wellcome Trust, grant number 092719/B/10/A. For the purpose of open access, the author has applied a Creative Commons CC BY public copyright license to any author accepted manuscript version arising from this submission.

Notes

1 Paul Gibbs, "The Evolution of One Health: A Decade of Progress and Challenges for the Future," *Veterinary Record* 174, no. 4 (2014): 85–91, 86.
2 Angela Cassidy, "Humans, Other Animals and 'One Health' in the Early Twenty-First Century," in *Animals and the Shaping of Modern Medicine*, eds. Abigail Woods et al. (Basingstoke: Palgrave, 2017), 193–236.
3 Jakob Zinsstag et al., eds., *One Health: The Theory and Practice of Integrated Health Approaches* (Oxford: CAB International, 2021).
4 Cassidy, "Humans, Other Animals," 193–236.
5 The best known of these accounts, which has informed all later iterations, is Calvin Schwabe, *Veterinary Medicine and Human Health* (Baltimore: Williams & Wilkins, 1984).
6 Nicolaas Rupke, ed., *Vivisection in Historical Perspective* (London: Routledge, 1990); Lisa Wilkinson, *Animals and Disease: An Introduction to the History of Comparative Medicine* (Cambridge: Cambridge University Press, 1992); Anita Guerrini, *Experimenting with Humans and Animals: From Galen to Animal Rights* (Baltimore: John Hopkins University Press, 2003); Keir Waddington, *The Bovine Scourge: Meat, Tuberculosis and Public Health, 1850–1914* (Woodbridge: Boydell Press, 2006); Neil Pemberton and Michael Worboys, *Mad Dogs and Englishmen: Rabies in Britain, 1830–2000* (Basingstoke: Palgrave Macmillan, 2007); Susan Jones, *Death in a Small Package: A Short History of Anthrax* (Baltimore: Johns Hopkins University Press, 2010).
7 Abigail Woods et al., *Animals and the Shaping of Modern Medicine* (Basingstoke: Palgrave, 2017).
8 Stephen Hinchliffe, "More than One World, More than One Health: Re-Configuring Interspecies Health," *Social Science & Medicine* 129 (2015): 28–35.
9 For a more expansive description of this approach and discussion of relevant literature, see Woods et al., *Animals*, 1–26.
10 Kathryn Kete, ed., *A Cultural History of Animals in the Age of Empire* (London: Bloomsbury, 2007).
11 Michael MacKay, "The Rise of a Medical Speciality: The Medicalization of Elite Equine Medical Care, 1680–1800" (PhD diss., University of York, 2009).
12 Andrew Cunningham, *The Anatomist Anatomis'd: An Experimental Discipline in Enlightenment Europe* (Farnham: Routledge, 2010).
13 Jessie Dobson, "John Hunter's Animals," *Journal of the History of Medicine and Allied Sciences* 17, no. 4 (1962): 479–486.
14 Abigail Woods, "Doctors in the Zoo," in *Animals and the Shaping of Modern Medicine*, eds. Abigail Woods et al. (Basingstoke: Palgrave, 2017), 27–69.
15 J. Baron, *The Life of Edward Jenner* (London: H. Colburn, 1838).
16 Thomas Henry Huxley, "A Contribution to the Pathology of the Epidemic Known as the 'Salmon Disease,'" *Proceedings of the Royal Society of London* 33, nos. 216–219

(1882): 380–389; Duncan George Forbes Macdonald, *Grouse Disease: Its Cause and Remedies* (London: W. H. Allen, 1883); Thomas Spencer Cobbold, *On the Destruction of Fish and Other Aquatic Mammals by Internal Parasites* (London, 1883).

17 J. N. Radcliff, "The State of Epidemic, Epizootic and Epiphytic Disease in Great Britain in 1861–1862," *Transactions of the Epidemiological Society of London* 1 (1863): 393–428; William Budd, "Variola Ovina, Sheep's Small-Pox: Or the Laws of Contagious Epidemics Illustrated by an Experimental Type," *British Medical Journal* 2 (1863): 141–150; Andrew Smart, *Reports to the Lord Provost and Magistrates of the City of Edinburgh on the Pathological Appearances, Symptoms, Treatment, and Means of Preventing Cattle Plague* (Edinburgh: MacLachlan & Stewart, 1866).

18 The annual *Reports of the Medical Officer to the Privy Council* provide many examples of these kinds of activities.

19 Graham Wilson, "The Brown Animal Sanitary Institution," *Journal of Hygiene* 83 (1979): 171–197.

20 *Catalogue of the Hunterian Collection in the Museum of the Royal College of Surgeons in London* (London: Richard Taylor, 1830).

21 *Transactions of the Pathological Society of London*, 1846–1910, *passim*.

22 *Third Report of the Commissioners Appointed to Inquire into the Origin and Nature, &c. of the Cattle Plague* (London: United Kingdom Parliament, 1866); *Royal Commission Appointed to Inquire into the Effect of Food Derived from Tuberculous Animals on Human Health* (London: United Kingdom Parliament, 1895).

23 H. R. Dean, "The Pathological Society of London," *Proceedings of the Royal Society of Medicine* 39, no. 12 (1946): 823–827.

24 *Transactions*, 1846–1910, *passim*.

25 *Transactions*, 1846–1910, *passim*.

26 Jones, *Death in a Small Package*.

27 Lise Wilkinson, "Glanders: Medicine and Veterinary Medicine in Common Pursuit of a Contagious Disease," *Medical History* 25 (1981): 363–384.

28 Michael French and Jim Phillips, *Cheated Not Poisoned? Food Regulation in the United Kingdom, 1875–1938* (Manchester: Manchester University Press, 2000); Waddington, *The Bovine Scourge*.

29 Pemberton and Worboys, *Mad Dogs*.

30 L. S. Jacyna, "Images of John Hunter in the Nineteenth Century," *History of Science* 21, no. 1 (1983): 85–108.

31 Anatomical Society, *Minute Book for 1833–1843* (held by Library of the University of Edinburgh).

32 Woods, "Doctors in the Zoo."

33 Charles Darwin, *On the Origin of Species by Means of Natural Selection* (London: John Murray, 1859).

34 Woods, "Doctors in the Zoo."

35 John Bland Sutton, *An Introduction to General Pathology* (London: Blakiston & Son, 1886), 376.

36 Dr. Cheadle, "Introductory Address: A Discussion on Rickets," *British Medical Journal* 2 (1882): 1145–1148.

37 Radcliff, "The State of Epidemic, Epizootic and Epiphytic Disease."

38 Abigail Woods, "From One Medicine to Two: The Evolving Relationship Between Human and Veterinary Medicine in England, 1791–1835," *Bulletin of the History of Medicine* 91 (2017): 494–523.

39 Delabere Blaine, *The Outlines of the Veterinary Art; Or, the Principles of Medicine: As Applied to the Structure, Functions and Economy of the Horse, the Ox, the Sheep and the Dog* (London: T. N. Longman and O. Rees and T. Boosey, 1802), viii.

40 Ibid.; Woods, "From One Medicine to Two."

41 Woods, "From One Medicine to Two."

42 William Budd, "Observations on Typhoid (Intestinal) Fever in the Pig," *British Medical Journal* 2, no. 239 (1865): 81–87, 81; William Budd, "The Siberian Cattle

Plague; or, the Typhoid Fever of the Ox," *British Medical Journal* 2, no. 242 (1865): 169–179; Charles Murchison, "On the Points of Resemblance Between Cattle-Plague and Small-Pox," *The Lancet* 86, no. 2209 (1865): 724–726; *Third Report of the Commissioners*; Smart, *Reports to the Lord Provost.*

43 William Youatt, *On Canine Madness* (London: 1830), 51–52.

44 Report from the Committee on the Bill to Prevent the Spreading of Canine Madness. *Parliamentary Papers* X (1830), 651: 685, 26–27.

45 St. Thomas's Hospital, "Glanders in the Human Subject," *Lancet* 19 (1833): 728–731.

46 J. B. Simonds, *A Practical Treatise on Variola Ovina, or Smallpox in Sheep* (London: 1848); J. F. Marson and J. B. Simonds, *Report of Experiments Made Under Direction of the Lords of the Council as to the Vaccination of Sheep and as to the Influence of Such Vaccination in Preventing Sheep Pox* (London: HMSO, 1864).

47 J. W. Ogle, "Drawing, Showing Atrophy of Muscles on One Side of the Larynx of a 'Roarer,'" *Transactions of the Pathological Society of London* 10 (1859): 339–347; Charles Murchison, "Intestines of a Pig," *Transactions of the Pathological Society of London* 10 (1859): 334; Budd, "Observations on Typhoid," 81; Budd, "The Siberian Cattle Plague."

48 Abigail Woods and Stephen Matthews, "'Little, If at All, Removed from the Illiterate Farrier or Cow-Leech': The English Veterinary Surgeon, c. 1860–85, and the Campaign for Veterinary Reform," *Medical History* 54 (2010): 29–54; Woods, "From One Medicine to Two."

49 Michael Worboys, "Germ Theories of Disease and British Veterinary Medicine, 1860–1890," *Medical History* 35 (1991): 308–327.

50 Keir Waddington, "'Unfit for Human Consumption': Tuberculosis and the Problem of Infected Meat in Late Victorian Britain," *Bulletin of the History of Medicine* 77, no. 3 (2003): 636–661.

51 Local Government Board: Supplement to Fifteenth Annual Report, containing Report of Medical Officer, 1885, *Parliamentary Papers* 1886 [C.4844-I]; Edgar Crookshank, "An Investigation into the So-Called Hendon Cow Disease, and Its Relation to Scarlet Fever in Man," *British Medical Journal* 2, no. 1407 (1887): 1317–1320, 1317; George Brown, *Report on Eruptive Diseases of Teats and Udders of Cows in Relation to Scarlet Fever in Man* (London: HMSO, 1888).

52 Abigail Woods, "From Co-ordinated Campaigns to Water-Tight Compartments: Diseased Sheep and Their Investigation in Britain, c. 1880–1920," in *Animals and the Shaping of Modern Medicine*, eds. Abigail Woods et al. (Basingstoke: Palgrave, 2017), 71–117.

53 Cheryl Logan, "Before There Were Standards: The Role of Test Animals in the Production of Empirical Generality in Physiology," *Journal of the History of Biology* 35, no. 2 (2002): 329–363.

54 Michael Bresalier, Angela Cassidy, and Abigail Woods, "One Health in History," in *One Health: The Theory and Practice of Integrated Health Approaches*, eds. Jakob Zinsstag et al. (Wallingford: CAB International, 2020), 1–14.

2

THE CASE FOR A ONE HEALTH APPROACH FROM A PHYSICIAN'S PERSPECTIVE

Laura H. Kahn

The One Health concept recognizes the inextricable linkages between human, animal, plant, environmental, and ecosystem health. This concept provides an important framework for examining and addressing complex health issues, such as emerging zoonotic diseases, food safety and security, water quality and sanitation, antimicrobial resistance, agriculture, and climate change. In addition, One Health encompasses comparative and translational medicine, recognizing that disease processes across species are shared. Humans and animals suffer from cancer, heart disease, asthma, diabetes, and many other diseases. This chapter will focus on the linkages between agriculture, health, and climate change. Examining them in a systematic, comprehensive way will allow users to identify the web of connections that humanity depends upon for survival. This work builds on the current One Health literature, including Planetary Health, by developing a three-dimensional matrix tool to examine the linkages between human, animal, and environmental health.

There are many ways to visualize the One Health concept. For example, the Centers for Disease Control and Prevention use intersecting circles representing interdisciplinary coordination, communication, and collaboration.[1] One Health Sweden and the One Health Initiative developed a One Health umbrella graphic that highlights the importance of zoonotic infections and comparative medicine.[2] The World Organization for Animal Health (formerly the Office International des Epizooties, OIE) uses one circle with humans, animals, and

FIGURE 2.1 A small bubble of contaminated seawater, analyzed by researchers at the Marine Science Laboratory, the Pacific Northwest National Laboratory. This small bubble contains pre-concentrated microorganisms, oils, fecal coliform bacteria, and other contaminants collected from the top 50 microns of the sea surface microlayer. US Department of Energy, 2013.

DOI: 10.4324/9781003294085-4

the environment visualized along the perimeter.[3] But none of these graphics are intended to be used as policy research tools.

A One Health three-dimensional cube provides a framework which can be used as a policy research tool to examine, analyze, and address complex health threats in a systematic, concise, and comprehensive way. The first dimension involves One Health factors that include humans, animals, plants, environments, and ecosystems. Environments encompass the abiotic (e.g., air, water, soil) aspects of a geographic area, and ecosystems involve the interactions between the biotic (e.g., microbes, insects, flora, and megafauna) aspects of a geographic area. The second dimension provides complexity and incorporates different biological levels such as microbial and cellular, the individual, and the population.[4] The third dimension focuses on the political, social, and economic factors and can be represented by political borders: local/regional, national, and international/global.[5]

This matrix will be used to examine the microbial connections between agriculture, health, and climate change. People cannot be healthy if they are starving, malnourished, or obese. Microbes are the most prevalent life forms on the planet.[6] Entire microbial ecosystems called "microbiomes" exist on and within humans, other animals, and plants and are important for normal metabolic function.[7] Agriculture and food security provide the essential nutrients needed for optimum health and wellbeing.

Implementing One Health requires the recognition that humans are a part of nature and are genetically related to other life forms. A One Health approach respects the welfare of all species and protects the world's natural resources, such as oceans and forests, that are essential habitats for wildlife. There are microbial connections between agriculture, food safety and security, and climate change. We live in a microbial world, and they made the Earth habitable. Over billions of years, cyanobacteria in the world's oceans produced oxygen which humans, plants and animals need to respire. A One Health analysis using the multi-dimensional framework reveals the microbial connections between climate change and agriculture. Recognizing these connections is a necessary step to developing policies to address them. But first, we must acknowledge our connections with the rest of life on the planet.

In her book, *Braiding Sweetgrass,* Robin Wall Kimmerer, Director of the Center for Native Peoples and the Environment at SUNY Syracuse, New York, writes that native peoples view plants and animals as "kin" rather than as objects to own.[8] Indeed, Indigenous peoples have long recognized the importance of respecting and protecting the plants, animals, and environments that they depend upon for survival.[9] This worldview is in contrast to an anthropocentric worldview in which humans alone have intrinsic value. An anthropocentric worldview posits that the planet and its flora and fauna exist solely for the benefit of humanity.[10] This belief system allows people to disregard the Earth's diminishing resources, species becoming extinct, and waste accumulation, which ultimately jeopardizes the future of humanity and civilization. We need to revise our belief systems to acknowledge that we are not separate from the other animals. Philosophically, the idea that we are part of life on the planet rather than superior to it might be difficult for some individuals to accept. Widespread refusal to recognize our many connections to the other animals, philosophically or religiously, might hinder our ability to develop and implement effective policies to address the many threats we face.

Domestication, Agriculture, and One Health

When the Ice Age ended and the planet warmed around 10,000 years ago, humans began domesticating plants and animals. The process of domestication fundamentally changed how humans viewed their relationships with other animals and with the natural world. This change is evidenced through artworks before and after domestication. For example, the Lascaux caves in the Dordogne region of southwestern France, painted during the pre-domestication Paleolithic era around 15-17,000 BCE, depicts thousands of lifelike animals including horses, stags, aurochs, bison, a bear, and others in bold, brilliant colors. The only anthropomorphic figure is a stick-like "birdman" with an erection and a broken spear by his feet. A large bison stands over him. The meaning of the artwork has been lost, but the artist drew the human in a submissive, defeated pose under the much larger animal.[11]

In contrast, one of the earliest images of Jesus Christ, as a good shepherd, is painted on a wall of the St Callixtus catacombs in Rome, Italy during the third century AD. In the painting, Christ carries a lamb on his shoulders with two small sheep, representing his flock, at his sides. A similar painting exists in the Catacombs of Priscilla in Rome. In both paintings, Christ dominates the scene, with the domesticated animals much smaller in size. These post-domestication artworks exemplify humans' changed perceptions of their relationships with other animals.[12] The animals need human care and protection.

Religious beliefs provided justification for humanity's elevated position in the world. In his book *Sapiens*, Israeli historian Yuval Harari writes that humans switched from gathering and hunting plants and animals to raising and owning selected plants and animals as reliable sources of food. These changes in food acquisition ushered in the Agricultural Revolution. Harari wrote: "This was the turning point where Sapiens cast off its intimate symbiosis with nature and sprinted towards greed and alienation." Humanist religions sanctified *Homo sapiens* as the most important beings on Earth; and all other beings—including plants and nonhuman animals—existed to benefit them. Indeed, many plants and animals became human property.[13]

Agriculture provided the foundation for civilization. For the past 10,000 years, a mild and predictable climate enabled agriculture to provide a relatively stable and secure food supply, allowing human populations to grow and establish towns, cities, nations, and empires.[14] While this development was largely beneficial to many humans, it came at considerable costs. Living in close proximity to animals facilitated their microbes to spread to new hosts, namely humans.

Throughout history, zoonotic diseases such as plague, influenza, and measles spread through densely populated cities, causing deadly epidemics.

In 1776, Adam Smith, the Scottish economist and philosopher, wrote, in *The Wealth of Nations*, that a division of labor in a well-governed society leads to growth in prosperity and wealth. Nations that relied on the labor of individuals to supply all of their food needs (i.e., hunters and fishermen) were miserably poor and hungry and occasionally had to abandon their infants, elderly, and infirm or perish from hunger. In contrast, nations that divided labor produced ten or a hundred times more food needed to feed people. Indeed, many people need not labor at all because they could consume the products from those that did.[15]

The Industrial Revolution, beginning in the eighteenth century led to the rise of new power sources, machines, factories, and the efficient production of goods. Large-scale industry provided affordable goods for the masses. Unfortunately, the benefits came with costs. Combined with an economic system based on profit-driven private ownership of factories (i.e., capitalism), the Industrial Revolution resulted in unregulated dumping of industrial wastes into environments, adversely affecting health. For example, coal-fired power plants belched black ash and greenhouse gases into the atmosphere contributing to air and water pollution, climate change, acute and chronic diseases, and preventable deaths. Because government regulations to protect the environment cost money to implement, corporations fight them. This adversarial relationship between corporations' needs for profits and governments' regulations to achieve environmental protection presents challenges to implementing global One Health.

Capitalism and the Industrial Revolution have been beneficial for humanity. As of 2021, the global population reached over 7.7 billion people.[16] Humans and their domesticated animals constitute approximately 96–98 percent of the global terrestrial mammalian biomass. Unfortunately, capitalism and the Industrial Revolution have been detrimental to the Earth's biosphere. Large-scale agriculture provides food security but comes at considerable environmental costs. The combined mass of broiler chickens exceeds the mass of all other birds on the planet. They are a sign of a reconfigured biosphere from human intervention.[17] Excessive hunting, fishing, and poaching have left few remaining wild species—an estimated 3 percent of wildlife remains on Earth.[18]

With a warming planet, diminishing natural resources, dwindling biodiversity, increasing wastes and pollutants, increasing zoonotic diseases, worsening storms and natural disasters, humans must ask: "Is civilization sustainable?" Although humans are highly adaptable to addressing short-term problems and well-defined threats such as the hole in the ozone layer, they are not as adept at addressing long-term, vague, systemic problems such as climate change.[19] Changing individual behavior is hard, but changing societal behavior is even harder. A One Health approach could facilitate many of the changes needed to protect civilization but requires broad political support.

Livestock, Fecal Matter, and Zoonotic Diseases

In 1996, the World Food Summit defined food security as existing when all people, at all times, have physical and economic access to sufficient, safe, and nutritious food that meets their dietary preferences for an active and healthy life.[20] Food security (aka "zero hunger") is so important that the United Nations listed it as their second sustainable development goal.[21] In 1948, the Universal Declaration of Human Rights (Article 25) declared food to be a human right.[22] But the *type* of food was not specified. For humans, high-fiber diets are important for gut microbial health and proper metabolic regulation.[23] But it's not clear how much meat and other animal proteins are needed for optimum human health. Evidence from the Blue Zones, the areas in the world where higher percentages of people live remarkably long lives, suggest that animal proteins should not constitute more than 5 percent of people's diet.[24] Should a meat-rich diet be a human right? Humans need proteins for health, but must these proteins come from animals?

There are pros and cons to eating animal-based protein diets. Meat provides micronutrients important for health, including iron and vitamin B12. There is evidence that early humans evolved into modern humans because they hunted, cooked, and ate meat.[25] Eating meat is an integral part of many religious and cultural traditions. The Abrahamic religions promote meat consumption through festive meals such as the Christmas ham, the Pesach lamb, and the Eid al-Adha sheep.

But animals did not evolve to live in densely packed conditions involving hundreds, thousands, or tens of thousands of animals in modern-day livestock production facilities. Animal wastes, particularly fecal wastes, contaminate environments and ecosystems. In addition, eating domestic or wild animals promotes the spread of zoonotic diseases which have been increasing in frequency since the mid-twentieth century.[26] People can be healthy eating plant-based proteins as long as they supplement their diets with vitamins and minerals, especially vitamin B12 that is primarily available in meat, eggs, and dairy products.[27]

In 2016, the world consumed almost 130 billion pounds of beef, averaging 17.4 pounds per capita. The countries with the highest beef consumption per capita included Uruguay (124.2 pounds per capita), Argentina (120.2), Hong Kong (114.3), the United States (79.3), and Brazil (78.9).[28] Food preferences are often influenced by religion, culture, and economic class. For example, India has one of the lowest per capita beef consumption rates and the highest proportion of vegetarians in the population, estimated at around 30 to 38 percent, in the world.[29] However, the distribution of vegetarianism in India varies dramatically across different states. The state of Rajasthan has the most vegetarians, almost 75 percent of the population, whereas West Bengal is only 1.4 percent vegetarian.[30] Rajasthan and West Bengal do not differ greatly in terms of GDP, ranking twenty-first and twenty-fourth out of 33 states, respectively.[31] Religions, particularly Buddhism and Jainism, influence Indian dietary preferences.[32] But even in India, meat and other animal protein consumption is increasing, following global trends.[33] Increasing affluence leads to increasing meat and milk consumption. Corporate agricultural interests, such as the meat and dairy industries, oppose policies such as dietary recommendations to reduce consumption of their products that might jeopardize their bottom lines.[34] Feeding billions of people meat and dairy products requires raising tens of billions of domestic livestock.

Taro Gomi, the famous Japanese author of children's books, wrote: "all living things eat, so everyone poops."[35] Berendes et al. estimated that, in 2014, over 7 billion humans and over 30 billion domesticated livestock produced almost 4 trillion kilograms of fecal matter.[36] Animals produced 80 percent of it, and the amount has been increasing by almost 52 billion kilograms each year. By 2030, it is estimated that the amount will increase to 4.6 trillion kilograms each year.[36] To visualize these massive amounts of fecal waste, the sum of all of it would cover the combined entire surface areas of the cities of Los Angeles, California and New York, New York under six feet of muck.[37] Sanitation systems usually process human fecal waste, not animal fecal waste. Animal waste is typically used as fertilizer for crops, but if it's not processed through heat or composting, the microbes in it mix with soil microbes and, after rainfall, they can be washed into nearby streams, causing eutrophication in downstream lakes and killing aquatic life.

Fecal contamination of crops contributes to foodborne and waterborne illnesses. According to the World Health Organization, there are 600 million cases of foodborne illnesses each year, from which 420,000 deaths occur. Of the 31 global hazards that caused foodborne illnesses, *Campylobacter* species and non-typhoidal *Salmonella enterica*, both of which come from animal fecal matter, were among the most frequent causes of foodborne illnesses.[38] Around 2.5 billion people lack adequate sanitation, and 780 million people lack access to safe drinking water. Around one billion people practice open defecation. A significant fraction of diarrheal diseases is due to water contaminated with fecal matter. Human and animal fecal microbes causing water contamination include rotavirus, *Escherichia coli*, *Cryptosporidium*, and *Shigella* species.[39]

Foodborne and waterborne illnesses lead to increased antibiotic use and worsening antimicrobial resistance. In many poor, low-income countries, over-the-counter antibiotics serve as substitutes for sanitation and hygiene. Fecal microbes mix with soil microbes and share resistant genes. Antimicrobial resistance genes are ancient and naturally occurring in the environment.[40] Combined, these factors worsen the crisis of antimicrobial resistance and threaten the future of modern medicines.[41]

In addition to contaminating soil, crops, and water, fecal microbes contaminate the atmosphere, producing nitrous oxide and methane, two extremely potent greenhouse gasses. In 2018, agriculture in the United States contributed 10 percent of greenhouse gas emissions. Of this 10 percent, methane and nitrous oxide constitute 10 percent and 7 percent of greenhouse gas emissions, respectively.[42] Methane is about 28 times more efficient at trapping heat than is carbon dioxide and lasts in the atmosphere for more than one decade. Nitrous oxide is about 300 times more potent a greenhouse gas than carbon dioxide and stays in the atmosphere for about a century, although carbon dioxide stays in the atmosphere for thousands of years.[43]

Beyond fecal matter, microbes generate methane through anaerobic fermentation in the digestive tracts of ruminants such as cattle. Cattle have four-chambered stomachs, one of which is called a rumen, hence the name ruminants. Through enteric fermentation, cattle release methane into the atmosphere through burping.[44] Enteric fermentation, combined with manure emissions, generate the largest sources of methane in the US[45] Reducing the size of domestic animal herds and flocks has the potential to reduce greenhouse gas emissions.[46]

Climate Change and One Health

According to the UN Environment Program Emissions Gap Report 2020, the world is headed for a temperature rise in excess of 3°C by the end of the twenty-first century, which is far above the desired goals of a temperature rise no greater than 2°C warming.[47] Agriculture is in the unique position of both being threatened by and contributing to climate change. Since agriculture is the foundation of civilization, climate change presents an existential threat to it.

To understand the impact of climate change on agriculture, the geological timeline of the Earth's temperature must be examined. As previously mentioned, for the past 10,000 years during the Holocene era, the temperature of the planet has been remarkably stable, relatively mild, and predictable, allowing humans to know when to plant crops and when to harvest them. Agriculture didn't develop during the Ice Age when the planet was much colder. There have been minor deviations below the baseline during the Holocene era varying between 1 and 2°C, known as the Late Antique Little Ice Age (AD 536–660) and the Little Ice Age (AD 1300–1850), which resulted in severe weather, crop failures, famine, death, and war.[48] The causes of these deviations is unknown. Philipp Blom, a German historian, found a correlation between crop failures, famine, and witch burnings in his research on the Little Ice Age. Poor, elderly women were typically accused of cavorting with the Devil, which caused severe storms and destroyed crops.[49] Blaming scapegoats is not a relic of the past. Blaming Asian-Americans for the emergence of COVID-19 has resulted in racially motivated violence.[50]

The three-dimensional One Health matrix reveals the microbial connections between agriculture, health, and climate change. Humans are adversely impacting upon the Earth's microbial ecosystems through the production of massive quantities of fecal matter which cause widespread environmental and ecosystem contamination and disease. Fecal microbes not only contaminate soils and waterways, but also contribute to climate change through methane and nitrous oxide emissions. Most sanitation systems only process human but not animal fecal matter. Animal fecal matter is typically used as a fertilizer in plant agriculture. Management of agricultural soils containing high levels of nitrogen contribute over half of the nitrous oxide emissions from the agricultural sector.[51] The One

Health analysis provides the information needed to address agriculture's greenhouse gas emissions which worsen climate change. The hard part is implementing policies to address them.

In September 2016, the United Nations General Assembly met in New York and agreed that antimicrobial resistance threatened global health. They recognized that a One Health approach was essential to addressing the problem and called upon the World Health Organization, the Food and Agriculture Organization, and the World Organisation of Animal Health to develop a global antimicrobial resistance framework to help countries develop national action plans.[52] The framework's third objective is to reduce the incidence of infection through effective sanitation, hygiene, and infection control measures. Unfortunately, the reference manual makes no mention of the issue of animal fecal matter contaminating environments and ecosystems.[53] This oversight seriously undermines a One Health approach to antimicrobial resistance.

Since the early 1990s, the United Nations Framework Convention on Climate Change (UNFCCC) has held annual climate change conferences, known as Conference of the Parties, or COP, to establish legally binding commitments by nations to reduce their greenhouse gas emissions. The Kyoto Protocol, adopted in 1997, expired at the end of 2020. One hundred and ninety-seven countries participate in these annual efforts.[54] In 2011, during COP 17 in Durban, South Africa, an *ad hoc* working group requested that the Subsidiary Body of Scientific and Technological Advice (SBSTA) work on issues related to agriculture. Two years later, workshops assessing agricultural issues related to climate change began to be held at COP meetings, focusing primarily on adaptation measures, improved soil health and fertility, improved manure management, improved livestock management, and contingency plans for severe weather events to ensure food security.[55] In November 2017, SBSTA 47 drafted the "Koronivia joint work on agriculture" which was adopted as decision 4/CP.23 at COP 23 in Bonn, Germany, which invited countries to submit their views on agriculture, food security, and climate change.[56] It did not address agriculture's greenhouse gas emissions or the targets needed to reduce them. Developing effective policies requires acknowledgement that the problem exists.

Conclusion: Connecting One Health and Climate Change

In December 2015, 193 states and the European Union adopted the Paris Agreement and agreed to limit the global average temperature rise to well below 2°C above pre-industrial levels, and to aim for 1.5°C.[57] Unfortunately, the Paris Agreement did not provide stimulus or guidance to develop agricultural greenhouse gas reduction policies.[58] Despite agriculture's sizeable greenhouse gas contributions, no country has developed policies to reduce them. In the United States, California and New York State passed greenhouse gas reduction policies that include agriculture.[59] In 2022, the US Congress passed the Inflation

Reduction Act, which includes investments of $21 billion for farmers, ranchers, and forest landowners to reduce greenhouse gas emissions.[60]

Internationally, policies to reduce greenhouse gas emission from agriculture remain aspirational. In November 2021, representatives of the 197 member-countries attended the COP 26 meeting in Glasgow, Scotland to deliberate on climate change. The conference had been delayed by one year due to the COVID-19 pandemic. The goal of the conference was to set targets to limit the warming of the planet to 1.5 degrees. Unfortunately, the targets were not reached,[61] nor were agriculture's greenhouse gas emissions discussed. Ignoring these potent, but short-lived, greenhouse gases missed an important opportunity to impact climate change.

For civilization to continue in the twenty-first century and beyond, it must learn to exist sustainably in a microbial world. With its sense of superiority to other animals and to nature, humanity has the hubris to continue producing massive amounts of wastes with inadequate regard for the consequences. Agriculture and civilization rely on a stable, mild, predictable climate for food security. Without it, civil society breaks down. Food safety, food security, and water quality are essential for health and wellbeing. Middle- and low-income countries are the most vulnerable to the excesses caused by wealthy nations. A One Health approach recognizes and highlights the connectedness of these issues, which are usually discussed in isolation, and presents them in a coherent narrative. Acknowledging that human health depends on healthy plants, animals, and a healthy planet would go a long way towards developing effective and equitable public policies that are necessary to achieve many of the United Nations' Sustainable Development Goals to ensure humanity's long-term survival.

Notes

1 "One Health," Centers for Disease Control and Prevention, April 19, 2022, https://www.cdc.gov/onehealth/index.html.
2 "The One Health Umbrella," One Health Initiative, November 29, 2019, https://onehealthinitiative.com/the-one-health-umbrella/.
3 "One Health," World Organisation for Animal Health, 2021, https://www.oie.int/en/what-we-do/global-initiatives/one-health/.
4 Peter M. Rabinowitz et al., "Incorporating One Health into Medical Education," *BMC Medical Education* 17 (2017): 1–7.
5 Laura H. Kahn, "Developing a One Health Approach by Using a Multi-Dimensional Matrix," *One Health* 13 (2021): 100289.
6 Paul G. Falkowski, *Life's Engines: How Microbes Made Earth Habitable* (Princeton: Princeton University Press, 2015).
7 Vincent B. Young, "The Role of the Microbiome in Human Health and Disease: An Introduction for Clinicians," *BMJ Clinical Research* 356 (2017): j831.
8 Robin Wall Kimmerer, *Braiding Sweetgrass: Indigenous Wisdom, Scientific Knowledge, and the Teachings of Plants* (Minneapolis: Milkweed Editions, 2013).
9 Linda Etchart, "The Role of Indigenous Peoples in Combating Climate Change," *Palgrave Communications* 3 (2017): 17085; Richard Schuster et al., "Vertebrate Biodiversity on Indigenous-Managed Lands in Australia, Brazil, and Canada Equals that in Protected Areas," *Environmental Science & Policy* 101 (2019): 1–6.

10 Helen Kopnina et al., "Anthropocentrism: More than Just a Misunderstood Problem," *Journal of Agricultural and Environmental Ethics* 31 (2018): 109–127.
11 Barbara Ehrenreich, "'Humans Were Not Centre Stage': How Ancient Cave Art Puts Us in Our Place," *The Guardian*, December 12, 2019.
12 Zelda Caldwell, "Three of the Oldest Images of Jesus Portray Him as the 'Good Shepherd,'" *Aleteia*, May 12, 2019.
13 Yuval Noah Harari, *Sapiens: A Brief History of Humankind* (New York: HarperCollins, 2015), 77–97, 228–232.
14 Laura H. Kahn, "Can We Remain Food Secure Amid Climate Change?," *Bulletin of the Atomic Scientists*, September 12, 2016.
15 Adam Smith, "The Wealth of Nations," Library of Economics and Liberty, accessed March 30, 2021.
16 "US and World Population Clock," United States Census Bureau, accessed March 31, 2021.
17 Carys E. Bennett et al., "The Broiler Chicken as a Signal of a Human Reconfigured Biosphere," *Royal Society Open Science* 5, no. 12 (2018): 180325.
18 Ulrich Zeller, Nicole Starik, and Thomas Göttert, "Biodiversity, Land Use and Ecosystem Services—An Organismic and Comparative Approach to Different Geographical Regions," *Global Ecology and Conservation* 10 (2017): 114–125.
19 Richard McKenzie et al., "Success of Montreal Protocol Demonstrated by Comparing High-Quality UV Measurements with 'World Avoided' Calculations from Two Chemistry-Climate Models," *Science Reports* 9 (2019): 12332; Feike Sijbesma, "Our Minds Are Wired to Fear Only Short-Term Threats. We Need to Escape This Trap," *World Economic Forum*, June 27, 2016.
20 "Food Security," FAO Agricultural and Development Economics Division, June 2006, http://www.fao.org/fileadmin/templates/faoitaly/documents/pdf/pdf_Food_Security_Cocept_Note.pdf.
21 "The 17 Goals," United Nations, accessed April 6, 2021.
22 "Universal Declaration of Human Rights," United Nations, accessed April 6, 2021.
23 Mari C. W. Myhrstad et al., "Dietary Fiber, Gut Microbiota, and Metabolic Regulation—Current Status in Human Randomized Trials," *Nutrients* 12, no. 3 (2020): 859.
24 Dan Buettner, *The Blue Zones* (Washington, DC: National Geographic, 2008).
25 Richard Wrangham, *Catching Fire: How Cooking Made Us Human* (New York: Basic Books, 2009).
26 Romain Espinosa, Damian Tago, and Nicolas Treich, "Infectious Diseases and Meat Production," *Environmental and Resource Economics* 76, no. 4 (2020): 1019–1044.
27 Alina Petre, "7 Supplements You Need on a Vegan Diet," *Healthline*, October 15, 2019.
28 Rob Cook, "World Beef Consumption Per Capita (Ranking of Countries)," Beef2Live, April 6, 2021.
29 Benjamin Elisha Sawe, "Countries with the Highest Rates of Vegetarianism," World Atlas, September 20, 2019.
30 "Distribution of Vegetarians Across India in 2014, by Major State," Statistica, accessed April 7, 2021.
31 "Indian States by GDP per Capita," Statistics Times, March 1, 2021.
32 Jamuna Prakash, "Chapter 6—The Dietary Practices and Food-Related Rituals in Indian Tradition and Their Role in Health and Nutrition," in *Nutritional and Health Aspects of Food in South Asian Countries*, eds. Jamuna Prakash, Viduranga Waisundara, and Vishweshwaraiah Prakash (London: Elsevier, 2020), 75–85.
33 Michael Pellman Rowland, "Demand for Meat Is Growing Rapidly in India. This Could Impact All of Us," *Forbes*, December 17, 2017.
34 Marion Nestle, *Food Politics: How the Food Industry Influences Nutrition and Health* (Berkeley: University of California Press, 2002).
35 Taro Gomi, *Everyone Poops*, trans. Amanda Mayer Stinchecum (La Jolla: Kane/Miller Book Publishers, 2001).

36 David M. Berendes et al., "Estimation of Global Recoverable Human and Animal Faecal Biomass," *Nature Sustainability* 1, no. 11 (2018): 679–685.
37 For volume, we assume that fecal matter has the density of water: 3.9×10^{12} kg $= 3.9 \times 10^{9}$ m^3. The combined surface areas of NYC + LA = 303 + 503 miles^2 = 806 miles^2 = 2087 km^2 = 2×10^{9} m^2. Height = volume ÷ surface area. Therefore, height = 3.9×10^{9} m^3 ÷ 2×10^{9} m^2 = 1.95 m = 6 feet. Therefore, fecal matter would cover NYC + LA to a height of 1.95 meters or 6 feet. For NYC alone, the height of the fecal matter layer would equal around 16 feet.
38 "WHO Estimates of the Global Burden of Foodborne Diseases," World Health Organization, 2015.
39 "Diarrhoeal Disease," World Health Organization, May 2, 2017.
40 Laura H. Kahn, "How Human and Animal Excrement Harm the Planet's Ecosystem," *Bulletin of the Atomic Scientists*, November 14, 2019.
41 Laura H. Kahn, "Antimicrobial Resistance: A One Health Perspective," *Transactions of the Royal Society of Tropical Medicine and Hygiene* 111, no. 6 (2017): 255–260.
42 "Sources of Greenhouse Gas Emissions," United States Environmental Protection Agency, April 14, 2022.
43 "Overview of Greenhouse Gases," United States Environmental Protection Agency, accessed April 14, 2022.
44 "Reducing Enteric Methane for Improving Food Security and Livelihoods," Food and Agriculture Organization of the United Nations, 2022.
45 EPA, "Overview of Greenhouse Gases."
46 Michael B. Eisen and Patrick O. Brown, "Rapid Global Phaseout of Animal Agriculture Has the Potential to Stabilize Greenhouse Gas Levels for 30 Years and Offset 68 Percent of CO_2 Emissions this Century," *PLOS Climate* 1, no. 2 (2022): e0000010.
47 Joeri Rogelj et al., "Emissions Gap Report 2020," United Nations Environmental Programme, 2020, https://www.unep.org/emissions-gap-report-2020.
48 Kahn, "Can We Remain Food Secure."
49 Philipp Blom, *Nature's Mutiny: How the Little Ice Age of the Long Seventeenth Century Transformed the West and Shaped the Present* (New York: W.W. Norton & Company, 2019).
50 Sam Cabral, "Covid 'Hate Crimes' Against Asian Americans on Rise," *BBC News*, April 2, 2021.
51 EPA, "Sources of Greenhouse Gas Emissions."
52 "Draft Political Declaration of the High-Level Meeting of the General Assembly on Antimicrobial Resistance," General Assembly of the United Nations, September 21, 2016, https://www.un.org/pga/71/wp-content/uploads/sites/40/2016/09/DGACM_GAEAD_ESCAB-AMR-Draft-Political-Declaration-1616108E.pdf.
53 "Antimicrobial Resistance: A Manual for Developing National Action Plans," World Health Organization, 2016, https://apps.who.int/iris/handle/10665/204470, 29.
54 "History of the Convention," United Nations, 2022.
55 "SBSTA Work on Agriculture," United Nations, 2022.
56 "Framework Convention on Climate Change," United Nations, February 8, 2018, https://unfccc.int/sites/default/files/resource/docs/2017/cop23/eng/11a01.pdf, 19.
57 "The Paris Agreement," United Nations, 2022.
58 Jonathan Verschuuren, "The Paris Agreement on Climate Change: Agriculture and Food Security," *European Journal of Risk Regulation* 7, no. 1 (2016): 1–4.
59 "Implementing California's New Methane Reduction Efforts," Dairy Cares, April 2017. Climate Leadership and Community Protection Act, New York State, July 2019. https://climate.ny.gov/Our-Climate-Act
60 "The Inflation Reduction Act Drives Significant Emissions Reductions and Positions America to Reach Our Climate Goals," US Department of Energy, Office of Policy, August 2022. https://www.energy.gov/sites/default/files/2022-08/8.18%20Inflation ReductionAct_Factsheet_Final.pdf
61 "COP26 Closes with 'Compromise' Deal on Climate, But It's Not Enough, Says UN Chief," *UN News*, November 13, 2021.

3

SPILLOVER INTERFACES FROM WUHAN TO WALL STREET

An Interview with Chris Walzer

Irus Braverman

***Chris Walzer** is Executive Director of Health at the Wildlife Conservation Society (WCS) in New York. He is a board-certified wildlife veterinarian, tenured professor of Conservation Medicine at the University of Veterinary Medicine in Vienna, Austria, and author of more than 120 peer-reviewed research publications and numerous book chapters. Walzer has an internationally recognized diverse One Health expertise. His work focuses on the nexus of emerging zoonotic-origin pathogens, environmental encroachment, and the commercial wildlife trade. Walzer is the recipient of several research and service awards, most notably the Distinguished Environmentalist Award from the Mongolian Ministry of Nature and Environment for contributions to the conservation of Mongolia's rare and endangered species.*

This interview was conducted on June 4, 2021, transcribed by Margaret Drzewiecki, and edited by the author, in communication with Chris Walzer.

IB: Please tell us in a few words your definition of One Health and why it is important.

CW: The idea of One Health is that the environment, animal health, and human health are all interrelated. Not only has our environment changed dramatically in the past decades but so has our understanding of health and disease. Today, health can no longer be viewed as simply the absence of disease, both disease and health being products of complex interactions and systems. Unfortunately, health appears in many instances entrenched in and focused on disease. Today, a consensus exists that health must be viewed beyond parasites and pathogens,

FIGURE 3.1 A palm civet (*Paradoxurus hermaphroditus*) at a wildlife market in Vietnam. This species is a popular delicacy throughout parts of Southeast Asia and the main ingredient of the dragon-tiger-phoenix soup prized by wealthy individuals in the Guangdong Province of Southeast China. The species acted as a bridging host in the spillover of SARS virus to humans in 2002–2003. Photo: WCS-Vietnam, used with permission.

DOI: 10.4324/9781003294085-5

incorporating economic, sociopolitical, evolutionary, and environmental factors, while also considering individual attributes and behaviors.

This is not a new idea, really. Hippocrates already pointed out that the environment was really important in his "On Airs and Water."[1] Obviously, the understanding of health and of infectious disease or disease in general was very rudimentary at the time, but Hippocrates did point out that the environment in which humans live had an influence [on their health] so this concept is very old. It wasn't acceptable [at the time] to go around chopping up humans and seeing why they died, so a lot of the knowledge was gained by dissecting animals and finding out how they actually work. Only much later, often illicitly, this was also determined in humans. So, there was a very close relationship [between human and animal health], and if you look at the history, you'll see that human doctors were treating animals and animal doctors were treating humans.

In the 1700s and 1800s [is when] you start seeing separation between human and animal medicine, [and both] often focused on disease rather than health. And then … you start seeing the specialization happening and clear divergence between veterinary medicine and medical science and, … within those disciplines, even more specialization which led to what I would call "hyperspecialization." If you go today to an expert for your fingers and you look yellow in the face because you're suffering from a liver condition, this specialist is most likely not going to comment on the fact that you look like death, … and too bad if you died two weeks later. [For this reason,] in the 1960s at UC Davis they came up with this idea of One Medicine.[2] Calvin Schwabe [then] talked about [this concept].[3] And that was sort of a precursor for One Health, although I have to say that One Medicine is really something very different: it's actually still driven by this idea that one can inform the other, but it doesn't realize the interconnections. So it is an interesting and important stage, but neglects the interconnections and the dependencies of the three: environment, human, and animal health.

In 2004, with the Manhattan Principles facilitated by the WCS [Wildlife Conservation Society, where CW works] and the Rockefeller University [as well as others]—that was the first time the public perceived One Health. And it gained traction. The World Organisation for Animal Health, or the OIE,[4] adopted it, word-for-word, in their mission, and other organizations started using the term. The 2004 Manhattan Principles are a set of 12 principles that are still very focused on animal health, human health, and infectious diseases.[5] The environmental component is not that strong and climate change is not even mentioned. It's not like in 2004 no one knew about climate change; it just wasn't seen as related.

IB: Isn't that also because of this hyper focus on medicine and the narrow understanding of the difference between medicine and broader health?

CW: Yeah, I think so. Along with the wider availability of antibiotics in the 1970s and 1980s there was also this idea that we were going to vanquish all infectious diseases. There was a big focus on infectious diseases at this time, at least in the West. When I went to vet school in the 1980s, there was this idea that we were going to get rid of infectious diseases. It was actually the practitioners and the medical experts from the southern hemisphere, the ones who are working on tropical diseases, often parasitic diseases, who realized that it's not that simple—there are intermediary hosts and the environmental conditions play an important role. For example, understanding the ecology of the snail in schistosomiasis, a parasitic disease of humans with significant morbidity and mortality, is really important. Additionally, if you change the landscape—specifically, the flow rate of water where snails and humans can interact—then you are going to change infection pressure. So if you go into standing water and bathe you will actually pick up this parasite from the snail.

FIGURE 3.2 *Biomphalaria* is a genus of air-breathing freshwater snails. The human disease schistosomiasis (snail fever), caused by all *Schistosoma* species, infects 200 million people annually. The fluke, which is found primarily in tropical areas, infects mammals (including humans) via contact with water that contains schistosome larvae (cercariae), previously released from the snail. Infection occurs via penetration by cercariae through the skin (from Wikipedia). Fred A. Lewis, Yung-san Liang, Nithya Raghavan & Matty Knight, the NIH-NIAID Schistosomiasis Resource Center. CC BY 2.5.

IB: And you're saying these realizations actually came from the Global South?

CW: Well, this happened in a colonial context, of course. Because those who were medically trained were mostly Europeans. This was in the 1970s and 1980s. Now we're seeing the Global South being the leader on many of these infectious diseases. Anyway, One Health came about initially with a strong focus on veterinary medicine, infectious diseases, and human health, and was sort of hijacked by the human health field between the Manhattan Principles of 2004 and the Berlin Principles of 2019.[6] Our motivation for the update of the 2004 Principles through the 2019 Berlin Principles was really that we felt that the term One Health had been negligently simplified by different actors. This was not done in a malicious way, they just used it as a public relations ploy. It was used a lot in veterinary public health, for example, [where] they would talk about One Health in the context of the farm-to-fork principle. And then there were the blatant misuses of the term. I often joke that [even] an aspirin is now labeled as a One Health product. That's not actually true, but there were products on the market that were developed as One Health products. The pharmaceutical industry uses that term broadly [for medicine that] had nothing to do with even the simplified Manhattan Principles.

And then there was a seminal shift as well with *The Economist* and [The] Rockefeller [Foundation] coming out with the Planetary Health concept in 2014.[7] Basically, they took One Health, [which at that point] wasn't very strong on the environmental side, but was very strong on the human medical side, and incorporated the environment, not just in its direct sense, but also the environment in the societal, political, and economic context. And that was a big change. The environment for human medical practitioners was "what's your house like?" The lead pipes, the heating, and things like that. So the definition of the environment changed considerably with Planetary Health.

IB: Does that also connect with environmental justice issues?

CW: Yes. In 2014, *The Lancet* [an influential medical journal] came out with a call to create a movement for Planetary Health, and in 2015 the Rockefeller Foundation, together with *The Lancet* and *The Economist* came out with that famous Planetary Health issue.[8] It really brought into central focus the socioeconomic and political components. Justice, as we understand it today in 2021, was probably not as at the forefront as it is now. But now everyone's become aware of health inequities across spatial entities. COVID-19 highlighted the multifactorial inequities across the five boroughs in New York City, for example. In the Bronx, where I live, there were far higher rates of hospitalization and death when compared to hospitals in Manhattan. Everyone knows that—but it has not necessarily transferred into policy. Now we're starting to see wider discussions.

IB: So what was the contribution of Planetary Health to the understanding of One Health?

CW: It brought into [One Health] the field of economics and sociopolitical drivers. But there's a difference between highlighting it and recognizing these factors and then addressing the injustice of the situation. For me, that's still a leap that needs to happen. We'll see people who will talk about the Global South, and who will say the resources and the access to resources in the Global South is not as good as it is in the Global North. But very few are saying that this is a global injustice.

The 2019 Berlin Principles really build strong[ly] on the concept of Planetary Health, [which] includes socioeconomic drivers and reinforces the foundational importance of an intact and functioning environment. Looking back, I would probably stress that aspect even further, [emphasizing] that intact and functioning ecosystems are critical and just foundations for health for all.

IB: So, is everybody today pretty much on the same page with realizing the foundational importance of intact and functioning ecosystems?

CW: No. So here's a good example and a very timely one [from] last week. The Global Health Assembly just met to discuss "Preventing the Next Pandemics."[9] [But] no one talked about prevention. Now it's all about preparedness. So we assume the next pandemic will happen anyway, and we might as well get prepared. The environment, and preventing and recognizing the importance of an intact environment, was not addressed at all. What is interesting, of course, is that the WHO [World Health Organization], [which] is a multilateral organization, absolutely understands the importance of the environment. They are also working a lot on climate change and the future impacts of climate change, and so on. But the member states that make up the WHO and that sit on the Global Health Assembly—all 194 of them—they obviously didn't get it and didn't want to get it. So that's a bit worrisome. It's the first pushback we've seen in 16 months against the idea that we need to conserve the environment. In the last 16 months [since COVID], everyone has been like, "Yeah, we need to do something; the planet is sick, we need to do something." This is the first time that there's been a reversal to "let's go back to what we were doing before and just prepare better for the next one."

IB: Isn't this because we have an exaggerated sense that technology might save us? This brings me to another question I have been meaning to ask you, and that is what you think about the term Anthropocene, and do you prefer it over the term the Capitalocene.[10] To clarify, some point out that blaming all humans for today's accelerated anthropogenic changes is problematic since it is mainly a certain group of humans—namely, the wealthier folks in developed countries—who are responsible for the situation.

CW: I took great pleasure in saying that COVID-19 [marks] the end of the Anthropocene. Welcome to the Capitalocene, in all its extent. And I enjoyed it because, obviously, I was trying to stimulate people to think a bit more than just about a virus, and to think more about where the spillover frontlines really are. Within a traditional thinking framework, the spillover frontlines would be wildlife markets and the edges of forests paired with behaviors and practices that increase spillover. But in reality, we must look a lot more closely at Wall Street, financial institutions in general, and corporations—as their decisions and practices significantly drive destruction and spillovers.

I believe quite strongly that COVID has given us an opportunity to highlight that not only are there climate change externalities—when you work in fossil fuel, for example—but there are also health externalities that someone needs to be paying [for]. You're cutting down forest in Brazil and [if] you do a partial deforestation, you change the rate of Zika infection in that area, [and as a result] the local hospitals will be treating that many more additional patients and there will be more and more neurological deficits in children, and so on and so forth. All those costs are not being considered.

So, I believe that we're completely in the Capitalocene. The Anthropocene for me is actually a terrible term because it implies that everyone is responsible. I'm responsible: I'm super privileged and I drive a car, and I drink wine and throw plastic out every now and then. Sure, I'm responsible, but, compared to Amazon or Exxon, my role is minute. We're always working on symptoms. But the root cause, for me, is definitely a growth-based economy that does not recognize and pay for and mitigate the externalities of what it is producing, consuming, and wasting.

IB: In addition to and alongside capitalism, there is the tight relationship between conservation and colonialism, right?

CW: In North America, there was genocide and [oppression of] Indigenous peoples and wildlife was wiped out. First, you had the colonial phase, the creating of nature and wildlife fortresses [by] keeping people out. Then, much later, you have the recognition that Indigenous peoples and local communities have rights to the land which need to [be] upheld and secured because their livelihoods, wellbeing, and cultural identities most directly depend on the sustainable use of natural resources.[11]

Therefore, today Indigenous peoples and local communities are often WCS's best partners and the strongest advocates for conservation. They are the guardians of the land. I mean, we work with hundreds of different Indigenous groups around the world and over a thousand local communities. [There] are also Indigenous groups which are not formally recognized as Indigenous peoples. The pushback is brutal, obviously in Brazil, but also in other places in Mesoamerica, where Indigenous leaders—environmental leaders—are assassinated by those who want access to the land and associated resources. It's a constant battle, really.

IB: Coming back to One Health. If the term One Health has been already so appropriated and so batted around in so many ways, would it make sense to perhaps coin a different term that will highlight the uniqueness of today's challenges?

CW: Let me indeed start from the challenges. My challenges are that I live in the real world of policymakers, decisionmakers, donors, and environmentalists—that's [who] I have to interact with. So that means I need to speak a language that they can understand. … It needs to be a common language that we speak. One Health has really become an accepted term, and it has made the interconnectedness of the three pillars very clear. However, for many governments and for some multilaterals and individuals, it is still a new term. The WHO had it in their documents, but most likely never gave it too much thought. … So, changing that term for something else makes no sense—it's embedded in legislation, it's embedded in policy, it's everywhere. My role is to expand One Health [further so as] to bring in these socioeconomic justice aspects. Everything can be in One Health, that's for sure. I love it when my boss says, "but everything is One Health!" And I respond, "Well, good, then you've got it, right?" … One Health is everything—every aspect is One Health.

Traditionally, One Health is illustrated with a Venn diagram: three static circles of animal and human health and the environment [see Figure 4.2 in John H. Amuasi's chapter, this volume]. But I prefer to use what's called the SDGs [the Sustainable Development Goals of the United Nations] wedding cake that you can see in Figure 3.3. You'll see the SDGs stacked up on top of each other.[12] You have the biosphere with the SDGs 6, 13, 14, and 15 at the bottom and then health is one layer above. … protecting the biosphere is foundational to the health of all.

[Although], at present, during a pandemic, we are focused on infectious diseases, we must necessarily also consider non-infectious One Health issues. My example is corn syrup. We produce corn and subsequently high-caloric corn syrup across vast stretches of North America, to the detriment of the soil and the environment. This production is subsidized, [which is] a political and economic decision. The corn syrup ends up in sugary sodas and practically all processed food and is associated with the obesity crisis and increasing rates of type 2 diabetes. Wellbeing is compromised and more than 35 million individuals per year in the United States end up requiring health care. These health costs of corn syrup production are not covered by the producers. In a simplified summary, taxpayer-subsidized corn syrup production outsources the health externalities to taxpaying individuals who are now often sick, while driving the profits in the production entities and the healthcare industry in the United States—all this while ignoring the environmental costs.

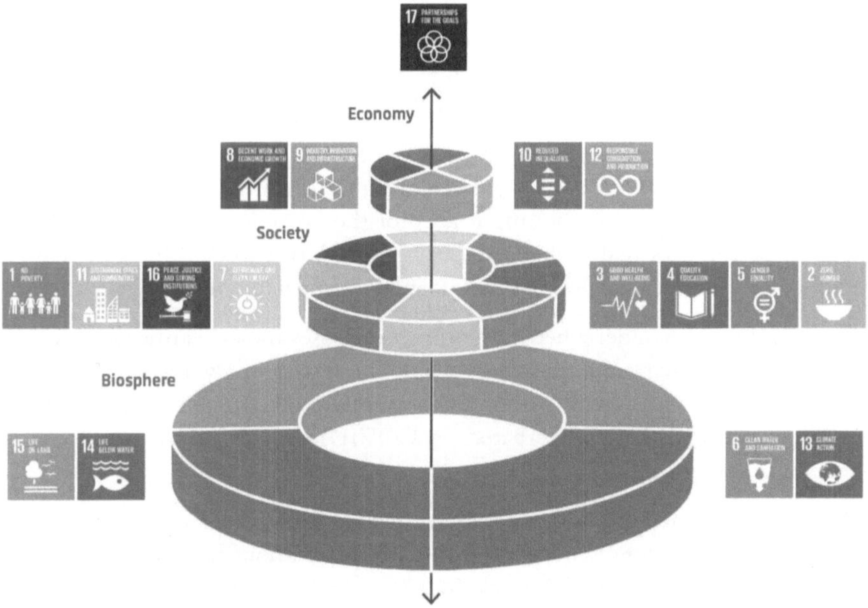

FIGURE 3.3 A novel way of conceptualizing the Sustainable Development Goals and their linkage to food, the illustration of the SDGs wedding cake "describes how economies and societies should be seen as embedded parts of the biosphere … [and] is a move away from the current sectorial approach where social, economic, and ecological development are seen as separate parts." Credit: Azote for Stockholm Resilience Centre, Stockholm University. CC BY 4.0.

IB: What you're pointing at is that a profit-driven healthcare industry, contrary to the state perhaps, has less interest in broad prevention. Making and keeping people unhealthy is good for business.

CW: In the past decades, huge gains in human health and wellbeing have been achieved. However, the broader costs to the environment have not been calculated. Unrestrained and continuous economic growth, paired with reduced social services, have dramatically increased health inequities across the globe. This leads back to the Capitalocene idea. There's a big difference between the United States and Europe: health is not an unregulated business opportunity in Europe—it is a tax-funded service provided by the state. Of course, corporations and service providers have to make money, but within tightly regulated boundaries. Health care is available to all and it's not simply a business opportunity. The United States has by far the most expensive healthcare system. And it is nowhere near the top ten in terms of quality. Maternal mortality in the United States is [the] highest amongst industrialized nations and, when you look specifically at BIPOC [Black, Indigenous, and People of Color] maternal mortality, it is even worse. If you take the wealthiest zip codes in the United States, the health care outcomes are still not in the top ten. And they have at least four times the costs and it's not available for all, it's crazy!

IB: Yes it is. And rather than changing this sick situation, many more countries are going down this path. But let me take you back to the Berlin Principles, and you've highlighted that One Health was aligning with the Planetary Health understanding by integrating economic and sociopolitical considerations. What has COVID added to your understanding of One Health and how differently do you see One Health now in comparison to how you understood it pre-COVID?

CW: COVID didn't change anything, fundamentally, but it helped clarify what's important. Up to 2019, I would always highlight the transdisciplinary ambition, but what I neglected was the importance of the cross-sectoral or intersectoral aspect, and that has really changed through COVID—the realization that all sectors of society and governance are affected or impacted in some way.

IB: Beyond this, I wonder whether focusing on viruses and other microbes doesn't erode some of these boundaries that we hold on to so strongly—for example, the boundaries between humans, nonhuman animals, and the environment. Does this focus on microbes, exacerbated by COVID, start to undermine some of these mega-categories?

CW: Well, there are two ways of looking at that. One is that we're all one. But clearly, our interactions with everything are not without massive costs, right? So, when we domesticated livestock 10,000 to 15,000 years ago this brought us humans as a species in very close contact on a constant basis with novel disease agents in animals—we brought the animals and their pathogens into our communities and interacted with them. It is thought that during this domestication period, the world population just didn't grow for about 5,000 years.[13] And one of the hypotheses is that this was because there were so many diseases spilling over [and] killing large proportions of the population during the domestication process. And some of those viruses and pathogens that we acquired from livestock during the domestication process, we've now made uniquely our own—measles, mumps, and others. Through a process of coevolving, we developed the immune system necessary to deal with these pathogens. So, on the one hand, yes we're all one and because [the human population is growing and needs ever-more resources], we're constantly losing natural barriers while creating new interfaces and interacting with a whole slew of new potential pathogens in a very interconnected fashion, so the protective barriers [are effectively] gone. Take the Ebola virus disease, discovered in 1976 on the banks of the Ebola River, as an example. Decades ago, an outbreak would wipe out a village, actually, more likely, half the village. The people who were impacted just moved away and left their dead behind. But now that every village is connected by road or airport (or both) to everywhere else in the world, the outbreaks have a different dynamic and constitute an immediate potential global threat. What we're seeing is a [human] encroachment into natural barriers and the establishment of new interfaces. It's nothing particularly new. Just in my own life[time], the world population has more than doubled and global GDP grew by an order of magnitude. I just can't wrap my mind around it—when you were born and when you went to high school there were half as many people on this planet as there are now. It's amazing.

IB: I know that a lot of ecologists like to say that the problem is population, but this statement is also problematic in many ways, isn't it?

CW: The problem is not the population per se. The problem is population *and* GDP. So it's the resources used. If you look at the line of population growth paired with GDP it's like drawing a vertical line. … We could have enough food for everyone if we shared equitably. [But] we're wasting 40 percent of our food on average across the globe. So the problem is the disproportionate use of resources in the North.

IB: Yes, and this circles back to the environmental justice point. But let me go back again to this idea that domestication was a point of contact. I have a layperson question here: don't our bodies come into indirect contact through air and water and soil, even before they come into direct physical contact?

CW: But they don't—that's a misconception. … Our immune systems mobilize to cope with the environments we are living in, both within the span of a single lifetime, and over the course of generations. So, as a human being [living] 20,000 to 30,000 years ago, our immune systems became ideally adapted to our surroundings in that era. Everything has a cost, so mobilizing an immune system to eradicate [germs with which we have no] contact is a waste of energy, and would come at the cost of reproductive output and thermal homeostasis. Our bodies only do what is absolutely necessary, for bioenergetic reasons. So, when our species domesticated animals and began living in close proximity to them for the first time, our immune systems took several generations to become optimally suited for protecting us under those conditions.

IB: So the problem of diseases is basically a geographic one?[14] In other words, is it this idea that wildlife and humans now encroach on each other's habitats? And, relatedly, does this also mean that successful conservation efforts could actually encourage pathogen "spillover" by bringing back species such as waterfowl and small mammals to areas where humans also now reside?

CW: That's an interesting way of looking at it. If you cut down the forest and put concrete over it, you're not going to worry about pathogens, okay? The worst place for pathogens and spillover are disturbed landscapes. Here, the full complement of species is lacking, barriers have been removed and we risk shaking out the pathogens. Beyond the simple fact of enhanced pathogen proximity, it is actually still pretty unclear how and why this happens. One hypothesis is the so-called "dilution effect,"[15] and it is thought to work in this way: in very diverse biodiverse landscapes, pathogen prevalence and the risk of pathogen spillover is low, and the reason that works is that there is less contact but additionally the environment contains a lot of additional hosts for potential pathogens and most of them are incompetent. Incompetent hosts can take up pathogens, but make poor petri dishes for them. Pathogens that cannot reproduce efficiently in a host will not be transmitted onwards to other hosts, or will be poorly transmitted, which "dilutes" the pathogen. Let's say you have a bowl of coffee grains which are the pathogen, and you have 50 people in the room, and everyone gets one, or you only have five people, but you need to share them all out and everyone gets six or seven—the dilution effect is stronger the more potential hosts you have. The dilution theory is great. [The idea is that] if biodiversity was high everywhere [then overall health would be better]. Conservation organizations could advocate for biodiversity and you're all good.

IB: But that's exactly what they say, right?

CW: Well, no—we certainly don't. There's a strong case for the dilution theory in specific contexts and it's got most of its drive because the one system where it does work is the northeast of the United States with Lyme disease. ... Unfortunately, really, it may not work in all systems, or at least it's been difficult to prove. If you think of a really biodiverse system, there are so many things going on at the same time that it's been impossible to prove the dilution theory in these complex systems; no one's been able to show them. To test the dilution theory, colleagues have created what are called experimental microcosms where they can manipulate the number, densities, and type of species in a microworld. Interestingly, they have clearly shown that, in the face of a parasitic infection in this microworld, a large diversity of species is protective—dilution works. The important point, however, is that dilution only works when the species assemblage mimics a natural diversity.[16] A random assemblage does not allow for pathogen dilution. That sort of points in the direction that these species and pathogens have coevolved together in a specific environment and this coevolution of traits mitigates pathogen prevalence and spillover. Once you start changing that assemblage with encroachment, human incursions, and deforestation, you may have the same biodiversity indicator and still have 25 species, but it's just not the right ones, because they haven't coevolved.

If you look at it from an epidemiological point of view, we exist and interact across multiple interfaces and compartments. And while these may seem to be hyperconnected, they're not necessarily [connected] from a pathogen immune response point of view. If you look at infectious disease, we still have distinct compartments along a gradient. We have an intact forest compartment, a forest edge compartment, a peri-urban and an urban compartment. A pathogen would have to slip across all these compartments in order to cause a pandemic. During this ongoing pandemic, we are uniquely focused on the pathogen side of the equation. But it is essential to understand the host side as well. A hunter-gatherer in the Congo Basin has a completely different immune system than mine, right? She is perfectly adapted to the pathogen landscape where she has evolved. When I move into her compartment—as a researcher, for example—I probably have to be vaccinated against 15 things. I'm looking out if I get diarrhea, I screen myself for parasites, I might have to take antibiotics after two weeks— because I'm not used to that pathogen environment. At the same time, my microbiome changes as I move into this novel environment and new feedback loops shift my immune responses. I am clearly not adapted. Pathogens can bind, replicate, and evolve in my cells as my immune system is completely naive in this environment. Similarly, our global interconnectedness has misaligned pathogen and immune landscapes. So, while we are one on our planet earth, movement between the compartments still engenders costs and consequences.

IB: Okay, I guess it does then become a geographical paradigm, highlighting the importance of space—but that almost sounds reductionist. And also going back to this idea of the dilution effect and its problems, maybe I didn't understand correctly, but what I understood you saying is that the dilution effect might be relevant only to those intact ecosystems where coevolution has happened, and not so much where humans have intervened and changed the environment. But isn't this idea that there is this pristine environment somewhere problematic because there is no such thing? And couldn't we then adopt a more gradational approach?

CW: Yeah, it is [problematic]. We always talk about thresholds: you remove three species from the assemblage, nothing happens; you remove the fourth one—something happens. There are some great studies in plant communities that have shown that, beyond the diversity, it's also about keystone species: when you remove that one specific species, the whole system collapses—a threshold.[17] A lot of similar work has been done on experimental aquatic environments that are somewhat easier to manipulate. You just remove one invertebrate or add in an extra fish, or something like that, and suddenly the entire system just collapses. Looking in from the outside, you'd think that wasn't such a big difference. But all was intertwined and, without that single linkage, it just doesn't work anymore.

IB: Okay, so are you saying that it is a question of intensity, rather than black or white?

CW: Exactly, it's always a gradient until it hits a threshold and things abruptly change. There [are] some really good examples with so-called vector-borne diseases where the dilution effect also works—for example, mosquito-mediated diseases like Zika and malaria. When you deforest or selectively remove trees, you'll change the habitat for mosquitoes and then certain mosquito species will increase—and it's often those that are the vectors for Zika and malaria. You cut down forest, and you will get an increase in Zika infections in the adjacent communities. This has actually been costed out in some areas. Cutting down the forest brings you $100 for the lumber, but the cost to the community for health services will be $200. I obviously invented those numbers, but that's the principle, which has been shown very robustly in Latin America and Mesoamerica for Zika and for malaria in Malaysia. However, attempts to show a similar effect for malaria in Africa have not succeeded. So yes, this effect exists with vector-borne pathogens but not everywhere and not in every context. Or at least it hasn't been proven everywhere.

IB: When you said that when everything was paved, we wouldn't have any parasites or disease problems anymore—

CW: I didn't say we wouldn't have problems. We wouldn't have infectious agents in those environments. We'd have a lot of other problems. If you cut down that forest, you immediately have all the climate mitigation problems of course. About 30 percent of all climate change effects can be mitigated by nature, so if you cut down that forest, you've got to go find some other way to do it. And then, amongst many other deleterious impacts, you destroy the pest controllers and pollinators—the bats. We simply can't live on this planet without bats, it's just not going to work.

IB: Which brings me back to One Health and to a question we touched upon earlier that I wanted to circle back to. What is your approach toward the involvement of Indigenous perspectives in One Health, and regarding the dynamic between One Health and Indigenous peoples generally?

CW: So, if we're working on the assumption that intact ecosystems with intact functions in nature are the basis of our own health then that aligns completely with what Indigenous knowledge [everywhere] in the world has told us for thousands of years: respect your environment, use it sustainably, and so on and so forth. If you take things from the environment, there's a cost involved and you'll need to give back—there's always been a trade-off. If I'm going to kill this

FIGURE 3.4 In Sabah, Malaysia, large areas of tropical forest and other ecosystems have been cleared and converted for oil palm plantations, destroying critical habitats for numerous endangered species and contributing to global warming. Deforestation is also tied to the emergence of infectious diseases. Photo: Chris Walzer/WCS.

animal for food, I have to honor it in some other way—it's a limited resource and there's a clear understanding of that very basic fact. Acknowledging that we are an integral part of nature and not outside and separate from nature is critically important. And the explanations and the worldviews that are part and parcel of that are a great framework for the One Health approach. But of course reconciling traditional knowledge with scientific method is no simple task and in many instances there is significant tension.

IB: Who and what exactly are in conflict here?

CW: This is not necessarily a conflict, but there are challenges in bridging foundationally different worldviews and overcoming transdisciplinary boundaries. Traditional knowledge encompasses know-how, skills, and practices that are developed, sustained, and passed on, mostly orally, from generation to generation within a community, often forming part of its cultural and spiritual identity. In contrast to the scientific method, which is evidence based, traditional knowledge is evidence informed. Now some Indigenous peoples—for example, the First Nations in Canada—are way ahead of the curve. They are well integrated in health monitoring and surveillance. Key is the "two-eyed seeing" approach as explained by Mi'kmaw Elder Albert Marshall: "To see from one eye with the strengths of Indigenous ways of knowing, and to see from the other eye with the strengths of Western ways of knowing, and to use both of these eyes together."[18] First Nations have made clear statements on the foundational importance of the environment and that their knowledge would be beneficial while also acknowledging the importance of vaccines, antibiotics, and modern diagnostics. One problem is that you often find individuals and groups that have misplaced, maybe even romantic, notions that [when there is] Indigenous knowledge then you don't need modern medicine for anything anymore and you don't need diagnostics. I certainly wouldn't want to go there. But I firmly believe that framing One Health in a holistic Indigenous understanding of nature and our place therein is very beneficial.

IB: So, despite the fact that One Health kind of emerged as a Western idea, now it is seen as something that is in line with Indigenous knowledge?

CW: Yeah. But don't forget that it's not like Western medicine was invented in 1960 and then from then on it was just a colonial enterprise—it's already been around for 2,000 years at least, and there's a lot of tradition there as well. We often forget that. It's not like only the Indigenous peoples have a long history of informing health—Western medicine does too.

IB: So you are basically suggesting that Indigenous knowledge is at least partially romanticized by Western approaches, maybe along the lines of what Edward Said referred to as Orientalism?

CW: Certainly, but it goes both ways. And sometimes it's really very confrontational. I was in a meeting where [we were] discussing One Health with Indigenous leaders present. And one individual held a very long speech about how One Health is an appropriation of ancient Indigenous knowledge. We've known this for years, she continued, and if you had listened to us, everything would be good and nothing would have happened. Well, while there is some truth in that, I'm also like, no, that's not how history works—without [Western] medicine, none of you would be around anymore; it's not like you lived in a vacuum. You also had smallpox and you also had influenza and all these other pathogens killing your communities. Spillovers happen all the time everywhere. And no place is so isolated today that it is not impacted by every other place and community on our planet. Constantly.

IB: Then why are they called spillovers? A spillover seems to imply that there is a containment and then something spills over. Whereas what you're describing is a constant flow.

CW: Good point! So a spillover is just a technical term to basically describe the movement of pathogens between two compartments and, in this case, between species or groups of species. Basically, it's just about germ movement. While we often think of humans as exceptional, we are all different species across a single environmental gradient. Spillovers describe germs spilling over and switching species. For example, the canine distemper virus spilled over from dogs into seals. That's a classic spillover with quite a big species jump, right? And similarly, spillovers from bats into horses and then from horses into humans, those are spillovers.

IB: So when the pathogens shift between taxa, then it's called a spillover? Does the spillover also apply to shifts between habitats and not only taxa—for example, when pathogens spill from one ecosystem to another, or does the term spillover just apply to species' bodies?

CW: Normally, it applies only to species because someone has to host and carry the pathogens from one ecosystem to another. But you also have spillovers through nonliving entities. A well-known example from the West Coast of the United States is the toxoplasma parasite. Domestic cats defecate in the gardens of houses on the steep shores of the Pacific and when it rains, the fecal matter with the parasite washes down into the sea and infects marine wildlife causing severe neurological disease in species such as the sea otter. That would be something like an environmental spillover. However, in reality it is an alien invasive—[the] domestic cat—[which is] contaminating an anthropogenically highly modified environment. And then there's the rain or water that transmits the pathogen.

Let me add as a side note here that the term "pathogen" simply refers to an organism that can cause disease and that there is a large diversity of them. The term becomes somewhat problematic when used in policy, as it is often perceived as existing in a static, clearly defined context. Many organisms, and especially those that are configured as a major concern to us humans, are actually potential pathogens that still need to evolve the necessary traits to elicit disease. Often, this will happen after spillover from wildlife and other animals. The uncertainty around this process is difficult to grasp and address. But clearly, we need to expand our understanding of pathogens to include evolution, recombination, and individual host attributes that allow disease to occur. And we can also talk a lot about the term "disease" and probe into that, but I'll flag that for another time.

You know, this is also a gradient. If you don't recognize things as gradients, you can get very hung up on terms. For example, zoonosis is a classically defined term which necessitates a reservoir. A pathogen circulates and is maintained in an animal species and then continuously or repeatedly spills over [rabies, for example]. What we see with pathogens like SARS-CoV-2, and HIV as well, is that a few spillover events lead to infections in novel species such as humans and subsequently evolve to become a predominantly human disease.

IB: That comes back to that question that we had in the workshop about whether or not the term "zoonotic" is a misconception, because it puts the blame on the animal. And humans are then configured as victims of wildlife. Does this criticism of the term zoonosis resonate with you?

CW: Yes. I mean this should not become a question of blame. We all carry our set of pathogens, you know. We potentially also transmit pathogens to wildlife if we come in contact with them. But that wouldn't be zoonotic, it would be "anthropozoonotic." However, the modern definition of zoonosis is that it moves between animals and human—and there is no required directionality for this movement. It's seen as going both ways.

IB: Oh, really? For something to be referred to as zoonotic, the origin doesn't have to be from animal to human?

CW: Right, it doesn't have to be in one direction or the other, it goes both ways. But the reality of the world is that there's human exceptionalism and all is viewed and discussed in this context. It's easier for me to simply accept human exceptionalism on a pragmatic level in order to interact with the rest of the world. If I portray myself as being equal to a badger when I talk to policymakers, this won't go very far, if you see what I mean. And it is not really going to help me co-create policy or legislation to mitigate the impacts from infectious diseases.

IB: Your life would be simpler. I mean, as a badger. I definitely want to be a cat [laughing]. Now seriously, I have a question coming back to our discussions in the workshop. It's already before COVID that we started hearing about herd immunity, and you were talking about that a lot in the various discussions you and I have had over the years, and also during the workshop. Herd immunity is another one of these loaded concepts that refers to spillovers from the nonhuman world into the human world and back. So I'm wondering if and how our conception of "herd" and of "herd immunity" has changed in light of the COVID pandemic.

CW: It certainly has. I mean, from the scientific point of view, there's been no change. It's just a term that now everyone thinks they understand. You know, everyone has become a hobby epidemiologist—everyone now knows what an R-value is and [what the meaning of] prevalence [is], and infection rates, and incidences. … Everyone sort of has a notion of that now. The concept of herd immunity is actually related to that geographic conception we spoke about earlier because it indicates a particular space where people or species have developed sufficient immunity as a cohort to limiting pathogen spread and impacts.

IB: And then, when globalization happens, that community changes, right?

CW: Exactly.

IB: I see. So where do you see One Health moving from here? How do you see the future?

CW: In the next two to three years we're going to see a lot of One Health in policy and in legislation. We now have the tripartite-plus One Health high level expert panel.[19] It's obviously highly political, but a clear and important recognition of the importance of One Health. And I think we're [also] going to see a One Health-framed pandemic prevention treaty. It's going to take a lot longer for One Health to be fully recognized in a socioeconomic context. We're seeing climate change being addressed in an economic context, so that if you produce and use a gasoline-powered car, there are certain climate costs. It's very simplistic still, but we do acknowledge the economic framing, and I think we will see health also being acknowledged in the economic context, but it's going to take some time. There are a lot of colleagues also working on non-infectious diseases and how society needs to address those. What changes in our individual lives and [in our] economic, social, political systems need to occur to decrease the impact and costs of these diseases? Cardiovascular diseases and diabetes are in orders of magnitude more important and costly than any infectious disease. But, as you pointed out, you can make a lot of money from such diseases. Drugs, tech, and other interventions actually keep multi-moribund individuals alive for a long time and somebody is making money all along until they die. Providing access to health care and, most importantly, framing health and wellbeing as a global commons appears, in my mind, to be a critical step as we move forward and hopefully beyond this pandemic.

IB: And when you say legislation that already incorporates One Health, did you mean international treaties or national legislation?

CW: So there's both. I've actually been involved in legislation efforts here in the United States, with One Health legislation being drafted and policies being made.[20] It's mostly focused around the recognition that the various disciplinary and agency silos and sectors need to work together and communicate.

IB: Intersectional, as you called it earlier.

CW: Yeah, but also more cross-sectoral—human health needs to talk to the agriculture sector and [they] both need to talk to the animal health sector. The EU has this interesting structure where animal health is managed at the EU level from Brussels, [whereas] human health remains national—so they already have an inherent One Health operationalization problem.

IB: It's embedded in the structure. And that, in a nutshell, is the problem of Western medicine, right? Differentiation between physicians who deal with humans and vets who deal only with animals and suddenly they come together and see that there are connections. Eureka!

CW: [Laughs.] With the EU, interestingly enough, the animal health sector and its sole focus on agriculture and livestock production is [indeed] outdated and definitely needs an overhaul. Granted, we see similar structures in many other countries. The idea of integrating the environment and wildlife health is really quite foreign to them still. We're seeing slow and cautious changes to this approach, and it often depends on who is the Minister of Agriculture and the agricultural lobby, how strong [it is], [and] how much space they have to change.

IB: And what about non-Western countries, is there a move toward One Health there?

CW: So you know many, many countries have developed national One Health plans. Nigeria has a good one. Tanzania. Kenya. You can find them online. Bangladesh has a pretty good one. The problem is [that while] these plans are quite easy to draft, they're much more difficult to operationalize and implement. And so, at the moment, this is the big discussion: how can we actually operationalize and implement One Health so that it makes sense and provides added value. We need ministries and administrations to talk with each other, we need shared information to flow between entities. But how does all that get translated in a way that makes sense and that can inform decisions? That's really quite difficult.

IB: This is why you are pointing out that it's so important to work across sectors, right? Because, again, the divides have been embodied in structures that are much harder to alter; they have been institutionalized and so now it's more difficult to make those connections. So you are talking about both distinct forms of institutional bureaucracy and distinct forms of expert knowledge, yes?

CW: Yes, siloed bureaucracy is always one of the potential problems. The other is politics. So let's say you have a coalition government and they split up their ministries. … And then the center-right will take the Ministry of Agriculture and the human health side will maybe go to the social democratic party. So there's already an inherent competition between these two ministries and it's often hard to reconcile between them. And one of the roles we embrace as a conservation organization is that we are great in bringing ministries together and facilitating collaboration. You will not believe how often we are asked, "but what's the other ministry doing?" And they're literally one kilometer down the road. And they say, "Oh, we haven't met them in two years," and you're like "yeah, but you're working on the same thing! This morning, I was at a meeting with them on the exact same issue!" So silos are obviously a huge, huge problem, and it's unfortunate because the administrators are mostly apolitical. You know, after a certain political level, the experts and the administrators basically stay in the institution when parties and governments change. But it's political influence and interventions that are stopping them from adequately collaborating.

IB: So we need spillovers [laughs].

CW: Yeah, that's where we need spillovers. Between administrative silos [laughs].

IB: Shift the spillovers from one institution to another rather than from one taxon to the other.

CW: I might steal that from you next time I talk to Congress.

IB: Go for it. Just don't forget to tell Congress that it was my idea [laughs]. I am realizing more and more that you're an optimist at heart. Whereas I like to dig under all these concepts and show how very problematic they are at their core, and this also pertains to One Health. At the same time, you are [also] a pragmatist, and look forward to the work that can be done using this framework rather than focusing on its problems—and that's really important, so I would like to highlight that. It was really interesting to see how you perceive One Health as much more complicated than just being this holistic health approach that everything is connected, like in the well-known butterfly wing example. And it is really important to understand that you still believe in categories, distinctions, and what you call compartments, where place and geography still matter. I guess there would be no meaning to One Health if this wasn't the case and if there weren't materialities to bridge, in a way?

CW: Exactly. Even in small areas like a city you have completely different prevalences of diseases in some area. Just think of diseases such as opioid addiction. Addiction will be high in certain zip codes, with the prevalence orders of magnitude higher than in other zip codes. And that's a consequence of the environment, the socioeconomic context, and political priorities. That's also the case with depression, maternal mortality, and many other wellbeing and health issues.

IB: It's mind-blowing, honestly.

CW: Yeah. I wanted to say that because you referred to it as "geographic."

IB: And I could sense that you didn't much like that, did you?

CW: No. And I'll tell you why. We do this a lot—we map things. It's a really good way of showing what's going on. But the problem of the map, even the best map, [is that] it's pretty unidimensional.

IB: Two dimensional.

CW: Yeah, sorry yeah, two dimensional, but, in my mind, it is just one dimension.

IB: I see, what you are saying is that it's certainly not three dimensional and surely not four dimensional, right?

CW: Yeah, exactly. You could add in the third dimension, but that wouldn't provide much additional data. What is creating these spatially explicit cohorts is so much more than just their geography. I think that's really essential to understand. Even though we use mapping tools to show where things are, the map only captures a subset of the complexity—and especially not all the drivers.

IB: I am not sure I understand. Could you give me an example?

CW: Sure, let's take a simple one: death after hospitalization in the Bronx, Queens, and Manhattan.

IB: I was thinking you'd give me an example that is more ecological, but okay, I guess humans are part of the ecosystem too [laughs].

CW: Well, it's just what I have in my head now. Admission to hospital and then dying at the hospital was four times higher in the Bronx than it was in Manhattan, right? If you just draw that spatially it gives you nothing. But if you then look at race, economics, income, density of housing, access to health care, and all the other factors, and start piling those on top—then it starts making sense. But there's only so much that can go into a map, even with all the layers.

IB: I understand your criticism of the map and your reluctant approach toward geography, but when I say geography, I approach it in a more complex way that can indeed take into consideration factors like race, class, colonial legacies, and so forth.

CW: Yeah, okay then, that's fine.

IB: It just struck me when you were talking that a lot of your considerations were spatial and geographic in nature. Not in the sense of mapping and GPS and technical coordinates, but more in the sense of the invocation of scales such as global and local, landscapes such as urban and peri-urban and, more generally, the critical importance of materiality. This played out really strongly when you were talking about spillovers. You're talking about these pockets and compartments—and this sounds very geographical at its core.

CW: You know, this is very interesting—so, for example, if you just take Lyme disease and look at a map of the five boroughs in Manhattan or any other city, Lyme disease can only exist in certain areas—they have to be green, they need to have deer, and they need to have mice. So that's not going to happen at the Union Square in New York City because there are no deer there and the mice don't walk that far.

IB: And things will change significantly in New York City when the water comes up because of sea level rise.

CW: Well, that's the other point—it's all dynamic, of course. We talk about infectious diseases and that they can only exist here and there. But when they spill over they can actually move between one spatially explicit area and another one, right? Especially if [the disease is] not too deadly. Diseases that kill everyone really quickly are not much of a problem, evolutionarily speaking. That virus or whatever pathogen is necessarily going to have to adapt its strategy over time and become less deadly.

IB: That image of spilling concrete all over the place to reduce pathogens has stuck with me. Because this was the settler colonial ideal that was used in so many contexts—for example, by the Zionists in Palestine. I'm now studying early Zionist poetry and there are stanzas promising to "dress the landscape with concrete." This poetry is filled with images of paving over the entire malaria-infested swamps of the Galilee, what's currently northern Israel. The unique ecosystem of the Hula Valley was paved over, with only tiny pockets left. It's a really powerful image—right?—of humans trying to get rid of these diseases by using concrete. [A few months after this interview, there was an outbreak of influenza among the birds in the Hula Valley, which caused the death of eight thousand cranes and more than one million chickens. Since the Hula is such a migratory hub, disease ecologists feared that the H5N1 disease would travel with the birds and spread across Europe and the world.[21]]Could there be a link between paving this ecosystem and disease outbreaks?

CW: Concrete and draining were the traditional tools used at the turn of the nineteenth century. Draining wetlands to get rid of malaria—I see that all through Mesoamerica, where the American army was in Panama draining and changing the environment.

IB: Wow, that's incredible. And of course we could go on and on about the relationship between colonialism and ecology and how the massive changes in landscape wrought by colonial projects has impacted the spread of diseases that in turn killed so many Indigenous people. Amitav Ghosh speaks about this in his newest book *The Nutmeg's Curse* but of course this is not a new idea. As far as I know, it was first recognized by Alfred Crobsy in his book *Ecological Imperialism*, and popularized by Jared Diamond. But I'd like to go back to the distinction between infectious and non-infectious—is this distinction also problematic in any way?

CW: It would be quite interesting to balance it out a bit—get away from just this singular pandemic infectious disease focus—to discuss the less acute but in many ways much more impactful context of non-communicable diseases in a One Health framing. A human physician told me that she really understands my frustration about mainstreaming One Health and talking to policy makers because she comes from the primary healthcare sector. She said that 30 years ago, when we were trying to get the medical community and legislators to understand that primary healthcare needs to be expanded across sectors, they were just laughed at. And it took decades for primary healthcare to actually do that. So there are clear parallels here with those who said in the context of human health that we need to think more about the community and about the environment where these people are living.

IB: Thank you so much, Chris, it was fascinating to speak to you.

Notes

1 Hippocrates, *Hippocrates on Airs, Water, and Places* (London, England: Wyman & Sons, 1881).
2 Jakob Zinsstag et al., "From 'One Medicine' to 'One Health' and Systemic Approaches to Health and Well-Being," *Preventive Veterinary Medicine* 101, nos. 3–4 (2011): 148–156.
3 Calvin W. Schwabe, *Veterinary Medicine and Human Health* (Baltimore: Williams & Wilkins, 1984). See also Zinsstag et al., "From 'One Medicine' to 'One Health.'" Calvin Schwabe was credited with coining the term, but, according to this article, there is controversy about this. See R. Scott Nolen, "LEGENDS: The Accidental Epidemiologist," American Veterinary Medical Association, June 19, 2013, https://www.avma.org/javma-news/2013-07-01/legends-accidental-epidemiologist.
4 The acronym is French: Office International des Épizooties (French: International Office of Epizootics; Paris). "One Health," OIE, 2021, https://www.oie.int/en/what-we-do/global-initiatives/one-health/.
5 "The Manhattan Principles," Wildlife Conservation Society, 2021, https://oneworldonehealth.wcs.org/About-Us/Mission/The-Manhattan-Principles.aspx#:~:text=In%20September%2C%202004%2C%20health%20experts,domestic%20animal%2C%20and%20wildlife%20populations.
6 Kim Gruetzmacher et al., "The Berlin Principles on One Health—Bridging Global Health and Conservation," *Science of the Total Environment* 764 (2021): 1–4.
7 "Planetary Health—Economist Intelligence Unit," The Rockefeller Foundation, 2014, https://www.rockefellerfoundation.org/report/planetary-health-economist-intelligence-unit/.
8 Ibid.
9 "Ending this Pandemic, Preventing the Next One: European Perspectives at the Seventy-Fourth World Health Assembly," World Health Organization, March 6, 2021, https://www.euro.who.int/en/health-topics/health-emergencies/coronavirus-covid-19/news/news/2021/6/ending-this-pandemic,-preventing-the-next-one-european-perspectives-at-the-seventy-fourth-world-health-assembly.
10 Jason W. Moore, *Anthropocene or Capitalocene? Nature, History, and the Crisis of Capitalism* (Oakland: PM Press, 2016).
11 "Securing Rights," Wildlife Conservation Society, 2021, https://www.wcs.org/our-work/communities/securing-rights.
12 Andreas Obrecht et al., "Achieving the SDGs with Biodiversity," Swiss Academy of Sciences (SCNAT), 2021, https://scnat.ch/en/uuid/i/d67fa591-6ef6-5e16-a770-f2347cb250f7-Achieving_the_SDGs_with_Biodiversity.
13 James C. Scott, *Against the Grain* (New Haven: Yale University Press, 2017).
14 It turns out that the importance of the geographic element in disease was already acknowledged in the medical health arena early on. See, e.g., Jacques C. May, "Medical Geography: Its Methods and Objectives," *Geographical Review* 40 (1950): 9–41.
15 Felicia Keesing and Richard S. Ostfeld, "Dilution Effects in Disease Ecology," *Ecology Letters* 24, no. 11 (2021): 2490–2505.
16 Pieter T. J. Johnson et al., "Community Disassembly and Disease: Realistic—But Not Randomized—Biodiversity Losses Enhance Parasite Transmission," *Proceedings of the Royal Society B* 286 (2019), https://doi.org/10.1098/rspb.2019.0260.
17 Bradley J. Cardinale et al., "Biodiversity Loss and its Impact on Humanity," *Nature* 486 (2012): 59–67.
18 Cheryl Bartlett, Murdena Marshall, and Albert Marshall, "Two-Eyed Seeing and Other Lessons Learned Within a Co-Learning Journey of Bringing Together Indigenous and Mainstream Knowledges and Ways of Knowing," *Journal of Environmental Studies and Sciences* 2 (2012): 331–340.
19 The panel is sponsored by four organizations: the Food and Agriculture Organization of the UN, the World Organization for Animal Health (OIE), the UN Environment

Programme, and the World Health Organization (WHO). "26 International Experts to Kickstart the One Health High Level Expert Panel (OHHLEP)," World Health Organization, 2021, https://www.who.int/news/item/11-06-2021-26-international-experts-to-kickstart-the-joint-fao-oie-unep-who-one-health-high-level-expert-panel-(ohhlep).

20 Advancing Emergency Preparedness Through One Health Act of 2019, H.R. 377, 116th Cong. (2019–2020).

21 "Grim Cleanup of Cranes that Died of Bird Flu Begins in Hula Valley Reserve," *The Times of Israel*, December 28, 2021, https://www.timesofisrael.com/grim-cleanup-of-cranes-that-died-of-bird-flu-begins-in-hula-valley-reserve/.

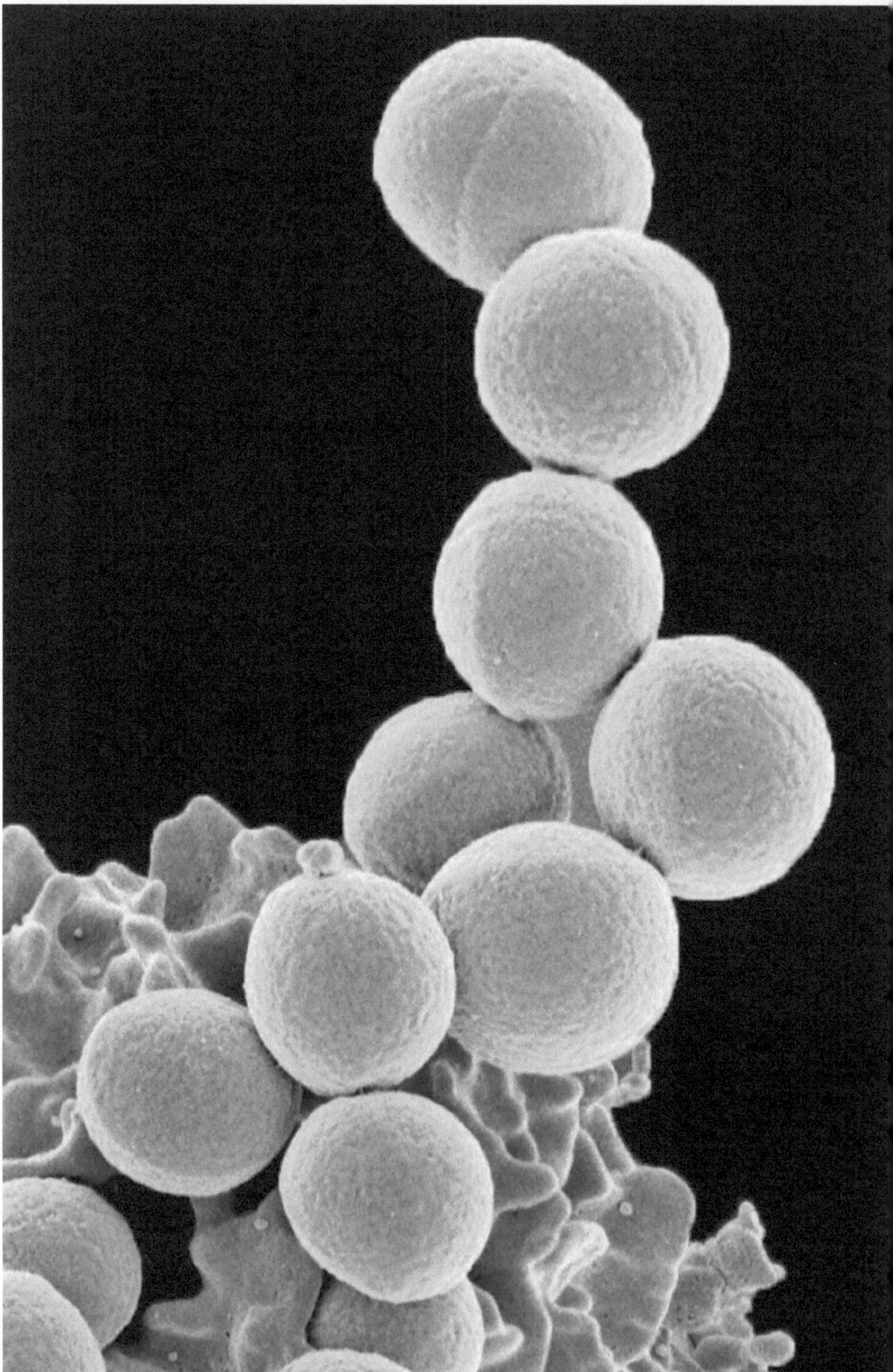

4

ONE HEALTH, SURVEILLANCE, AND THE PANDEMIC TREATY

An Interview with John H. Amuasi

Irus Braverman

***John H. Amuasi** is faculty at the Global Health Department of the School of Public Health, Kwame Nkrumah University of Science and Technology (KNUST), Kumasi, Ghana and Group Leader of the Global Health and Infectious Diseases Research Group at the Kumasi Center for Collaborative Research in Tropical Medicine. He is also the Executive Director of the Secretariat of the African Research Network for Neglected Tropical Diseases (ARNTD). Amuasi trained as a physician at the KNUST School of Medical Sciences and later graduated from the University of Minnesota School of Public Health with a PhD in Health Research and Policy. He served as Head of Research and Development at the Komfo Anokye Teaching Hospital in Kumasi. Amuasi's work focuses on improving health systems, including policy analyses and using both primary and secondary data in low- and middle-income countries. His research currently involves field epidemiologic studies on cryptosporidiosis, malaria, snakebite, and other neglected tropical diseases and conditions. Amuasi also serves as an Executive Committee member of the African Coalition for Epidemic Research, Response and Training (ALERRT) and is co-chair of The Lancet One Health Commission.*

This interview was conducted via Zoom on April 28, 2022, transcribed by Margaret Drzewiecki, and edited for clarity by the author.

IB: What can One Health do for our world, and why is it important?

JA: There have been several perspectives to One Health and it has evolved quite a bit over the years. There has also been so much debate on what One Health is and a lot of jostling for space in One Health even across different disciplines. In December 2021, the One Health High Level Council, the OHHLEP, published its new definition of One Health:

FIGURE 4.1 The methicillin-resistant strain of *Staphylococcus aureus* is being recognized by a white blood cell responsible for destroying foreign cells. Antimicrobial resistance is becoming a central concern of One Health. National Institute of Allergy and Infectious Diseases, National Institutes of Health, CC BY-NC 2.0.

DOI: 10.4324/9781003294085-6

> One Health is an integrated, unifying approach that aims to sustainably balance and optimize the health of people, animals, and ecosystems. It recognizes [that] the health of humans, domestic and wild animals, plants, and the wider environment (including ecosystems) are closely linked and inter-dependent. The approach mobilizes multiple sectors, disciplines, and communities at varying levels of society to work together to foster well-being and tackle threats to health and ecosystems, while addressing the collective need for clean water, energy, and air, safe and nutritious food, taking action on climate change, and contributing to sustainable development."[1]

This was the culmination of quite a lot of work with different stakeholders and I think it's the one definition that we could all live with for now.

One Health has traditionally been the space of veterinarians. It's a bit unfortunate, because it is really not a veterinary issue, but it certainly is not a human health issue either. So, the vet community of old sort of kept One Health to themselves and were not very open to working with others on this subject—the big grants on One Health were largely in the vet space until perhaps the last couple of years. COVID changed all that, or [perhaps that happened even earlier] with the Ebola crisis in West Africa.

With COVID, One Health has quickly become a subject for everybody. It is no longer recognized as an exclusive space for the vets and now we have the quadripartite. The quadripartite partnership is formed by the UN Environment Program (UNEP), which joined in March 2022 with the pre-existing tripartite partnership of the Food and Agriculture Organization of the United Nations (FAO), the World Health Organization (WHO), and the World Organisation for Animal Health (WOAH, formerly known as OIE). These four bodies have the formal commitment to work on One Health issues globally, [and are] all now working together on One Health. I mean, this has never happened before, so it's a very important sign of the times and the prominence that One Health now has globally. This is definitely not changing anytime soon—it has come to stay.

Now, let me circle back a little bit. When One Health first started, it was couched as the intersection between animal health and human health and then finally the environment also had some role to play. Now, this framing of One Health has really been thrown out because it only looks at areas where animal health and human health interact. In other words, it focuses unduly on zoonotic diseases and doesn't address the fundamental interconnectedness of humans, animals, and other life forms within an environment which we all share. Someone described it like this: instead of drawing those three circles and showing One Health as the intersection of those three circles, One Health is in fact the larger circle that encapsulates these three circles. They may still intersect, but One Health is really the encapsulation of these three rings.

IB: So, if you were to visualize it, it wouldn't be the space in between humans, animals, and the environment—

JA: No.

IB: It would be a broader circle that contains all three realms.

JA: Yes.

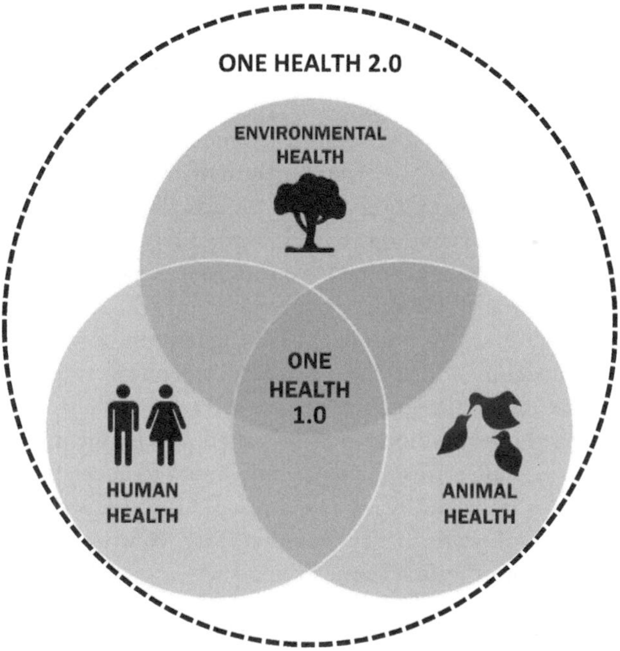

FIGURE 4.2 The three circles of One Health, surrounded by a larger circle to illustrate Amuasi's vision. Original three circles by Thddbfk, CC BY-SA 4.0, with addition by the author of the One Health 2.0 circle (using a broken-line).

IB: Oh, that's a really helpful illustration. The three circles have been so central to One Health that they have almost come to represent this vision.

JA: Yeah, [but] we're really moving away from this now to think about that which [it] encapsulates. Otherwise, what you're saying is that when something does not fit into that intersection, then it's not One Health. But if the framing is that we are fundamentally interconnected humans, animals, and other life forms and we share an environment—and that this environment is both a natural and a built environment—then our understanding of One Health expands. And that's what makes space for non-communicable diseases in One Health, and it very, very quickly makes space for a lot more understandings of our fundamental interconnectedness. At the same time, let me say upfront that this notion of fundamental interconnectedness really needs to be thought through very carefully, because, if it's a false premise, then everything that's built on it is a problem. It's easy for everyone to accept that we are fundamentally interconnected—humans, animals, and other life forms—and that we share an environment. But although people say, "yeah, yeah, yeah," people don't think about it deeply enough. What that really means is that, whether you recognize that connection or not, it is there. So, it is really for us … to find out what these interconnections are and what they imply for us.

Now beyond that, there is the environment which we share. There is the natural environment, and there is also the built environment. The built environment and the natural environment interact so much with each other. I mean, just tonight I was riding on the train from Greifswald, which is close to the Baltic Sea in Germany, back to Berlin. We were going through thick pine forests and sometimes there were lakes and I could see some deer and some other wildlife here and there. But then this is a train, a train with electric lines, metal, and stones—and it's right there, cutting through a natural environment. [So, we are moving toward an] understanding that the built environment cannot be separated from the natural environment. What this does, for me, is it recognizes this fundamental interconnectedness in the environment, which is shared.[2]

In One Health, a lot of practitioners have suggested that we have a problem with anthropocentricism, which is an over-emphasis on the role and importance of humans and so on. But I argue for a positive framing of anthropocentricism. For me, the current state of the world's so-called development is unsustainable and is driving various life forms, including humanity, to extinction. The extinction of so many species have already taken place—and so it's only a matter of a couple of thousands of years, maybe less, that if we continue along the present trajectory, human beings [will become extinct].

IB: Wow, you're giving us a lot of time. Climate change scientists count in the hundreds, not in the thousands.

JA: Right, yeah. I'm being very conservative here. I'm just trying to point out that sooner or later this will happen. The term Anthropocene depicts this notion exactly: if human beings are largely responsible for this accelerated deterioration of the planet which we share, then it only makes sense that human beings should be responsible for fixing it. And this is where my positive framing of anthropocentrism comes in.

IB: You mean using the focus on humans in terms of responsibility and stewardship.

JA: Exactly the point. I think that the spotlight is clearly on human beings as responsible for fixing the mess. Now, that being said, it also means that the whole ethos of One Health and the solutions that we propose to One Health problems largely have humans front and center, because we are doing [this] for ourselves, while recognizing that it is not possible without doing it for other life forms and the environment we share. But I wanted to make my personal position clear, because there are other people in the One Health space who really take a very different view of this. For some, human beings as a life form exist on an equal footing with other life forms. They do not see the preservation of humanity as a goal in itself, but more the preservation of life—and if human beings have to go, then too bad.

IB: How would these different approaches actually matter when it comes to practical policy decisions?

JA: Right. So, when it comes to practical decisions, a lot of the debates are currently happening within the space of food systems. There are people who believe that it is fundamentally wrong to eat other animals and think that this should not happen and that human beings should be able to live well enough on a plant-based diet. But if you look at the world as it is currently, and from a food security perspective, certainly shifting people to a plant-based diet would be a catastrophe in terms of nutrition in many parts of the world.[3]

IB: Especially in low-income countries.

JA: Exactly the point. So, this is certainly not the time to be moving in that direction at least for LMICs [low- and middle-income countries]. Notwithstanding, there's evidence of people living very healthy, long lives, who are not necessarily vegetarian or vegan. In fact, the longest living populations in the world are in Japan and in Italy and France—people who enjoy their meat and wine, and you know they live normal, long lives.

IB: It could also be their environment that contributes to this longevity.

JA: That's right, it's really a balance. Interestingly, one of the points that they raised in terms of a key to long life is never taking themselves too seriously and trying to avoid getting angry and that kind of stuff. I found that very interesting.

IB: This adds mental health to the One Health story, doesn't it?

JA: Exactly the point. So, this goes back to the understanding of this fundamental interconnectedness and its implications for quality of life and longevity. Now let me come back to your question about the implications of the debates over what is One Health. I think the strongest example of where these debates have implications is with regard to food and food systems. We have pastoralists—people who live with their animals, breed them, and then eat them—this has been their lifestyle for centuries. I find it odd and, quite frankly, disrespectful that we would come along today and tell them that no, that's not the best way for human beings to live and human beings should not really be eating animals.

IB: Which comes back again to this question of context: is One Health one platform with one morality and ethics for all? Or maybe we can have one ethical approach in the context of the United States, which is high up there in terms of how many farm animals it produces for slaughter. So here in the United States we might tell people to limit their meat intake but then in certain other places where there are pastoralists and others who eat meat for their sustenance—in those contexts maybe it should not be the message. I'm just wondering if there is enough flexibility for a One Health approach that can actually differentiate itself in a variety of locales rather than being a single global approach.

JA: I agree with you. The importance of reducing meat intake is very clear, it has implications for health.

IB: And climate.

JA: Right—and climate. You can think about it either at the population level or at the individual level. Either way, reducing meat intake is a great idea. This is undisputed. Where I have a problem is with those in One Health who advocate for the move toward removing meats from diets entirely. I respect people's personal choice along those lines—that's fine—but I'm talking about global movements that are against the consumption of animals. The only place where I agree with this approach is in environments where certain animals are close to extinction and need to be preserved. This is very clear in the context of the ban against whaling, which was very timely at the time it was introduced, though I think Japan continues to argue that currently there are enough whales to go around for food. It's a very interesting proposition and I struggle with it a bit myself.

IB: Let me just say from my perspective that I have a problem with rigidity. The problem with certain animal rights people is this rigidity, which has brought about lot of tensions between them and conservationists as well, for example in discussions about whether to protect species from extinction by bringing their conspecifics from the wild into captivity to save them. I guess what I'm hearing from you is that, in One Health, some proponents of the animal rights discourse take a myopic approach. I understand that you want to enable a more flexible approach where the specific context matters, right?

JA: Exactly the point. In fact, I have couched this as is the "W-4," which is the When, the Who, the What, and the Where. The W-4 is critical in addressing One Health problems. And these four are changing all the time as they interact in very interesting, unusual, and sometimes unpredictable ways, because "who" is one thing in one setting but then the same "who" [is different] with a different "where." This has implications for health. The clear example for me is that we've seen the distinctly lower morbidity and mortality from COVID in Africa: I am very clear and convinced that there is something we do not understand about COVID in Africa—and it's not about a poor reporting of deaths, it definitely is not. I have my finger on the pulse with the COVID figures because of a couple of clinical trials and other studies I'm leading in Ghana and so I get the numbers. And we have no patients now—it's zero, zero, zero. People are just not getting COVID at this present time, and when the samples come through, even for people who are traveling out of the continent—because, until recently, it was still a requirement to have a negative PCR—samples are all negative. So COVID is just not there currently as much as it used to [be].[4] There's still transmission going on, I'm certain, but it's very much on the low side presently.

The point I'm making is that we saw this lower mortality in Africa. [But at the same time], there was a disproportionate impact of COVID, both morbidity and mortality, for people of African origin or descent in the UK and also in the US, because these are where large populations of Africans outside of Africa live.[5] And when you think about it, you wonder: [are] these are the same Africans who maybe, had they been in Africa, [their] situation would have been different? So, this is what I meant when I said that the "who" behaves differently depending on the "where." There are people [who] have much worse socioeconomic conditions in Africa, but they weren't so badly hit.

IB: You're pointing to the context—that context is important.

JA: Exactly. What this means is that One Health approaches cannot be a one-size-fits-all. They can, or they should, have the same end objective, which is to promote the health of life forms, with a recognition that it is in our collective benefit. And, if one wants to be selfish about it, one can say that if human beings are to really live healthy, full lives then we must be fully pursuing One Health, which is also the health of other species and the shared environment. So, whether you look at it from what could be called an anthropocentric viewpoint or not doesn't change anything because what you need to do is just about the same thing. If your aim is to give human beings a better quality of life and live longer—fine; if it is to help make life better for the animals, that's fine, too; but it doesn't change what you have to do.

IB: So, what is health?

JA: Well, the definition put forward by the WHO has not been without contention, but there is no other definition that has stood the test of time. The WHO defines health as "a state of complete physical, mental and social wellbeing and not merely the absence of disease or infirmity."[6] Many, many people have written a lot of articles on this and I actually teach this to my students. This definition may be considered by some as less than optimal, but I think it's better than any others that have been proposed.

IB: What are the important components of this definition, from your standpoint?

JA: The social part of it—the mental, physical, and social wellbeing. And the fact that the absence of disease or infirmity does not necessarily [indicate] health.

IB: One Health is so intuitive and so inclusive that it seems to lose significance or meaning, especially when you're trying to bridge all these different groups. I'm saying two different things here. Number one, that such a broad definition of health doesn't leave out anything, really, and so I wonder how operative it is. And number two is whether, with these massive institutions pulling in different directions, there is much left there in common in the shared space between the circles.

JA: There are certainly things that are more in the middle than others—zoonotic diseases, for example. I keep referring to COVID because I think it teaches us so many amazing lessons. COVID started in horseshoe bats and spilled over to human beings, but now we know that it also infects ferrets and other kinds of animals.

IB: Yeah, my cat had COVID.

JA: And the argument that this could be undergoing some mutations in these animals and could then be transmitted back to human beings just tells you that a disease like that is right there in that middle space—because it's going back and forth, it could morph into anything, it could hide for a while and come back later. You think about the West African Ebola epidemic which started in Guinea, and bats were the culprits again. But human beings interacting directly with these animals who were just carrying the virus and doing their own thing—no problem—until we came along and got it from them and then it becomes a problem.

IB: My work on corals highlighted for me that symbiotic relations are really significant. Moreover, it highlighted that there isn't such a clear-cut distinction between the environment and individual bodies, especially when we understand our bodies as a conglomerate of different more-than-human cells—as holobionts. The separation between bodies and their environment seems to be even more questionable when it comes to oceans and life forms within water. Since we are in fact mostly composed of bacterial cells, is there a separation between our bodies and others' from which to work toward interconnectedness?

JA: It's a great point, but the transmission and the movement of pathogens either among humans or interspecies has mostly emerged as a result of lifestyles that human beings have adopted. This is why I made mention of the built environment and its interaction with the natural environment. So, this is where we are now: we have developed this kind of interaction and we have developed the speed of this interaction. I am here in Germany and tomorrow by this time I will be in the middle of Ghana and would carry anything that I might have picked up here to the middle of Ghana by tomorrow. This situation was developed by us.

IB: Yes, but birds travel, too.

JA: Exactly, yes, but one could argue that the relationship between birds and the environment is a natural relationship, whereas the relation between human beings moving across the continent in a space of ten hours is not a natural activity—so what happens is that it changes the nature and the speed of relationships between species and with their shared environment.

IB: I see what you're saying. The microbiome develops slowly. And as part of an ecosystem, right? So if I travel to the other side of the world then it's not something that my microbiome is used to interacting with—

JA: Well, yeah, exactly. So, this change has been introduced by human beings. We will not go unpunished, quote unquote, for there's a price to pay. If we want to mitigate that price, then we need to put in place [certain strategies]. So we have drugs, and we have the treatments and interventions which are aimed at fix[ing] ourselves and to pay for the changes in the nature and the speed of interactions that we have created. Even in the so-called natural world, animals have their own ways of dealing with the negative aspects of interactions. I mean, some animals are known to chew different plants at different times, which have antibiotic properties that shield them. So, there's this intrinsic shielding that species have from what would be harm or danger. Even so-called nature tries to shield itself from interactions that can be critical to its survival. That's how I think of it, and this is why I started off by talking about the natural and the built environment, and how they interact with each other in ways that we are yet to fully understand.

IB: And, of course, many have questioned that very distinction between nature and culture. I mean to call it a "built" environment and distinguish it from a "natural" one separates humans from nature, as if whatever we do is not natural. But when bees build their homes and beavers create their dams that's not really referred to as a built environment—

JA: That's right! [laughs] For me that is the point. There is the fundamental interconnectedness, but there are also cases where this interconnection is much stronger and has implications very directly for the health of different species. This is perhaps why, interestingly, zoonotic diseases that hardly have an impact on animals would affect human beings so badly. COVID is not causing deaths among animals, as far as we know.[7] They can carry it. But it's causing death among human beings. By contrast, the foot and mouth disease, which can decimate herds of livestock, can be transmitted from animals to human beings, but hardly has any impact on human populations.[8] Those diseases highlight the differences between human and nonhuman species. Despite the importance and the function of interaction, they affect one species much more negatively or even catastrophically than they would others.[9]

IB: I agree with you that, on the whole, COVID has been far more dangerous to humans than most other animal species. Although occasionally we do hear news stories about zoo animals getting sick and even dying from COVID, like the three snow leopards in a Nebraska zoo.[10] But I would like to come back to the institutional collaborations. What's going on with One Health and why does it seem so crucial to have this quadripartite collaboration that you said never happened before? What does it matter that it's happening? And how is it going to help us?

JA: Well, as I said, the concept of One Health has existed largely within the space of the veterinary sciences. If you look carefully at funding, One Health grants have largely been for veterinary type of activities. Big funders like the Bill & Melinda Gates Foundation, the Wellcome Trust Foundation, and even USAID—although USAID was a bit ahead of the curve when they started funding One Health big time a little over a decade ago—but the other big ones like Wellcome and Gates until recently hardly mentioned One Health at all.

Now, with One Health becoming mainstream and these four major UN bodies agreeing to work together, it sets the stage for funding and for global policy at a whole different level. There has even been talk of the need for a pandemic treaty of some sort. And even talk about a possible fund which would address pandemics. This is unlikely to happen anytime soon, but having this quadripartite [partnership is itself] a first step toward a global recognition of the need for fundamental changes to the way the world functions at different levels, including the world economic system. In fact, the major culprit, which is so antithetical to One Health, is the global financial system or the prevailing global economic system. You think about how a tree in the forest is less valuable in hard economic terms than when it's a log in a container on the high seas going somewhere. It is just the way that the global economy functions. This will call for change, because if we continue with this way of making money then we'll surely be heading very quickly toward self-destruction.

IB: So, is this a critique of capitalism at large?

JA: Not necessarily, no. It's beyond capitalism, because even communism still puts the same value on things in terms of commodities.

IB: But that is much enhanced in the capitalist approach. Some have even referred to the Anthropocene as the Capitalocene. This is an attempt to shift the blame from the entirety of humanity to highlight that a particular part of humanity should be blamed and, more importantly, that a specific group of humans should assume the largest responsibility to fix [the problem].[11] In climate change discussions, there's a lot of talk about developing countries not being as responsible because they did not use [carbon] as much as developed countries.

JA: Right. What do you say to a country that discovers stockpiles of coal today? Can you now tell them that, sorry you can't touch that because it's not good for our environment?

IB: Especially after we, the developed countries, have already developed using our coal!

JA: They'll say you've been mining coal for hundreds of years and made a whole ton of cash. Now it's our turn!

IB: In One Health, there seems to be a tippy-toeing around criticizing the liberal and neoliberal economic systems, which to me are a fundamental part of the story and critical to our considerations when thinking about health. This is where an intervention by humanists and social scientists can help move One Health forward by bringing some of the criticism explored in these disciplines toward capitalism and colonialism. You can't really understand the entire picture if you don't see the social and economic processes that have led us to view a tree as a resource. Otherwise, you're mostly taking care of the symptoms and not really addressing the core problem.

JA: So the point I'm making is the need for what I call "radical slow change." It needs to be radical, but it cannot be fast—it's just not possible. A third aspect of this change is that I'm chasing a moving target, which makes it even harder. An immediate change would just throw the world into chaos, because of the way the world functions. But continuing in this direction is heading toward disaster. So, making a clear move toward changing the way the world economic system functions is the way to go.

IB: Can you give me examples of things that you are advocating for?

JA: I think it goes beyond advocating for a change in one thing or the other. Instead, it's a deep change in our perceptions and our value systems. This is why starting in one place in the world is just not enough and this is why I'm excited about the quadripartite [WHO, WOAH, FAO and UNEP] possibly being a starting point for more global action. Because if only one part of the world does it, they would put themselves at a disadvantage and this imbalance could actually turn this world into something worse off than it is already. So one [party] says, "Okay, I will stop mining coal," while the other says, "No, I'll keep on mining my coal."

IB: Which is exactly what is happening.

JA: Exactly my point. So it will have to be everybody agreeing that "okay; this is our new way of doing things." But everybody agreeing to the new way of doing things still puts some people at a disadvantage. So how do we fix that inherent disparity and disadvantage that is placed on people on account of this change that we need to make? And then the moving target bit is more of understanding that some of the interventions that would be One Health-worthy today, or interventions that would have been considered to be global goods, could actually be harmful to One Health. The example I often give is mechanized farming. Large-scale farming is what brought India out of famine and has been touted as what will deliver Africa out of economic dependence and poverty. But if Africa should go the way of Europe in terms of large-scale mechanized farming, that's a lot of forest disappearing. That's huge—massive—in terms of greenhouse gas emissions. At the same time, look at what's happening in Ukraine. This is going to have serious implications for food security in Africa, which means greater focus on self or local food security, which puts pressure on the environment.

So it's very complicated to work with a moving target, because what you're aiming at today moves so much that doing that same thing could actually be harmful rather than helpful if you did it later. And this is why I used the image of a slow radical change to capture moving targets. That means that we always have to be thinking together and moving together to address issues based on a common understanding of how things are going and how things are changing.

IB: Can you possibly give me an example of a success of the One Health approach? I will say about the global climate change context that international law doesn't usually have the same power as national laws, unless it's ratified as such, of course. But then there is always this idea that countries can always pull away, which, as we know, happened with the United States and the Paris Agreement. And so I wonder about these collaborations on the international level and through bodies such as the United Nations—which look very nice, the whole world is collaborating—but are they more symbolic than anything else? What kind of implementation and enforcement power does a body like that have?

JA: Yeah, those bodies do not have any enforcement power at all. This is very clear. But the expectation is that [following] a member state's agreement or participation, some of these treaties will translate into local legislation, which will then be enforced at the local level. Now, what would allow for adherence to these acts of legislation, and how would these treaties translate into such legislative acts? The important thing is that member states will be seen as moving in the same direction. People are unwilling to participate in these treaties when they see themselves at a fundamental disadvantage. I mean, why should Côte d'Ivoire, for example, put the brakes on some mining activity which is harmful to the environment, and next door in Ghana we're doing it and raking in all the foreign currency? That's just not going to happen, especially if we have the capacity to do more than them. This is why we need to truly be together in this, which is where the [international] bodies come in, otherwise it just will not happen as a matter of course.

IB: Can you give me a practical example for where this kind of international collaboration was successful?

JA: Well, this quadripartite collaboration is very recent. The four international organizations working together started barely three or two years ago, so the very collaboration at this level is pretty new in itself—it has never been performed at this level.

IB: That sounds exciting. What would be your vision of how a healthy One Health could manifest in the world?

JA: Hmm. One good example is the issue of antimicrobial use and antimicrobial resistance. AMU and AMR.[12]

IB: There are so many acronyms in public health!

JA: Yeah. Antibiotics are used in animals, they are used in human beings, they are sourced in some way from the environment, the air, and they are released into the environment through waste. This has implications for the morphing of pathogens, which may be found in the environment and then jump into different species, leading to resistance. They are also closely linked with food and food systems. So, for me, that's one area which is very ripe for this deep level of collaboration.

IB: Alright, so what is the One Health path forward with regard to antimicrobial resistance?

JA: The One Health way forward would be to have a truly integrated surveillance system, which would track the development of the mutation of these pathogens in animals and in human beings, which will be looking at the genetic linkages, whether these are developing independently or they are linked to each other, [and] the speed at which this is occurring.[13] The question is whether or not the use of antimicrobials in animals poses an existential threat to their use in human beings. So far, the evidence of this in high-income countries suggests that the connection is not as strong as we thought. Hypothetically, this could have been really bad. But the evidence shows that it really takes a lot more for resistance to develop than just the use of antibiotics in animals, and then us consuming that meat. So, it's not as bad as we thought it would be. But in low- and middle-income countries no one has really looked at this in sufficient detail. So, having this truly One Health surveillance approach with a strong genetic sequencing backbone to it, which calls for a lot of technology and gigabytes of storage and trawling through tons of data—that's a classic One Health approach to addressing antimicrobial use and resistance.

IB: Okay. Is the central contribution of One Health the shared data among institutional bodies that used to be separate?

JA: Exactly. So the governance aspect would be very tricky. But if all the four institutions aim toward the same objective, this will ultimately make a difference. Whereas the concept of One Health in itself is shared, the way it will translate on the ground and in different contexts might be different, as in the example of rabies, which illustrates how different countries govern rabies differently. As part of our *The Lancet* One Health Commission, we have been debating an illustration that would show the interconnections between One Health, Planetary Health, and Eco-Health. While theoretically, the three might be thought of as having some distinct elements, ultimately we are aiming for the same thing. We have a lot more visuals coming out in the final Commission report, and that illustration of the three converging circles will change a bit, as our focus really is the outer circle beyond the three intersecting ones. So we stayed away from the human, animal, and environment three-circle image [because] we're really trying to highlight the avenues via which we can pursue One Health. Each of those parts of the outer circle are major avenues via which we can promote One Health—our food systems, shared medicines, and diagnostics. Thinking of issues such as gender and big data will be the avenues through which we can pursue One Health goals, rather than this traditional human–animal–environment intersection.

IB: Right, you're moving away from the fraught distinctions that divided One Health rather than bringing people together. It almost seemed like each of the three in the tri-circle approach was fighting for their own turf, so moving away from framing it like that allows people to come together for a particular end goal. So I think it's very pragmatic—or should I say very healthy—to think about it in this way. But aside from declarations that come from political pressure to do more preparedness toward the next pandemic, I am wondering, again, what is the importance of this kind of institution building? Would it lead to anything other than more conferences? Has anything meaningful emerged out of these collaborations, like a different approach than what we've taken before or a different path forward? And has COVID shaken up some of these realizations and changed how we see things?

JA: Well, we are recognizing that a lot of the data and information that we have, which has informed policy for several years, may actually be wrong. And it is wrong because we only have the tip of the iceberg, and we have not invested in trying to determine the true size of the iceberg and what that really means for interventions. For example, when the WHO finally decided to do the mapping of neglected tropical diseases, they realized that several neglected tropical diseases were overestimated in terms of their prevalence, while several others have been underestimated in different parts of Africa. That is one area that would need closer attention to the data.

A second area was highlighted during COVID. We know for sure that the estimates of infection rates in Africa are not reflective of the reality. And they're wrong because a lot of countries hardly did any surveillance at all. Those who did some semblance of surveillance stopped very early. In Ghana, for example, contact tracing lasted for a couple of months, and then it stopped entirely. Now, if you're going to base your disease estimates and planning on those figures, you couldn't be more wrong. … This is just an example for how the data that we have regarding these diseases is terribly wrong. And it has implications for how we design One Health and what the successes could be.

IB: How do we get deeper into that iceberg?

JA: It's about designing these truly well-thought-through surveillance systems. And it's difficult because surveillance entails a lot of looking. I think people get tired of looking—why are you looking when there's nothing? But that's exactly the point: we must be integrating these surveillance systems into life as it is. This is the way we live life—surveillance needs to be a part of it.

IB: What do you mean by surveillance?

JA: Surveillance in terms of understanding what [are] the implications of our interaction with each other, with other species, and [with] the environment moving forward. Is there anything that we can put in place to mitigate the potential negative impacts of that?

IB: When you say surveillance, that requires some sort of technology, right? And so what exactly do you have in mind?

JA: Some of it will be active surveillance and some of it will be passive. Passive surveillance in the sense that if you have a true vital registration system where you're capturing all births, all deaths, and all significant health events in between—that is a big start already. Significant health events mean that there's a way of capturing when somebody is ill, when they don't feel well, when there's a health issue.

IB: Human health?

JA: Human and animal, for that matter.

IB: You're talking large animals, not microorganisms, right?

JA: Well yes, that's what I mean. For bacteria that will be more of an active surveillance. This has already been done. For example, there's active surveillance for polio where sewage and other environmental locations are swabbed to check if there is wild polio around us. This has been ongoing in parts of Africa.

IB: Who is being constantly checked?

JA: Well, the environment. Waste. The occasional detections of wild polio in the environment called for various interventions, which have been helpful. Actually, without this kind of active surveillance, there would have been outbreaks.

IB: So, if the bats in Wuhan were tested regularly, we might have caught COVID before it triggered an outbreak?

JA: There is that possibility, but in that case it would have been like searching for a needle in the haystack.

IB: Yes, and there are so many needles. And furthermore, these are needles that we can't even see. I mean, there are so many of those diseases around, and so few of the viruses are actually known, that it's almost surprising that they so rarely develop into epidemics, isn't it?

JA: There's so much that must happen before there's a real spillover which would be of concern. But then [there] are small things which can suddenly tip the balance. I mean all of a sudden, one little thing just moves everything.

IB: If it's so happenstance, can we actually predict and prepare for it?

JA: There is an element of luck for sure, but there's also clear evidence that the kind of interventions that were put in place in many African countries during the Ebola scare certainly helped to a large extent [to] mitigate the early period of COVID. This goes to show that, yes, we may not be able to prevent spillovers, but maybe that's not where the greatest focus should be—and it should [be] more on early detection and putting measures in place. So, for example, temperature scanning, hand washing stations at ports of entry, the kind of health information that was collected at ports of entry—many of these really didn't stop at all [after Ebola] in many African countries—at least in Ghana they didn't.

IB: And that was also the case in Asian countries after SARS, right?

JA: After SARS 1, the wearing of a mask became normal for them. And one of the most interesting things is how that little change in lifestyle during the pandemic has significantly impacted the transmission of other infectious diseases, even in places that are a little bit different, such as countries in Africa. So, it just shows how little modifications aggregated together do make a significant blip in the risk of disease transmission. In the same way, if surveillance is augmented to certain level, it will also make a difference. It may not stop spillovers, but [it] would possibly reduce the rate of occurrence or risk and certainly put us in a better place for early detection.

IB: I see your point. I'm just always concerned when I hear surveillance because, at least where I come from—which, as you know, is Israel—surveillance can sometimes be used as a means of population control. So first we're going to put on masks and then we're going to constantly check temperature—and then what? Another issue to consider is how will this enhanced and centralized surveillance impact our mental health and our relationships to one another. I am not an anti-mask person by any means, I just want to get your sense about the way that certain governments might take advantage of these powers to control their populations in ways that are not necessary for pandemic management.

JA: Yeah, that is a strong concern. But I worry more, actually, about the safety of the data and who has access to it and how it will be used. I don't have the answer to that [and] I am very worried about it. I really don't have an answer—it's a "drink deep or taste not [issue]." We must invest equally in data protection and in all other data safety measures.

IB: Israel is very big on data surveillance and protection.

JA: Exactly. But you see how that has helped the country to be fairly successful in many ways.

IB: Is that where we want to see the world heading?

JA: We will have to make our choices as to which is the necessary evil, but less interaction between human beings and animals is not even good for our own health. I also don't think that the rate at which we move around the planet, physically, is going to change drastically. Those are things that are just not going to change. Given that they will even accelerate, these are the trade-offs we'll have to make.

IB: Maybe we can end with some reflections from your work as director of the Secretariat of the African Research Network for Neglected Tropical Diseases (ARNTD)?

JA: Well, One Health tended to be more Eurocentric and [at that time] conservation was mostly about nonhumans. So those two tracks have been pursued very, very, very separately, which was problematic in itself. I would like to speak a bit about neglected tropical diseases. These are, essentially, diseases of poverty—diseases that disproportionately affect the poorest of the poor and end up exacerbating or accentuating disparities. These diseases would not exist and tend to almost disappear when the socioeconomic status improves. [This is] because the lifestyles of individuals are what bring them into contact with some of the vectors that transmit these diseases and when the socioeconomic status changes, the lifestyle changes and these people are brought into less contact with these vectors, so automatically these are less transmitted and so they don't manifest as much. And this is why they're called diseases of poverty. There is a strong relationship between NTDs—neglected tropical diseases—and One Health. In fact, the WHO published an important write-up on this about a month ago, because a lot of these neglected tropical diseases are infectious diseases that involve vectors, most of which are animals and insects—and so that's where the zoonotic component comes in. For example, rabies is a neglected tropical disease.

IB: Yes, according to the WHO, rabies remains one of the most lethal viruses on our planet.[14] Deborah Nadal and Susan McHugh's chapters in this volume discuss this, and so does Stephen Hinchliffe in the Foreword. Each discusses rabies from a slightly different perspective, but these contributions at the same time strongly complement each other.

JA: Right. Another example is the risk of leishmaniasis with black flies.[15] The *Aedes aegypti* mosquito will transmit pathogens, causing Zika and lymphatic filariasis, as well as a bunch of other [diseases].[16] So this vector alone accounts for a couple of infectious diseases, including some neglected tropical diseases. People's lifestyles, largely on account of poverty, puts them in closer contact with these vectors. The black fly, for example, lives in abundant termite mounds and can also propagate in mud walls. So people who live in mud houses obviously would have more contact with these. But if their lives improve and they are able to plaster their houses—away with the black flies [and so] the disease will go away, too. And this is why, in the One Health approach, neglected tropical diseases are closely tied to socioeconomic development. At the same time, when we focus so much on the built environment, it's separating human beings from animals and from the natural environment, which then comes back in the form of non-communicable diseases and mental health issues and so on and so forth. So improvement in socioeconomic indices cannot be the sole solution.

IB: What do you mean by non-communicable diseases?

JA: I mean decreased mobility, poor dietary choices, such as processed food, which then will lead to an increase in cardiovascular diseases, or hypertension and diabetes and so on and so forth.

IB: So one way or the other—we're basically doomed?

JA: Essentially.

Notes

1 FAO, OIE, and WHO, "Tripartite and UNEP Support OHHLEP's Definition of 'One Health,'" World Health Organization, December 1, 2021, https://www.who.int/news/item/01-12-2021-tripartite-and-unep-support-ohhlep-s-definition-of-one-health#:~:text=One%20Health%20is%20an%20integrated,closely%20linked%20and%20inter%2Ddependent.

2 Emma L. Bird et al., "Built and Natural Environment Planning Principles for Promoting Health: An Umbrella Review," *BMC Public Health* 18, no. 1 (2018), https://doi.org/10.1186/s12889-018-5870-2.

3 Marco Springmann et al., "Health and Nutritional Aspects of Sustainable Diet Strategies and Their Association with Environmental Impacts: A Global Modelling Analysis with Country-Level Detail," *The Lancet Planetary Health* 2, no. 10 (October 2018): e451–e461; Tina H. T. Chiu and Chin-Lon Lin, "Ethical Management of Food Systems: Plant Based Diet as a Holistic Approach," *Asia Pacific Journal of Clinical Nutrition* 18, no. 4 (2009): 647–653, "Food Security," World Preservation Foundation, accessed May 21, 2022, https://worldpreservationfoundation.org/environment/food-security/.

4 Stephanie Nolen, "Trying to Solve a Covid Mystery: Africa's Low Death Rates," *The New York Times* (March 23, 2022), https://www.nytimes.com/2022/03/23/health/covid-africa-deaths.html?smid=url-share.

5 Eboni G. Price-Haywood et al., "Hospitalization and Mortality among Black Patients and White Patients with Covid-19," *New England Journal of Medicine* 382, no. 26 (2020): 2534–2543, https://doi.org/10.1056/nejmsa2011686; Katherine Mackey et al., "Racial and Ethnic Disparities in COVID-19–Related Infections, Hospitalizations, and Deaths," *Annals of Internal Medicine* 174, no. 3 (2021): 362–373, https://doi.org/10.7326/m20-6306; Elizabeth J. Williamson et al., "Factors Associated with Covid-19-Related Death Using OpenSAFELY," *Nature* 584, no. 7821 (2020): 430–436, https://doi.org/10.1038/s41586-020-2521-4.

6 "Constitution of the World Health Organization," World Health Organization, 2022, https://www.who.int/about/governance/constitution.

7 Salleh N. Ehaideb et al., "Evidence of a Wide Gap Between COVID-19 in Humans and Animal Models: A Systematic Review," *Critical Care* 24 (2020): 1–23.

8 "Causes & Transmission of Hand, Foot, and Mouth Disease," Centers for Disease Control and Prevention, February 2, 2021, https://www.cdc.gov/hand-foot-mouth/about/transmission.html.

9 Alla Katsnelson, "How Do Viruses Leap from Animals to People and Spark Pandemics?," *Chemical and Engineering News*, August 30, 2020.

10 Reis Thebault, "A Zoo's Three 'Beloved' Snow Leopards Die of Covid-19," *Washington Post*, November 14, 2021.

11 Jason W. Moore, *Anthropocene or Capitalocene? Nature, History, and the Crisis of Capitalism* (Oakland: PM Press, 2016).

12 Scott A. McEwen and Peter J. Collignon, "Antimicrobial Resistance: A One Health Perspective," *Microbiology Spectrum* 6, no. 2 (2018), https://doi.org/10.1128/microbiolspec.arba-0009-2017; Markus Huemer et al., "Antibiotic Resistance and Persistence—Implications for Human Health and Treatment Perspectives," *EMBO Reports* 21, no. 12 (2020): 1–24.

13 See, e.g., Cécile Aenishaenslin et al., "Evaluating the Integration of One Health in Surveillance Systems for Antimicrobial Use and Resistance: A Conceptual Framework," *Frontiers in Veterinary Science* 8 (2021): 1–12.

14 "WHO Rabies Fact Sheet," World Health Organization, May 17, 2022, https://www.who.int/news-room/fact-sheets/detail/rabies.

15 "About Leishmaniasis," Centers for Disease Control and Prevention, May 19, 2020, https://www.cdc.gov/parasites/leishmaniasis/gen_info/faqs.html.

16 "*Aedes aegypti*—Factsheet for Experts," European Centre for Disease Prevention and Control, December 20, 2016, https://www.ecdc.europa.eu/en/disease-vectors/facts/mosquito-factsheets/aedes-aegypti.

PART II

Expanding One Health: Beyond the Human-Animal-Environment Triad

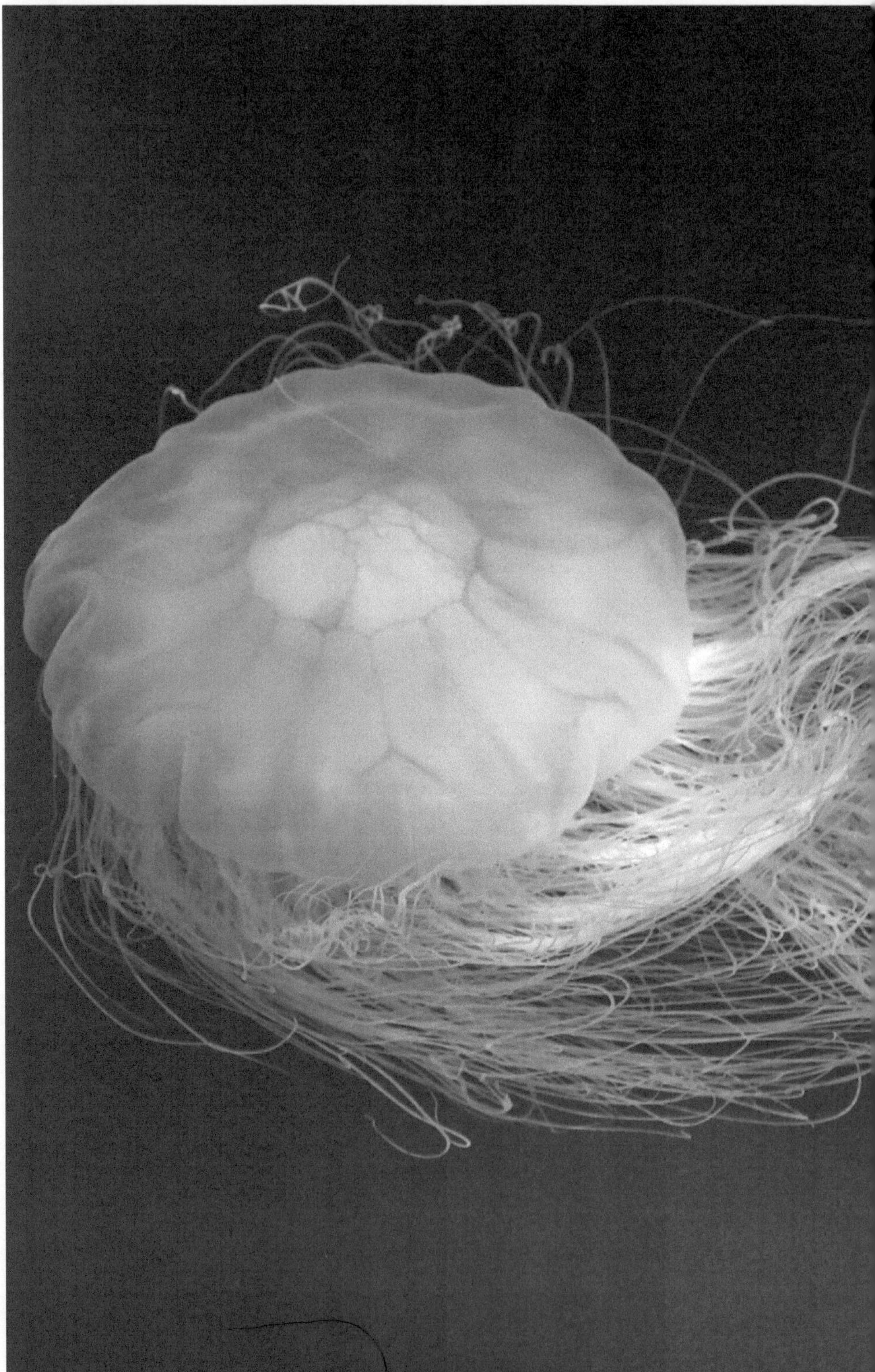

5

BETWEEN HEALTHY AND DEGRADED OCEANS

Promising Human Health through Marine Biomedicine

Elizabeth R. Johnson and Hannah Dickinson

One Health for the Anthropocene

The One Health paradigm's past and present rest on a pathogen-based framework of healthcare for more-than-human bodies. As noted in Irus Braverman's Introduction and elsewhere in this volume, the One Health movement gained ground over the past half-century primarily because of the emergence and spread of viral zoonotic diseases. As Abigail Woods' contribution to this volume makes clear, the history of One Health has long focused not on the root causes of these diseases, but on the entanglement of animals, humans, and ecological systems. By avoiding the demonization of particular disease agents or vectors, One Health approaches promote a more ecological perspective.[1] Rather than expelling pathogens or their animal carriers, many of which are domesticated and fully integrated within human social and economic networks, One Health planning enables a more expansive and nuanced approach to human–nonhuman relations that recognizes the world as immanently interconnected.

In this vein, One Health initiatives are distinctly more-than-human and responsive to the rapidly changing world ecologies that encapsulate the geologic era now commonly referred to as the Anthropocene.[2] One Health promises to extend epidemiological analysis beyond the study of nonhuman vectors to develop an understanding of ecological relations—healthful or otherwise. As Thomas Lovejoy and others have noted, climate change and ecological degradation contribute to the ill health of nonhumans and humans alike. Accordingly, markers of ecological health that typically fall outside of healthcare provisions are incorporated within the paradigm. This includes the effects of climate change

FIGURE 5.1 A single lion's mane jellyfish, *Cyanea capillata*, in Gullmarn Fjord at Sämstad, Sweden. Photo credit: W. Carter, Creative Commons.

DOI: 10.4324/9781003294085-8

as well as extinction rates and measures of biodiversity. As Ilaria Capua, former Vice President of the Science, Culture, and Education Commission of the Italian Parliament and current director of the University of Florida's One Health Center of Excellence, put it: biodiversity is crucial to development of multispecies health; it is "the elasticity and flexibility of the planet."[3] For Capua, biodiversity creates the conditions for multiple pathways toward collective good health, as the narrowing of species diversity forecloses on some of those pathways, making both human and nonhuman systems more rigid and thus more exposed to risks. The One Health paradigm is positioned to produce solutions to twinned health and ecological crises fueled by over-exploitation.

Social scientists adopting a more-than-human perspective (including the authors of this chapter and several others in this volume) have long advocated for environmental ethics and politics based on multispecies interconnection rather than environmental mastery. For us, the One Health approach seems promising. It suggests an anti-essentialist take on organisms and their processes that denies deep-seated nature–society divides. Rather than "good" or "bad" organisms, the entanglements that make up more-than-human relations can become healthful or pathological, depending on composition.[4] One Health perspectives accordingly appear to offer an antidote to the predominant capitalist and colonial regimes of extraction, as well as biopolitical exclusion.

This chapter explores two case studies that draw humans together with the ecological concerns of the Anthropocene via initiatives that echo the ideologies of the One Health agenda. In it, we interrogate—and ultimately question—the expansive possibilities of One Health through the lens of ocean governance, where marine organisms and ocean ecologies intersect with human health concerns and imagined futures—both healthful and pathological.

The first case study involves the extraction of jellyfish biomaterials for the dual purpose of environmental sustainability and biomedical innovation. A rising frequency of jellyfish blooms in recent decades has fueled fears of depleted biodiversity, destroyed fisheries, abandoned beaches, and ruined livelihoods. Previous marine management policies have focused on the control or elimination of blooms. More recently, however, jellyfish are being reimagined as the food and pharmaceuticals of the future and revalorized as potentially useful rather than damaging. The second case study focuses on how the shrimping industry is working to redirect its own indigestible byproduct, shrimp shells, for a wide array of sustainable innovations. A developing market for shrimp-derived chitosan promises that its applications are nearly limitless. Valued as a fertilizer, potential pesticide, and the basis for bioplastics and other sustainable materials, chitosan's uniquely multifaceted properties promise to simultaneously transform material production, the trajectory of planetary ecology, and the wellbeing of human life.

Both jellyfish and chitosan join a collection of other marine biomaterials in the pursuit of healthier—rather than diminished—more-than-human futures. Yet thinking with these and other marine biomaterials raises several questions about the assumptions of the One Health paradigm. The co-authors collectively analyzed scientific research and policy writing on these case study biomaterials from the late nineteenth century to the present. We also conducted a series of interviews with bioscientists and innovators working with jellyfish and chitosan.

Through our analysis, we question what innovators and policymakers envision when they attempt to ensure the emergence of one "correct" definition of a collective, more-than-human health. Focusing on assumptions made in policies surrounding threats and benefits to health, we suggest that these case studies reveal that what constitutes a "healthy" ecosystem for humans and nonhumans is not singular. Rather, we find that definitions of health in ecosystems and humans alike are plural and heavily contested: scientists and other producers of ecological knowledge are not at all unified in their understanding of what constitutes "good" health. This suggests that the conditions of the Anthropocene require not only a "more-than-one world" approach to ecological health,[5] but also critical and informed analyses of the scientific knowledge practices, epistemic viewpoints, and ideologies that make claims about what constitutes "good" and "ill" health.

One Health in Blue

As they stand, emergent One Health initiatives have focused primarily on the more-than-human health of terrestrial ecosystems. But as the United Nations Ocean Decade of Science ramps up, many are beginning to ask how ocean governance and One Health strategies are aligning intersections of marine biodiversity, oceanic ecologies, atmospheres, and human life.[6] As Lora Fleming et al. note, degraded, hypoxic, or toxic ocean environments cause ill health to humans through exposure to the water as well as ailing fisheries. And the salubrious benefits of healthy beaches and sustainable fisheries are increasingly recognized as necessary components of a livable and socially just future in the era of climate change.[7] The United States National Oceanic and Atmospheric Administration (NOAA) already incorporates a One Health model across its operations to reduce health risks for both humans and nonhumans. The work involves predicting, planning, and prevention around issues including harmful algal blooms, marine zoonotic diseases, and forecasting the links between infectious diseases and their atmospheric-ecological drivers.[8]

Like NOAA, much of the current conversation surrounding intersections of marine management and the One Health paradigm has focused on disease vectors like harmful algal blooms, infectious diseases, and toxic effluents. But a growing number of scholars and policymakers have emphasized the importance of marine fishery resources to health.[9] Take, for example, the "One Health Aquaculture" approach spearheaded by researchers working with the Centre for Environment, Fisheries and Aquaculture Science (CEFA) of the British government. Their work asserts the importance of developing the aquaculture sector around One Health principles, focusing on achieving environmental protection and equitable food access to ensure the healthy protein requirements needed for a global population expected to reach 10 billion by 2050.[10] The One Health framework has also been attached to a growing literature on ecosystem services. As the health of marine populations, like coral reefs, oyster beds, and other forms of "green infrastructure," are recognized as being vital to the maintenance of resource ecosystems, carbon sequestration, and coastal infrastructure, they too have been drawn into One Health initiatives.[11]

Finally, as scholars and public policymakers increasingly recognize the importance of marine environments as both a threat to and potential resources for human health, One Health initiatives have also been brought into conversation with the policy framework of the blue economy. Blue economy strategies aim to expand the economic potential of ocean resources—including fisheries, aquaculture, biopharmaceuticals, and deep-sea minerals—often in alignment with sustainable development values. While acknowledging the need for conservation and care in ocean governance, such policies often also highlight the abundance of well-managed oceans. In doing so, blue economy rhetoric and policies promise the concomitant emergence of large-scale (even industrial-scale) economic productivity with expanded ecological and social wellbeing.[12] The connection between One Health and the blue economy is exemplified most clearly in the economic pursuit of marine pharmaceutical resources, which have become one of the most promising elements in the blue economy narrative. This is the case, for example, in the EU-funded *Seas, Oceans and Public Health in Europe* (SOPHIE) project (2017–2020) which explored the economic potentials of developing marine biodiversity for medicine and disease prevention, while simultaneously protecting marine environments to ensure the provision of ecosystem services for future human health.[13]

In what follows, we explore some of the particularities of these alignments between good health, oceans, and economic productivity via our two case studies of jellyfish biomaterials and shrimp-derived chitosan. The healthful futures imagined through the applications of jellyfish and chitosan are, we argue, inescapably tied to histories of resource extraction and colonial narratives of abundance. To this extent, we consider how ostensibly progressive efforts to expand marine biotechnology in the pursuit of human and ecological health may serve to further marginalize particular "unhealthy" human bodies and "undesirable" nonhuman organisms. This leads us to question how paradigms of good and ill health are defined across human and marine ecologies alike.

The Promises of One Health I: Jellyfish Management for Human Health

Jellyfish blooms are understood by policymakers and scientists as a "wicked problem" and an emblem of anthropogenic changes on ocean ecologies. Scientists have documented an increase in jellyfish blooms since the turn of the twenty-first century, which have positioned them as a growing ecological and economic threat. "Jellyfish" is an imprecise umbrella term often used to encompass two phyla (Cnidaria and Ctenophora), several classes, and thousands of species of gelatinous zooplankton, many of which metamorphose into their motile adult stage *en masse*, resulting in blooms that can stretch for miles. The emergence of such blooms can be a problem for several sectors, including tourism, electric power plants (where they can disrupt cooling towers and shut down electricity generation in some regions for days), and the fishing industry. Large blooms can sink fishing boats and outcompete the fish that humans consume. For example, the expansion of the indigenous species of true jellyfish *Pelagia noctiluca* in the early 2000s led to the complete collapse of several Atlantic caged salmon operations.[14] Finally, jellyfish are also considered by many scientists to be an issue for wider ecological biodiversity, as rising jellyfish populations may cause a wide range of other species to decline.

Such far-reaching problems associated with blooms are compounded by the complexity of their ecologies. The supposed uptick in blooms seems to be indicative of a syndrome of ecological stressors caused by commercial shipping, terrestrial runoff, climate change, overfishing, and the expansion of underwater infrastructure like wind turbines, sea walls, and dykes, which create space for colonization by jellyfish polyps and promote blooms.[15] Like all wicked problems, technological solutions have not been forthcoming.[16] When scientists and policymakers have simplified the problem of blooms and made jellyfish themselves the target of change, they have encountered failure. In 2013, for example, South Korea famously innovated a fleet of unmanned robots known as the Jellyfish Elimination Robotic Swarm (or JEROS). They expected that the machines would combat blooms by shredding them.[17] But scientists discovered that shredding jellyfish only more effectively mixes their reproductive cells. Although these tactics eliminated existing blooms, they only served to promote future ones.[18] On French and Italian coastlines, municipalities erected nets to keep jellyfish away from ocean-going tourists. But even these are often worse than doing nothing as they tend to injure the jellyfish as they brush up against them, distributing tentacles in the water—and detached tentacles can sting just the same as live ones.[19] Accordingly, engaging in tactics that geographers Stephen Hinchliffe et al. have referred to as a "will to closure" in response to biological threats seems quite ill advised.[20]

In the wake of these realizations, a completely different paradigm of governance has taken hold. At an Expo Milano 2015 conference on nutrition for the Italian National Research Council, jellyfish were featured alongside algae and insects as the foods of the future. Incorporating "future foods" is a growing part of Italy's "slow food" movement.[21] Jellyfish have been consumed as a delicacy for over 1,500 years in East Asia. Recently, in the United States, more frequent blooms of cannonball jellyfish in the Atlantic South have blossomed into a thriving trade with China, with jellyfish processing plants being built in the Carolinas and Georgia. The Italian National Research Council hoped that jellyfish would be brought out of their "ethnic niche to become a mainstream solution to the unprecedented depletion of the oceans."[22]

What do these changes in jellyfish management have to do with One Health? To date, there has been scant scholarship on jellyfish in the One Health context. Although jellyfish can be toxic to humans, no zoonotic diseases have been found to pass between these marine organisms and our mammalian bodies. But they have nevertheless been implicated in the history of the One Health movement. Jellyfish appear in the canon of One Health literature, beginning in the late twentieth century in the work of veterinary pathologist and "father of modern epidemiology," Calvin Schwabe. In his 1964 monograph, *Veterinary Medicine and Human Health,* Schwabe formalized the concept of One Health and, following Rudolf Virchow, elevated the importance of veterinary science from its role as a lesser medicine.[23] Although Schwabe is best known for his medical research and writing, he also wrote a 1979 cookbook *Unmentionable Cuisine*, that promoted One Health ideals.[24] Among unconventional cuts of more conventional meats, the book includes recipes for dog, cat, and insects. It also features five recipes for jellyfish from East Asia and the South Pacific. The book, in Schwabe's own words, was meant to be "a practical guide to help us and our children prepare for the not-too-distant day when the world's growing food/population problem presses closer upon us and our overly restrictive eating habits become less tolerable."[25] Although he went on to emphasize that expanding food cultures could also expand culinary pleasures, Schwabe's cookbook was an early attempt to connect food cultures, consumption patterns, and human health directly to planetary environmental management.

Recent paradigms of jellyfish management have expanded on Schwabe's ideas, similarly linking the future of jellyfish to a future of expanded human health and wellbeing through direct consumption. To that end, a project funded by the European Union published the *European Jellyfish Cookbook* in 2020. It is, the authors argue, the first-ever publication of "western style" jellyfish recipes. To develop the cookbook, scientists teamed up with Michelin star chefs around Italy to develop new preparations. These include a jellyfish tagliatelle, jellyfish carpaccio, and jellyfish marinated with fennel, purslane, and grapefruit.[26]

Italian marine biologist and noted jellyfish researcher, Stefano Pirano, wrote the preface. In it, he linked population growth and future consumption patterns. In his words:

> With a world population that grows at [an] exponential rate (we will be just under 10 billion in 2050), and with a much slower food production (that grows only thanks to continuous technological innovations and the use of mainly non-renewable energy sources), the only way out is to find different, new, sustainable food resources. The biological cycle of jellyfish and their growing diffusion and abundance in coastal seas (also due to ongoing climate change) allows us to think of these fascinating creatures as a possible new food source, whose favorable biochemical and nutritional characteristics are now well recognized.[27]

Pirano and the cookbook's authors are not alone in their attempt to make palatable these strange bodies, otherwise associated with stings and slime. Other projects have similarly attempted to transform jellyfish bodies into gastronomical delights, in the form of crisps and relishes.[28]

The extraction of jellyfish from the oceans for the health and wellbeing of humans extends beyond their direct consumption. The authors of the cookbook project are indeed part of a wider, interdisciplinary, funded research consortium that set out to identify a diversity of products that can be made from the masses of gelatinous medusae. As part of that project, jellyfish bodies are being tested as a fertilizer additive, while jellyfish collagen and antioxidants are being extracted for biopharmaceuticals and cosmetics.[29] A separate field of anti-aging research is exploring jellyfish life cycles as part of advanced stem cell biomedicine and even radical life extension.[30] In the United States, supplements, including one called KollaJell, promise that the ingestion of jellyfish will delay the onset of cognitive decline.[31] Finally, jellyfish mucus is also being developed into wastewater treatment filters capable of capturing microplastics—yet another emblem of the Anthropocene.[32]

Jellyfish are therefore poised to transition from a "trash animal" that overburdens fisheries,[33] to a "sustainable protein," biopharmaceutical, industrial-scale infrastructure, and fishery in its own right.

The Promises of One Health II: The "Obesity Crisis" and the Shrimp-Shell

The exponential growth of shrimp aquaculture since 1950 is one of the key Earth system trends marking the Great Acceleration characterizing the Anthropocene. Shrimp shells represent the material excesses of accelerating aquaculture: they are indigestible and a potentially hazardous, yet abundant, global mass. Typically, 45–48% of shrimp raw material is discarded as waste.[34]Although some shrimp shells are recycled and turned into rudimentary fertilizers, most are destined for landfill, or are burned, or dumped into coastal waters with possible harmful ecological effects and potential endangerment to human health.

Since the 1970s, the United States Environmental Protection Agency (EPA) has regulated the dumping of fish waste—including crustacean shells—via the Marine Protection, Research and Sanctuaries Act. This Act details how bacterial decomposition of (shell)fish waste in coastal waters can lead to reduced oxygen levels in seawater, the smothering of living organisms, and the introduction of non-native and invasive species into oceanic ecosystems. Accordingly, as wild shrimp catches and shrimp aquaculture continues to grow globally, the question of how to sustainably deal with proliferating shrimp waste streams becomes ever more pressing.

Increasingly, scientists and biotechnology start-ups have diverted waste-shrimp shells away from the seafloor and toward laboratory settings, as they seek to re-appropriate and re-valorize crustacean biowaste in pursuit of a circular blue economy. This has seen the transformation of previously indigestible shrimp-shell waste into the biomaterial chitosan. Chitosan is produced via the chemical deacetylation of chitin. Chitin is the second most abundant polysaccharide in nature after cellulose, and is the primary structural component in the exoskeletons of crustaceans, insects, spiders, and nematodes, as well as fungal cell walls. Chitin and chitosan were discovered in the late nineteenth century, although their uses were limited until the 1970s, when the biomaterials and their socio-ecological and commercial potentials were "rediscovered."[35] By the twenty-first century, over 2,000 applications of chitosan have been developed, spanning agriculture, biomedicine, pharmaceuticals and food sectors.[36]

Chitosan boasts an extraordinary number of material properties that make its application potentials seem limitless. It is antifungal, antimicrobial, antibacterial, biodegradable, non-toxic, biocompatible in humans, hemostatic, has metal-chelating properties, and much more. Due to its ubiquity and versatility, chitosan has been heralded as a panacea for a staggering range of socioecological ills which tarnish the Anthropocene.

Chitosan is centrally positioned in an upcoming "biomaterial age"[37] which will align with One Health agendas by seeking sustainable biotechnological innovations for housing, nutrition, and medicine to achieve healthy, interconnected multispecies futures. The transition toward a "biomaterial age" will harness chitosan to target the accumulation of materials deemed socio-ecologically harmful at a range of scales. Through processes of chitosan-augmented digestion, the bodies of humans, nonhumans and wider planetary ecologies become refigured and made "healthier."

The obesity "crisis," first postulated in the 1990s, is now considered to be a growing global-health challenge in the Anthropocene, and is seen as a leading cause of ill health worldwide.[38]The narrative of the obesity crisis has primarily focused on measures of Body Mass Index (BMI)—with medical professionals and epidemiologists historically linking statistics that indicate increasing BMIs to the over-consumption of highly processed, unhealthy foods. But epidemiologists have recently linked the global obesity crisis to Anthropocene geologic and environmental transformation. Urbanization, biodiversity losses, industrialization, and land-use changes are said to have impacted human gut microbiomes in ways that bear directly on body mass.[39] Whereas responses to the so-called obesity crisis have long targeted individual practices of eating and exercise, biotech innovators have ironically attempted to solve the "problem" of obesity by producing further consumables for humans to digest. Chitosan dietary supplements are one such "solution" that purports magical weight-loss results. By binding with negatively charged fat molecules and bile acids, chitosan interferes with the emulsification of neutral lipids, such as cholesterol, by binding to them with hydrophobic bonds.[40] Put simply, chitosan can inhibit fat transfer by preventing absorption of fat in the gut, instead excreting fat molecules as waste from the body. Chitosan can, it seems, render dietary fat as waste.

Promises about the fat-fighting potentials of chitosan were first enshrined in self-help diet books in the 1990s, which sung the praises of the biomaterial as a "natural fat blocker,"[41] and a "revolutionary discovery that removes fat naturally."[42] In William J. Hennen's 1996 book, *Chitosan: Natural Fat Blocker,* the author positions chitosan as a "solution" for those seeking "safe and effective ways to lose weight, manage dietary fat intake, and stabilize factors leading to heart disease, cancer, and other conditions."[43] Similarly, in *The Fat Blocker Diet,* Arnold Fox and Brenda Adderly promise that chitosan is "completely healthy" and allows you to lose weight with "absolutely no known side effects."[44] Ultimately, these books present chitosan supplements as an inexpensive "magic bullet." Seemingly, chitosan is the perfect biomaterial for the Anthropocene: it promises to transform environmental waste into a solution to an ostensible public health crisis.

Biomaterials like chitosan appear perfectly suited to an ecologically minded One Health movement. Indeed, One Health practitioners, such as John Amuasi in his interview for this volume, positions food systems management as a key area of current debate and activity in the global One Health arena.[45] Relatedly, the varying food management interventions, made through both chitosan and jellyfish biomaterials, offer possibilities for simultaneously solving ecological issues and a constellation of human health problems, including obesity, high cholesterol, and the protein requirements of an ever-expanding global population.

Organisms and Biomaterials Between Health and Illness

As it turns out, ecological systems do not always play along with the fantasy of a well-plated, right-sized, and well-managed Planetary Health. These two dreams, of managed health networks in a shared, more-than-human world, are riddled with both practical and conceptual problems. In this section, we explore some of those practical and theoretical problems and their challenge to the One Health paradigm.

Although Western governments, researchers, and media outlets tout jellyfish as a low-calorie, low-fat food that is also high in collagen, B12, and protein, there is currently no regulatory infrastructure for processing food-grade jellyfish within the EU. In East Asia, where jellyfish are regularly harvested and packaged for consumption, they are processed and dried with aluminum salts, or alum. This leads not only to an impressive sodium content; it has been argued that alum could increase the risk of neurological degeneration and Alzheimer's.[46] Processed jellyfish, at least, may not be the healthful, anti-aging product that many have had in mind. As alternative processing strategies are being developed, many EU researchers are developing products using jellyfish biomaterial imported from East Asia, rather than extracted from the Mediterranean or North Seas. Accordingly, the notion that new resource markets will solve the ecological problem of excessive blooms in the immediate region remains a pipedream.

And processing is not the only issue. As noted at the outset of this paper, jellyfish blooms are thought to be indicative of a syndrome of ecological stressors caused by circulations of capital and petrol-economies and the effects of climate change. In popular press pieces and aquariums, jellyfish are presented as stress amplifiers, further degrading already-degraded ecosystems. But, as several researchers have reported, there is so little known about how jellyfish function in ecosystems—and their role in biodiversity—that these claims are speculative at best. In his report on jellyfish to the United Nations Food and Agricultural Organization in 2013, Fernando Boero referred to the "problem" of blooms as a "lottery" of timing.[47] The lottery is thought to involve the coincidence of planktonic and marine fish breeding cycles. Jellyfish become a "problem" when blooms emerge simultaneously with plankton blooms in a season in which fish hatches are delayed. Under these conditions, jellyfish can outcompete vertebrate fish in a given year. But this is not necessarily a cumulative problem. Rising jellyfish populations in one year do not necessarily mean rising populations in subsequent years. And given that vertebrates live in the adult stage for several years, rather than a single season, means that fish can recover from the effects of these annual ecological lotteries over multiple year periods. Some researchers are now questioning the narrative of increased blooms altogether, given how little we know of long-term cycling of marine jellyfish.

Considering these challenges, Boero and others have speculated about future seas dominated by jellyfish. If the energy lottery favors jellyfish over several years, Boero has written that a regime shift or tipping point is possible that would shift the quantities and kinds of biodiversity in a given marine ecology. But, importantly, Boero emphasizes that the results of such a shift are speculative, possibly not any less stable or less flexible than the current ecological system. It is, merely, otherwise. Other scholars I've spoken with, including those who are part of these circular health initiatives in Europe, fully recognize this as a problem for the harvesting they have proposed: creating a fishery of jellyfish not only risks a host of unintended ecological consequences, but there remain too many unknowns about the timing of blooms to ensure that future markets, if created, would be met in any sustainable way. Moreover, the emergence of these "potential otherwise" ecologies would, as Boero has written, be a "problem," not necessarily for the stability of ecologies as such, but only for existing human economic systems. Defining existing ecological arrangements as "healthy" and their alternative potentials as "ill" is a distinctly anthropocentric view that challenges the presupposition that a more-than-human "One Health" is really "more-than" at all. Indeed, few are asking questions about the health of jellyfish populations themselves. The dream of a circular jellyfish economy is predicated on bracketing off the "health" of jellyfish themselves, assuming that their bodies and populations are indicative of ill health overall—and entirely expendable.

Shrimp Shells: The New Snake Oil

There are reasons to be similarly skeptical about the promises of chitosan as a palliative to both waste streams and the so-called "obesity crisis." By the start of the new millennium, annual sales of the top three chitosan diet pills in the United States exceeded $6 million.[48] But despite being presented as a "magic bullet," scientific studies of chitosan supplements have largely failed to produce magical weight losses. In fact, almost immediately after promises about chitosan supplements garnered popular attention, there has followed decades of contested debate about the impact of these "health products" on obesity in humans.[49]

As with other common dietary supplements, chitosan pills have not undergone extensive clinical trials, are not prescribed, and are largely unregulated by the US Food and Drug Administration. Accordingly, their efficacy is contested. Although some scientific experiments have shown chitosan to decrease bodyweight gain in rats[50] and pigs,[51] the effect on humans is less conclusive. One study reported a 3.3 kilograms greater weight loss for those taking chitosan supplements *versus* those taking placebos in a short-term treatment for obesity.[52] However, a paper measuring fecal fat excretion after taking chitosan supplements concluded that chitosan had no effect on fat-absorption among healthy males.[53] The long-term effect of chitosan supplements remains inconclusive, and, since 2004, some scientists have discouraged the use of chitosan for weight loss, branding it ineffective.[54]

Nevertheless, popular enchantment surrounds the promise that chitosan can transform dietary fats into waste. A 2011 article in UK newspaper *The Independent*, entitled *All The Fat You Can Eat,* acknowledged the ongoing debate surrounding chitosan supplements, yet continued to glamorize the "ultimate fantasy" that chitosan pills "let you pig out without piling on the pounds."[55] While accepting these promises might be an "impossible dream," the report fuels hope that chitosan could be the magic solution for obese and overweight individuals. Over ten years later, chitosan supplements are still widely sold, with product reviews on health-store websites seeing consumers extol the virtues of ingesting chitosan supplements for magical weight loss without dietary changes. However, rave user-reviews stand in stark contrast to recent reports in *The Guardian,* which stress that the capacity of chitosan to support weight loss appears minimal, and certainly not at levels considered clinically significant.[56] All of this suggests that chitosan as a weight loss supplement is little more than twenty-first century snake oil.

Even more, the narratives that accompany weight loss supplements like chitosan perpetuate the equation of low or moderate weight with good health—and higher weights with ill health. As many scholars have argued, such narratives are more rooted in a Eurocentric fat-phobic culture rather than well-documented health research.[57] Indeed, the narrative of a global obesity crisis and the use of BMI as a measure of overall health, have been widely critiqued for well over a decade. In a 2010 paper, Rachel Colls and Bethan Evans painstakingly debunk several of the assumptions made by the obesity crisis discourse, including that: BMI is a measure of fatness; BMI is an indicator of health and co-morbidity; children with high BMIs become overweight adults; and high-BMI individuals should engage in diet and exercise to reduce weight and lengthen their lives.[58] The debunking of the obesity crisis presents a further issue for the One Health paradigm, illustrating that public health approaches to the definition of "good health" are at times riddled with social prejudices, poor research, and bad faith arguments. Ultimately, public health initiatives can reinscribe existing inequalities between social groups— thus, as Chris Walzer discusses in his interview for this volume, One Health projects should become better attuned to tackling, rather than reinforcing, health inequities.

Not One, But Many Healths

So, what does our critique of the underlying ideologies of these "health-giving" marine biotechnologies mean for One Health and its notion of a holistic, more-than-human collective health? When paired with biodiversity and environmental conservation, One Health strategies seem overridingly progressive: they necessitate an awareness of the indistinction between material and social processes, an obliteration of nature–society divides. But there are risks embedded within this idealized conceptualization of interconnected Planetary Health. As Stephen Hinchliffe writes, an understanding of infinite connectivity can support an ideology where the "one world" imagined is a Western-centric one, in which a panoply of systems and relations are approached with a singular set of epistemic solutions.[59] As above, these narratives also work to buttress existing assumptions not only about "good health" but the shape and direction of the future.

Apocalyptic narratives, like the one envisioned by Charles Schwabe, are still evident in some popular science writers' descriptions of jellyfish blooms, which frequently focus on the catastrophes of destroyed fisheries, ruined livelihoods, depleted oceans, and widespread famine.[60] But these warnings are met with a seemingly new and paradoxical alter-narrative: a vision of sustained abundance and a world in which collective health is enhanced rather than degraded. This is a narrative found in both case studies, of jellyfish biomaterials and chitosan. It encourages us not to worry about coming apocalypses or contemporary injustices: we will eat our way out of these problems—on white tablecloths and with all the trappings of a tasteful future in which we are all forever young and optimally sized.[61] This may be more libidinal fantasy than scientific endeavor toward collective health. Indeed, circular economy visions are consistent with much of the history of capitalism, in which ostensible limits are merely barriers to be overcome by technology or innovation, financialization, or the creation of new markets.[62]

That these narratives converge on the bodies of jellyfish and the waste of the shrimping industry indicate that marine organisms are uniquely powerful figures within the dream of a sustainable and circular blue economy. Jellyfish and shrimp both often appear not as one of many fragile marine fisheries requiring better regulation or more intensive surveillance, but rather, as *The New York Times* wrote of jellyfish in 2017, a potentially "inexhaustible" food resource.[63] Like jellyfish, shrimp and their shells are viewed as a resilient, almost limitless marine resource. Louisiana shrimp processors explained how short shrimp life cycles mean that a "bad crop" in one season is not necessarily indicative of damaged ecologies or a harbinger of poor future harvests. Rather, shrimp populations display remarkable resilience with large bounties of shrimp—and their shells—showing little signs of disappearing. Indeed, in discussions of jellyfish, shrimp and other "future foods," we find echoes of eighteenth and nineteenth century discourses around other seemingly infinite resources like buffalo or beavers in North America, or herring in the North Sea. In the biomedical applications of jellyfish biomaterials and chitosan, we also find dreams of the fountain of youth and immortality alive through potential biotech and cosmetic applications. These visions of "health" draw on the seeming "excesses" of jellyfish life and shrimp-shell waste to hope for the extension of certain forms of life on land—ones that continue to be tied to high levels of consumption and production.

We might also consider how some formulations of One Health create the conditions for what postcolonial author Sylvia Wynter has critically described as "the incorporation of all forms of [life] into a single, homogenized descriptive statement" whose health—indeed, whose life—can be managed with a transparent "science."[64] Wynter wrote those words as a critique of appropriations of land and racialized bodies through histories of colonization and the enslavement of Africans. That history of a universal or homogenized vision of life finds echoes in the One Health movement. This version risks perpetuating a colonial vision of collective health, undermining not only the plural forms of nonhuman life on the planet, but also non-Western voices and non-Western ways of living and knowing. Whereas the worlds of bioscience innovation, shrimp, and jellyfish seem perhaps distant from Wynter's postcolonial critique, they risk producing a similarly homogenized vision of human and nonhuman life alike. Look to chitosan diet pills and their associated marketing, which continues to perpetuate a Eurocentric vision of "healthy bodies." This imaginary builds upon and feeds into an ongoing biomedical preoccupation with BMI and other inherently racialized "health" measures and metrics, deployed as part of what Syad calls "obesity surveillance" of racialized bodies.[65] To combat such singular and potentially damaging visions, One Health might need to better accept the plurality of forms and expressions of life on land and sea—and risk breaking its vision of a singular and shared planet.

Acknowledgements

Portions of this chapter were published as Hannah Dickinson and Elizabeth R. Johnson, "Digesting Planetary Harms: Ocean Life, Biomaterial Innovation, and Uncanny Ingestions of the Anthropocene," *Journal of the History of Science and Technology* (forthcoming). We are both grateful for the time, patience, and incredible thoughtfulness of our research participants involved in shrimping, chitosan research, and jellyfish science.

Notes

1 See Meredith Barrett et al., "Integrating a One Health Approach in Education to Address Global Health and Sustainability Challenges," *Frontiers in Ecology and the Environment* 9, no. 4 (2011): 239–245; J. Zinsstag et al., "From 'One Medicine' to 'One Health' and Systemic Approaches to Health and Well-Being," *Preventive Veterinary Medicine* 101, no. 3 (September 1, 2011): 148–156.

2 The term, of course, is heavily contested. For a thorough contextualization of the racial histories of geoscience that gave birth to the term, see Kathryn Yusoff, *A Billion Black Anthropocenes or None* (Minneapolis: University of Minnesota Press, 2020).

3 Matthew Taylor and Ilaria Capua, "Rethinking the Concept of Health," The RSA, May 11, 2021.

4 We see this perspective emphasized in the social science writing on parasitic hookworms. Hookworms can cause anemia and digestive issues in adults and have severe negative effects on the growth and development of children. Medical professionals typically prescribe anthelminthic medications—which *de*worm mammalian bodies—to those with known infections. However, hookworm has also been lauded as a niche treatment for autoimmune disorders by a minority in the West. As geographers Jamie Lorimer, Will McKeithen, and Skye Naslun describe, an increasing number of individuals suffering from overactive immune systems have actively *wormed* themselves to modulate their body's hyper-active response. Thus, hookworm and other helminths are increasingly understood by medical communities as both the cause of disease and a health care treatment, depending on the system to which it is introduced. See Jamie Lorimer, *The Probiotic Planet* (Minneapolis: University of Minnesota Press, 2020); Skye Naslund and Will McKeithen, "Brave New Worms," in *The Routledge Handbook of Critical Resource Geography*, eds. Matthew Himley, Elizabeth Havice and Gabriela Valdivia (Abingdon: Routledge, 2021), 68–78, 68; See also Linda J. Wammes et al., "Helminth Therapy or Elimination: Epidemiological, Immunological, and Clinical Considerations," *The Lancet Infectious Diseases* 14, no. 11 (2014): 1150–1162.

5 Stephen Hinchliffe, "More than One World, More than One Health: Re-Configuring Interspecies Health," *Social Science and Medicine* 129 (2015): 28–35.

6 Lora E. Fleming et al., "The Ocean Decade—Opportunities for Oceans and Human Health Programs to Contribute to Public Health," *American Journal of Public Health* 111, no. 5 (May, 2021): 808–811. See also: Meredith A. Barrett et al., "Integrating a One Health Approach in Education to Address Global Health and Sustainability Challenges," *Frontiers in Ecology and the Environment* 9, no. 4 (2011): 239–245; Donald C. Behringer and Elizabeth Duermit-Moreau, "Crustaceans, One Health and the Changing Ocean," *Journal of Invertebrate Pathology* 186 (November 1, 2021): 107500. Ronan Foley, "'One Blue,'" *Dialogues in Human Geography* 7, no. 1 (March 1, 2017): 32–36; Michael Sweet, Alfred Burian, and Mark Bulling, "Corals as Canaries in the Coalmine: Towards the Incorporation of Marine Ecosystems into the 'One Health' Concept," *Journal of Invertebrate Pathology* 186 (November 1, 2021): 107538.

7 Lora E. Fleming et al., "The Ocean Decade—Opportunities for Oceans and Human Health Programs to Contribute to Public Health," *American Journal of Public Health* 111, no. 5 (May, 2021): 808–811.
8 For more on NOAA's approach to One Health see: https://cpo.noaa.gov/Serving-Society/NOAA-One-Health
9 Andrew D. Turner et al., "Marine Invertebrate Interactions with Harmful Algal Blooms – Implications for One Health," *Journal of Invertebrate Pathology* 186 (November 1, 2021): 107555.
10 G. D. Stentiford et al., "Sustainable Aquaculture Through the One Health Lens," *Nature Food* 1 (2020): 468–474. See also: Jess Vergis et al., "Food Safety in Fisheries: Application of One Health Approach," *The Indian Journal of Medical Research* 153, no. 3 (March, 2021): 348–357; Donald Behringer and Elizabeth Duermit-Moreau, "Crustaceans, One Health and the Changing Ocean," *Journal of Invertebrate Pathology* 186 (November 1, 2021): 107500.
11 Paul Sandifer and Ariana Sutton-Grier, "Connecting Stressors, Ocean Ecosystem Services, and Human Health," *Natural Resources Forum* 38, no. 3 (2014): 157–167, 163.
12 S. Smith-Godfrey, "Defining the Blue Economy," *Maritime Affairs: Journal of the National Maritime Foundation of India* 12, no. 1 (January 2, 2016): 58–64.
13 For more details on the SOPHIE project see: https://sophie2020.eu
14 M. Marcos-López, S. O. Mitchell, and H. D. Rodger, "Pathology and Mortality Associated with the Mauve Stinger Jellyfish *Pelagia noctiluca* in Farmed Atlantic Salmon *Salmo salar* L.," *Journal of Fish Diseases* 39, no. 1 (2016): 111–115.
15 Karl Mathiesen, "Are Jellyfish Going to Take Over the Oceans?," *The Guardian*, August 21, 2015, https://www.theguardian.com/environment/2015/aug/21/are-jellyfish-going-to-take-over-oceans; Jane Qiu, "Coastal Havoc Boosts Jellies," *Nature News* 514, no. 7524 (2014): 545; Lauren Frayer, "Jellyfish Invade Mediterranean Beaches," DW, August 26, 2013; Lily Whiteman, "Jellyfish Swarms: Bellwethers of Environmental Change," *Live Science*, September 11, 2014, https://www.livescience.com/47788-jellyfish-signal-climate-change-nsf-ria.html.
16 Kristy de Salas et al., "The Super Wicked Problem of Ocean Health: A Socio-Ecological and Behavioural Perspective," *Philosophical Transactions of the Royal Society B: Biological Sciences* 377, no. 1854 (July 4, 2022): 20210271.
17 Donghoon Kim et al., "Development of Jellyfish Removal Robot System JEROS," (conference paper, 2012 9th International Conference on Ubiquitous Robots and Ambient Intelligence (URAI), Daejeon, South Korea, 2012), 599–600. See also Delf Rothe, "Jellyfish Encounters: Science, Technology and Security in the Anthropocene Ocean," *Critical Studies on Security* 8, no. 2 (2020): 145–159; Elizabeth R. Johnson, "Governing Jellyfish: Eco-Security and Planetary 'Life' in the Anthropocene," in *Animals, Biopolitics, Law*, ed. Irus Braverman (London: Routledge, 2016).
18 Rebecca Helm, "Jelly Killing Machine Tested in Korea," *Deep Sea News*, October 3, 2013, http://www.deepseanews.com/2013/10/jelly-killing-machine-tested-in-korea/.
19 A company named EcoBarrier has designed nets that mitigate this problem, but scientists still express concerns about the effectiveness of these measures.
20 Stephen Hinchliffe et al., "Biosecurity and the Topologies of Infected Life: From Borderlines to Borderlands," *Transactions of the Institute of British Geographers* 38, no. 4 (2013): 531–543.
21 Dany Mitzman, "Invasion of the Jellyfish: Is it Time to Get Frying?," *BBC News*, August 12, 2017, https://www.bbc.com/news/magazine-40899378.
22 Carola Traverso Saibante, "In Praise of the Jellyfish, from Menace to Resource," *Fine Dining Lovers*, November 25, 2015, https://www.finedininglovers.com/stories/eating-jellyfish/.
23 Calvin W. Schwabe, *Veterinary Medicine and Human Health* (Baltimore: Williams & Wilkins, 1984).

24 Calvin W. Schwabe, *Unmentionable Cuisine* (Charlottesville: University of Virginia Press, 1988).
25 Ibid., 1.
26 Antonella Leone, *The European Jellyfish Cookbook* (Rome: Locopress Industria Grafica, 2020).
27 Ibid., 5.
28 Toki Mie et al., "The Microscopic Structure of Crunchy and Crispy Jellyfish," *Biophysical Journal* 114, no. 3 (2018): 538a; Lou Del Bello, "Jellyfish Chips Are the Future of Junk Food," *Futurism*, February 20, 2018, https://futurism.com/jellyfish-chips-future-junk-food; Natasha Frost, "Jellyfish May Be the Snack Food of the Future," *Atlas Obscura*, July 31, 2017, http://www.atlasobscura.com/articles/jellyfish-chips-climate-change-snack-food-science.
29 Iraj Emadodin et al., "Assessing the Potential of Jellyfish as an Organic Soil Amendment to Enhance Seed Germination and Seedling Establishment in Sand Dune Restoration," *Agronomy* 10, no. 6 (2020): 863; Antonella Leone et al., "The Bright Side of Gelatinous Blooms: Nutraceutical Value and Antioxidant Properties of Three Mediterranean Jellyfish (Scyphozoa)," *Marine Drugs* 13, no. 8 (2015): 4654–4681; Birgit Hoyer et al., "Jellyfish Collagen Scaffolds for Cartilage Tissue Engineering," *Acta Biomaterialia* 10, no. 2 (2014): 883–892. See also S. A. Trim, F. Wandrey, and C. Trim, "Beauty from the Deep: Cnidarians in Cosmetics," in *The Cnidaria: Only a Problem or Also a Source*, eds. G. L. Mariottini, N. Killi, and L. Xiao (Hauppauge: Nova Science Publishers, 2020); Alina Sionkowska et al., "Collagen Based Materials in Cosmetic Applications: A Review," *Materials* 13, no. 19 (2020): 4217, 1–15; Dong Wook Kim et al., "Moisturizing Effect of Jellyfish Collagen Extract," *Journal of the Society of Cosmetic Scientists of Korea* 42, no. 2 (2016): 153–162.
30 Sosuke Fujita, Erina Kuranaga, and Yu-Ichiro Nakajima, "Regeneration Potential of Jellyfish: Cellular Mechanisms and Molecular Insights," *Genes* 12, no. 5 (2021): 758, 1–18; Paul C. Guest, "Of Mice, Whales, Jellyfish and Men: In Pursuit of Increased Longevity," in *Reviews on Biomarker Studies in Aging and Anti-Aging Research*, ed. Paul C. Guest (New York: Springer, 2019), 1–24; Ronald S. Petralia, Mark P. Mattson, and Pamela J. Yao, "Aging and Longevity in the Simplest Animals and the Quest for Immortality," *Ageing Research Reviews* 16 (2014): 66–82.
31 "KollaJell," Longevity by Nature, last accessed May 1, 2022, https://www.longevity-bynature.biz/product/kollajell/.
32 Sarah Lawrynuik, "Slime Time," *New Scientist* 240, no. 3208 (2018): 44–45.
33 Kelsi Nagy and Phillip David Johnson II, eds. *Trash Animals: How We Live with Nature's Filthy, Feral, Invasive, and Unwanted Species* (Minneapolis: University of Minnesota Press, 2013).
34 Kandra Prameela, Murali Mohan Challa, and Hemalatha Kalangi Padma Jyothi, "Efficient Use of Shrimp Waste: Present and Future Trends," *Applied Microbiology and Biotechnology* 93 (2012): 17–29.
35 Grégorio Crini, "Historical Review on Chitin and Chitosan Biopolymers," *Environmental Chemistry Letters* 17 (2019): 1623–1643.
36 Tuyishime Philibert, Byong H. Lee, and Nsanzabera Fabien, "Current Status and New Perspectives on Chitin and Chitosan as Functional Biopolymers," *Applied Biochemistry and Biotechnology* 181, no. 4 (2017): 1314–1337.
37 Javier G. Fernandez and Stylianos Dritsas, "The Biomaterial Age: The Transition Toward a More Sustainable Society Will Be Determined by Advances in Controlling Biological Processes," *Matter* 2, no. 6 (2020): 1352–1355.
38 See R. Colls and B. Evans, "Challenging Assumptions: Re-Thinking 'the Obesity Problem,'" *Geography* 95, no. 2 (2010): 99–105; see also Walter Willett et al., "Food in the Anthropocene: the EAT–Lancet Commission on Healthy Diets from Sustainable Food Systems," *The Lancet Commissions* 393, no. 10170 (2019): 447–492.
39 Cecilie Torp Austvoll et al., "Health Impact of the Anthropocene: The Complex Relationship Between Gut Microbiota, Epigenetics, and Human Health, Using

Obesity as an Example," *Global Health, Epidemiology and Genomics* 5, no. e2 (2020): 1–10.

40 Ritva Ylitalo et al., "Cholesterol-Lowering Properties and Safety of Chitosan," *Arzneimittelforschung* 52, no. 1 (2002): 1–7.

41 William J. Hennen, *Chitosan: Natural Fat Blocker* (Orem: Woodland Publishing, 1996).

42 Arnold Fox and Brenda Adderly, *The Fat Blocker Diet: The Revolutionary Discovery that Removes Fat Naturally* (New York: St. Martin's Press, 1997).

43 Hennen, *Chitosan*.

44 Fox and Adderly, *The Fat Blocker Diet*.

45 Amuasi is, of course, not alone in recognizing the importance of food and eating in the One Health agenda, as Schwabe's earlier cookbook indicates. See also, David Tilman and Michael Clark, "Global Diets Link Environmental Sustainability and Human Health," *Nature* 515, no. 7528 (November, 2014): 518–522.

46 Linping Wang, "Entry and Deposit of Aluminum in the Brain," *Advances in Experimental Medicine and Biology* 1091 (2018): 39–51; J. R. Walton, "Chronic Aluminum Intake Causes Alzheimer's Disease: Applying Sir Austin Bradford Hill's Causality Criteria," *Journal of Alzheimer's Disease* 40, no. 4 (2014): 765–838.

47 Ferdinando Boero, "*Review of Jellyfish Blooms in the Mediterranean and Black Sea*, Volume 92 (Food and Agricultural Organization of the United Nations, General Fisheries Commission for the Mediterranean, 2013).

48 M. D. Gades and J. S. Stern, "Chitosan Supplementation Does Not Affect Fat Absorption in Healthy Males Fed a High-Fat Diet, a Pilot Study," *International Journal of Obesity* 26 (2002): 119–122.

49 Prior to being available as a dietary supplement on the mass-market, biomedical researchers had explored—and continue to explore—the potential use of chitosan for therapeutic treatment of diseases and dysfunctions of the human digestive system. Although much debated and with often conflicting results, researchers have ultimately suggested that chitosan treatments can offer promise and symptomatic relief to those suffering from chronic health conditions, including Crohn's disease, irritable bowel syndrome (IBS), high cholesterol, ulcerative colitis, and celiac disease.

50 Jiali Zhang et al., "Dietary Chitosan Improves Hypercholesterolemia in Rats Fed High-Fat Diets," *Nutrition Research* 28, no. 6 (2008): 383–290.

51 Ann M. Walsh et al., "Multi-Functional Roles of Chitosan as a Potential Protective Agent Against Obesity," *PLoS ONE* 8, no. 1 (2013): e53828.

52 E. Ernst and M. H. Pittler, "Chitosan as a Treatment for Body Weight Reduction? A Meta-Analysis," *Perfusion* 11 (1998): 461–465.

53 Gades and Stern, "Chitosan Supplementation."

54 Robert B. Saper, David M. Eisenberg, and Russell S. Phillips, "Common Dietary Supplements for Weight Loss," *American Family Physician* 70, no. 9 (2004): 1731–1738.

55 Glenda Cooper, "All the Fat You Can Eat," *The Independent*, October 23, 2011, https://www.independent.co.uk/life-style/health-and-families/health-news/all-the-fat-you-can-eat-1314916.html.

56 Robin McKie, "Herbal and Diet Supplements Have No Effect on Weight Loss," *Guardian*, 9 May, 2021, https://www.theguardian.com/lifeandstyle/2021/may/09/herbal-and-diet-supplements-have-no-effect-on-weight-loss.

57 Julie Guthman, *Weighing In: Obesity, Food Justice, and the Limits of Capitalism* (Berkeley: University of California Press, 2011).

58 Rachel Colls and Bethan Evans, "Re-Thinking 'the Obesity Problem,'" *Geography* 95, no. 2 (2010): 99–105. See also Bethan Evans and Rachel Colls, "Measuring Fatness, Governing Bodies: The Spatialities of the Body Mass Index (BMI) in Anti-Obesity Politics," *Antipode* 41, no. 5 (2009): 1051–1083.

59 Stephen Hinchliffe, "More than One World, More than One Health: Re-Configuring Interspecies Health," *Social Science & Medicine* 129 (2015): 28–35.

60 Lisa-ann Gershwin, *Stung!: On Jellyfish Blooms and the Future of the Ocean* (Chicago: University of Chicago Press, 2013).
61 Or, as Elsbeth Probyn writes, "The idea that you can solve such intricate and complicated human–fish relations by voting with your fork is deluded narcissism." Elsbeth Probyn. *Eating the Ocean* (Durham: Duke University Press Books, 2016), 10.
62 David Harvey, "Globalization and the 'Spatial Fix,'" *Geographische Revue* 2 (2001): 23–30.
63 Jason Horowitz, "Jellyfish Seek Italy's Warming Seas. Can't Beat 'Em? Eat 'Em," *New York Times*, September 17, 2017, https://www.nytimes.com/2017/09/17/world/europe/jellyfish-climate-change-italy.html.
64 Katherine McKittrick, *Sylvia Wynter: On Being Human as Praxis* (Durham: Duke University Press, 2015).
65 Iffath Unissa Syad, "In Biomedicine, Thin is Still in: Obesity Surveillance among Racialized, (Im)migrant, and Female Bodies," *Societies* 9 (2019): 59.

6

MORE-THAN-ALMONDS

Plant Disease and the Politics of Care

Emily Reisman

Introduction

What role does plant disease play in a One Health approach? The One Health High-Level Expert Panel, an advisory group representing the Food and Agriculture Organization of the United Nations (FAO), the World Organisation for Animal Health (WOAH, formerly the Office International des Epizooties or OIE), the United Nations Environment Program (UNEP), and the World Health Organization (WHO), explicitly includes "plants," "ecosystems," and "the wider environment" in its most recent definition.[1]

Yet, beyond calling for a holistic consideration of all life, the significance of plants, bacteria, fungi, or other organisms for a One Health approach remains vague. Zoonotic disease, in which humans are impacted by disease originating in other animals, understandably motivates the move towards One Health. Plants, not carrying the risk of disease spilling over to human populations, thus often appear as a backdrop against which animal interactions unfold. At times, they are considered for their role as primary producers along a food chain which can impact animal nutrition or introduce contaminants, such as lead,[2] into animal bodies. Novel plant diseases might also be considered as threats to biodiversity, which holds the promise of yet-undiscovered healing capacities for human ailments.

As this chapter illustrates, plant disease also offers an opportunity to consider broader conceptualizations of health which extend beyond a narrow focus on physical wellness and to spotlight the inherently political contours of disease management. As the Wildlife Conservation Society makes clear, One Health

FIGURE 6.1 Sheep graze among almond trees that are likely suffering from the bacterium *Xylella fastidiosa*, Mallorca, Spain. Photo by author.

DOI: 10.4324/9781003294085-9

must go beyond pathogens and parasites to address the socioeconomic and political factors impacting the wellbeing of all life.[3] And, as the term "More-than-One Health" underscores, we should be warry of presuming universally shared goals of health interventions as they are embedded in power relations and often require trade-offs.[4] In keeping with these concerns, I offer a feminist theorization of care to highlight the social origins and political implications of a rapid plant disease outbreak.

The socioecological devastation wrought by plant disease is well known. Plant disease famously provoked mass migration and mortality during the Irish Potato Famine, forever changed the ecology of North America during the chestnut blight, and nearly eliminated European wine cultivation. As global transportation networks become increasingly fluid, resistance to agrochemicals becomes more common, socioeconomic change refashions landscapes, and climate change alters the ranges and populations of organisms, plant disease is expected to be even more frequent and possibly more destructive.

In Mallorca, the largest of Spain's Balearic Islands, almond tree health was declining so swiftly that researchers in 2018 expected nearly every rainfed almond tree on the island would be dead within as little as five years. An introduced bacterium, *Xylella fastidiosa*, enabled by its spittle-bug vector and emboldened by climate change, had flooded the xylem of these rainfed trees, impeding the flow of fluid and nutrients to the point where the trees could no longer survive. As almond farmers, farm advisors, government officials, and scientists grappled with this new reality, often with sharp disagreement and blame, their words shared a common fundamental concern: care, or *cuidado* in Spanish.

What does it mean for farmers to care for trees? For a government to care for farmers? For trees to care for a diverse agroecological landscape? What does it mean to fail to care in these contexts? Or to be careless? Puig de la Bellacasa offers a theorization of care which does not entirely correspond to its vernacular usage. Rather than benevolent concern, care is a form of maintenance work. Research on care follows in the feminist tradition of drawing attention to unrecognized labors, such as Silvia Lopez Gil's description of care as largely unseen work "without which life does not function."[5] It is embedded in everyday practice,[6] so seamless a process that it lacks a clear beginning or end. Consonant with "new materialist" approaches which decenter human agency and emphasize the biophysical aspects of social worlds, care for Puig dela Bellacasa includes not only human acts, but the activities of an assemblage that promote ongoingness, irrespective of scale, aliveness, or species. As Annemarie Mol elaborates, "articulating 'good care' is an intervention rather than [a] factual assessment."[7] Feminist scholars introducing a special issue on the politics of care argue along these lines that "practices of care are always shot through with asymmetrical power relations."[8] "Care organizes, classifies and disciplines bodies," these authors note, with colonial governance as an instructive illustration. They encourage a critical approach to care which situates care practices within power relations. Care is non-innocent. It is a form of more-than-human biopolitics, in that it privileges the maintenance of certain lives over others. Care is, to quote Puig de la Bellacasa "a thick, *impure* involvement in a world where the question of how to care is posed."[9] She suggests that, as researchers, we look for where this question is not easily answered.

Epidemiological accounts of plant disease used widely by scientists and policymakers tend to reify boundaries, positioning pathogens as an outside threat or invasion. Here, I want to draw attention to how pathogens, no matter their origin, are produced from within. In keeping with a more-than-one-health approach that considers the disease assemblage inextricable from its social world, causes and effects are not reducible to the activities of a single species. Pathogenicity describes "the relational ways in which infectious diseases are made."[10] Spain's *Xylella* epidemic is produced not only by a bacterium, but also by the conditions of possibility created by tourism, unstable land tenure, histories of marginalization, and retreat of government from farm advising. Theorizing a plant epidemic through care demonstrates that the pathogen is only part of the picture. It invites a level of political engagement well beyond trade restrictions or government-funded tree removal and replacement. It broadens the scope of dialogue to reflect long-term commitments to the land. Understanding landscape care as non-innocent and laden with power relations avoids a good–bad dichotomy, asking difficult questions about how to care. Answers to this question may be incommensurable, as political projects often are, and carry uneven effects. Care is not a shorthand for universal values. It is a call to give relational maintenance work the political weight it is due.

Working with Care

How does one observe and classify care? One way is by listening for the multiple interpretations of care articulated by interlocutors. I did not begin this research seeking to examine care, nor did I anticipate encountering an epidemic of such dramatic proportions. It was the voices of my companions repeatedly deploying the word *cuidado* and detailing the decline in rural maintenance work inherent to the *Xylella* outbreak that drew me to revisit and deepen my engagement with theories of care in order to make sense of an unexpected phenomenon. The politics of care are a politics of relational maintenance: what relations are being sustained and to what ends?

This chapter derives insights from 24 confidential interviews conducted between January and July of 2018 with farmers, managers of almond grower cooperatives, government officials, and scientists in Mallorca identified via key informant and snowball sampling. These interviews were complemented by participant observation at farmer organizing meetings to discuss the epidemic, and analysis of archival records and secondary sources relevant to Spain's almond ecology. Through these engagements, I drew the landscape itself into walking interviews,[11] seeking to understand the multiple ways my interlocutors read the landscape for relations between people, trees, bacteria, insect vectors, climate, and more. A landscape, rather than land, is an assemblage of more-than-human relations rich with history, economy, cultural practices, and aesthetic values.

In examining the intricacies of landscape care, I build on multispecies ethnographic approaches,[12] and more-than-human geographic thought,[13] fields deeply indebted to, though often lacking engagement with, Indigenous scholarship.[14] Doing so serves to decenter human beings as exclusive protagonists, and instead bring relations among beings—plant, animal (humans included), mineral, and otherwise—to the foreground. I find this approach particularly necessary because disease is inherently relational across species categories. *Xylella fastidiosa* alone is simply a bacterium. *Disease*, as a phenomenon, exists as a pathogen–host–vector–environment complex.[15] It is those relations, not the bacterium nor the almond tree as discrete organisms, which demand a more explicit politics of landscape care.

El Oro De Mallorca

Xylella fastidiosa is a bacterium named for its habitat, the xylem of plants, and for its "fastidiousness": it is notoriously difficult to culture.[16] Whereas scientific convention calls for referring to the bacterium as *X. fastidiosa*, I will use the name "*Xylella*" as it circulated among my interlocutors. Although *Xylella* dislikes laboratories, it finds a plethora of comfortable homes within the xylem of vascular plants; it has 359 known plant hosts from 75 different plant families.[17] Like most bacterial pathogens, it is asymptomatic in the rainforests of South and Central America, where it has coevolved as an amiable endophyte which inhabits but does not kill its host. This makes sense as tropical plants benefit from abundant water and a few bacteria are unlikely to restrict their flow. Killing the host is generally bad for business. Once introduced to a new environment, *Xylella* rather easily finds a xylem-feeding insect whose mouth and foregut it can colonize, thus hitching a ride to the next juicy xylem the insect seeks out.

Almond trees are unlike the plants of *Xylella*'s tropical origin. They are adapted to arid lands, with wild relatives native to the region stretching from Central Asia westward to the Levant.[18] For the past few millennia, almonds have accompanied farmers throughout the Mediterranean basin, as they are particularly well suited to the long, hot, dry summers characteristic of the region. Although almonds have long been present in Mallorca, they did not gain a prominent place in the landscape until the turn of the twentieth century. At that time, a global grain glut and new industrial substitutes for olive oil made agriculture less profitable for aristocratic landlords while their political power gradually eroded.[19] Under these conditions, landlords began selling off parcels and an emerging class of merchant capitalists bought the land at bargain prices before subdividing and reselling it to peasants through long-term annuities.[20] Wine grapes were initially the preferred cash crop of this emerging class of diversified peasant farmers, as a global shortage caused by the *Phylloxera* wine blight in continental Europe promised spectacular profits.

The boom was quickly followed by a bust. A surplus of grapes drove a steep decline in prices and *Phylloxera* eventually made its way to Mallorca. Many chose to replace vineyards with a polyculture of tree crops in which almonds held a prominent place. The land which peasant farmers had been sold by reluctant landlords and speculating middlemen was often at the agricultural margins, much of it formerly forested and on steep slopes, and waged work was an important complement to subsistence production.[21] Almonds were an ideal choice: a drought-tolerant tree able to thrive in dry, rocky, low-nutrient soils, easily intercropped with grains and legumes, producing fuelwood, fodder, and fertilizer in addition to marketable nuts, and requiring little maintenance while the farmer engaged in waged labor elsewhere. While large estates also benefited from the profitability of almonds,[22] these nuts were an indispensable lifeline for peasant producers.

In 1930, the Balearic Islands commanded a greater area under almonds than any other Spanish province, despite their relatively small size. Almond trees served as pillars within a sophisticated, diversified rainfed farming system. A typical farm would include a mix of trees (often almonds, olives, carob, and figs) pruned at chest height to allow sheep grazing in the understory which was seeded with a rotation of winter grains, legumes, and fodder crops. Limited agrochemical inputs, such as synthetic fertilizers and pesticides, due to embargoes placed on the Francoist regime by countries fearing the spread of fascism, maintained a largely organic, agroecological system by necessity.[23] As a result, almond plantings grew increasingly popular. By 1975, according to government records, 15 percent of the Balearic Islands' total surface area was planted with almonds, with an additional 206,365 individual trees scattered across the landscape along field edges, roads, or hillsides.[24] Although other regions on Spain's Mediterranean coast grew almonds, nowhere else were they quite as economically significant as the Balearic Islands, where they earned the title *el Oro de Mallorca*, the gold of Mallorca. Whereas grains, olive oil, and meat provided sustenance for small farmers, almonds were primarily sold for export. As one farmer explained, almond harvest was the time of year for buying new clothes and gifts. It was the crop that put money in their pockets.

When commercial airline travel brought an influx of tourism to Mallorca, beginning in the 1960s, the landscape of almonds in bloom served as a stunning visual spectacle attracting visitors.[25] Most tourists came for the beaches, however, not the almonds. Fueled by substantial German investments, tourism has come to dominate the island's economy. Every year, an estimated 10 million tourists visit Mallorca,[26] whose resident population remains less than one million. Mallorcans I spoke with often repeated a striking statistic: in the summertime, an airplane leaves or takes off from the island every minute. Almond farmers I spoke with felt the landscapes they tend are valued as photo fodder while they themselves are forgotten. "*Xylella* didn't kill the almonds," one man told me, "tourism did."

The shift towards coastal tourism in the late twentieth century swiftly reoriented the maintenance practices of Mallorca's landscape, from facilitating agricultural reproduction to enabling a lucrative industry of transitory leisure.

Care operates as part of a political economy, prioritizing certain relations over others as a new *oro de Mallorca* takes the place of the old.

A Crisis of Care

When I arrived in Mallorca in January of 2018, I was told *Xylella* had been detected only 15 months previously. Once it had been identified, the Consejería de Agricultura, Pesca y Alimentación (Ministry of Agriculture, Fishing and Food) studied the issue internally while trying to calm farmers who feared they might lose their livelihoods. Not only were almonds declining, but according to the European Union's directive (Directive 2000/29/EC), all trees within a 100-meter radius of any site infected by the bacterium were to be removed. A team of plant pathologists soon found that infected trees spanned the entire island.[27] Although almonds suffered most visibly, the bacterium was also found in olives, wine grapes, and several shrubs. To follow the directive would be to denude the island. Naturally quarantined by the island's surrounding waters, the Mallorcan government pleaded their case to the EU Commission and received an exemption.

Xylella was first confirmed to be present in a cherry tree in a plant nursery, and many believed the international nursery trade was the route by which the bacterium was introduced. Most farmers suspected that it arrived from California, their almond-growing rival, in the 1990s when government officials had traveled there to learn about improved almond varieties and intensive growing techniques. When asked, a ministry official quickly dismissed the possibility that a scientist would carry plant material from overseas. A long-time farm advisor remembered returning with a small sample of the "Texas" variety from California in the 1990s, but nothing since. Another agronomist described pocketing a few almonds on research visits and propagating them back home, as a matter of course, inherent to scientific curiosity. Regardless of whether such research trips were to blame, the rumor resonated with farmers' perception of the ministry as careless and naïve, underestimating a single plant's infectious power. Further complicating matters, plant pathologists identified three distinct subspecies of the bacterium, suggesting multiple independent introductions.[28] As Natalia Gutkowski notes, unwanted organisms are often described as blind to borders, yet their enactment is profoundly territorial.[29]

The first thing I encountered when I entered the ministry's office was a large vertical banner illustrating the symptoms of *Xylella*. Plant quarantine notices for airports and the cruise ship and ferry docks were swiftly circulated. I was shown boxes awaiting distribution of freshly printed color pocket guides to identifying and managing the disease. The booklets illustrated best practices, recommending that farmers plow their fields or use herbicides to remove any understory vegetation which might harbor the insect vector.

The ministry was in a flurry of activity to address this *new* disease, but the almond farmers scoffed at the claim that this was anything new. They told me they had noticed abnormalities for over 15 years but lacked the support to address the issue. Since the 1980s, there has been no agricultural extension service providing agronomic advice. No one at the island's only university was actively studying almond diseases. In fact, an enthusiastic plant pathologist working at the airport did a great deal of the early investigation into *Xylella pro bono* in his spare time. A lack of official employment as a researcher at a university or research institution limited his credibility, potentially delaying action on the issue. When consulting the ministry, farmers were either told, incorrectly, that the problem was a fungus or that they had not properly cared for their trees. According to many at the ministry, the real problem was *falta de cuidado*, a lack of care.

This accusation of lack of care felt deeply unjust to many. "The administration comes back saying '*son almendros mal cuidados* (they are poorly-cared-for almond trees).' Obviously when your trees are dying and people have told you there's no solution, you stop taking care of them. Why would you invest?" The ministry's assertion of a lack of care seemed out of touch, as farmers were quick to note that the man who first identified the seriousness of the problem with almonds was himself a trained agronomist, praised for his carefully tended land. It also put a heavy burden on individuals who had been swimming against the tide of the tourism-driven economy. Even the most meticulously cared for orchard could be surrounded by farms that had been abandoned because economic opportunity was elsewhere, or turned into vacation rentals, prioritizing aesthetics over tree health. How could more pruning, weeding, or pest treatments possibly compete with "strangulation" by neighboring fields, they asked.

> We have a 33-hectare planting, rainfed, old, traditional, organic, that I'm wondering … this year should I pull it all out? … For the last two years I don't know whether to prune, not to prune, what to do, because no one gives you anything, no breaks … Now they're saying off the cuff that what you should do is tear everything out because if you hadn't had that single weed, because if you hadn't had that single insect, because whatever. Right. But in Mallorca, you can have a half-hectare of perfectly cared for almonds, without a single weed, but the neighbor left you with their farm abandoned. You'll be in the same situation. That's what we told the Ministry.

Abandoned orchards signaled a shift in more-than-human relations with implications spilling over beyond property boundaries. Spittle bugs carrying *Xylella* could proliferate freely in the leafy understory of a neglected orchard. Trees harboring the bacterium undetected would act as perpetual reservoirs. The disease assemblage itself proved the limits of an individualist approach to controlling the spread of *Xylella*.

Almond farmers fumed at the notion that they had not cared for their trees, because it seemed to negate the systemic disruptions of rural life. Farmers had found their economic base gradually dissolving as hotels multiplied. Their regional government had long touted tourism as the saving grace of an island once known for its antiquated, agrarian ways. Their children had chosen more comfortable and secure urban lives, often catering to visitors at hotels or restaurants. Their neighbors had abandoned the land, leaving trees that were uncared for and giving passersby the impression that agrarian lifeways had already disappeared into the history books. Agriculture, which had been cared for by a community not so long before, found itself excluded from the cares of most Mallorcans. As a long-time almond farmer described it:

> The image of those almonds, many of them already dead, or those trunks, the big ones on the side of the highway, farms, some abandoned, all the dry wood that's dying. For me, this is the graphic image of our agricultural society, of rural society, or what's left of it.

While the ministry's pamphlets detailed tactics to triumph over a novel bacterial invader, farmers were confronting a crisis much broader in scope. In the grand scheme, the contested approach to the care work of maintaining almond trees—to prune or spray or mow or not—felt nearly insignificant in the face of widespread collective failure to maintain agrarian lifeways. The *Xylella* outbreak merely caused those more distanced from agriculture to take notice and perhaps share momentarily the experience of grief. The large, decaying trunks, yellowed leaves and lifeless limbs in such an iconic species, the materiality of the disease, demanded attention in ways that a dwindling group of aging farmers did not. The faltering trees grabbed a few fleeting headlines in the local press, which insisted that the ministry show some kind of effort. The desiccated bodies of the trees, unexpected and unsightly, posed questions of landscape care much more publicly than such matters would have been discussed if healthy almond trees had continued their gradual fade with economic change.

"People Would Care If it Affected Tourism"

Almond production went from being the island's agricultural jewel to a husk of its former splendor in less than 30 years, an almond cooperative manager reflected. Farmers I spoke with felt that agriculture had become an afterthought for public officials. In the words of one almond grower:

> You can't compete with the quantity of money generated by an activity as powerful as tourism, which requires investments in infrastructure, airports, sea ports, highways, hotels, streets, sidewalks, [compared] with an agricultural sector where people are aging, where there's no relief, where there's not much interest in change.

A few farmers I met had begun incorporating agritourism into their activities, enticing visitors with the authenticity of rustic lodgings or freshly prepared lamb. They had always juggled multiple roles, often selling their labor as a tractor driver or in a factory to make ends meet. "You have to be polyvalent here. You can't live from one thing," one farmer explained. Most farmers felt the wealth generated by tourism remained highly concentrated. "For all this tourism, it seems like we should be the wealthiest place in Europe. But only three or four people are really earning … the hotel owners. We work very hard now in the summers [in the tourism industry] and we're still just as poor."

Political economic change had shifted relations of care. Tourism's economic heft now demanded wide-ranging forms of infrastructural maintenance and planning, from macro-level public works to household-level chores. The ongoingness of activities like agriculture had become almost incidental. Frustration towards tourism reflected concerns not only over the decline of agriculture, but over the concentration of wealth in Mallorcan society. The inability of agriculture to "compete" for attention was also an inability to challenge a pattern of deepening inequality. What counts as good care among public officials, be it smooth roads or plentiful harvests, echoed shifting power relations and social values. Care, which lays the foundations for responding to sudden shocks like a disease outbreak, is thus constituted within and through the political economy. Maintenance work, or the lack thereof, reflects and shapes social configurations.

No One Left to Consult

A steady decline in institutional support for crop health due to shifting political configurations had laid the groundwork for the devastation of the *Xylella* outbreak. Despite their frustration with how the *Xylella* epidemic was being handled, most farmers did not actively display their discontent or make demands upon the government. Franco's dictatorship likely contributed to a sense of resignation towards administrative dismissiveness. Despite 40 years of democratic governance since then, fascism was still a living memory for many Mallorcan farmers. The regime had cared for farmers, providing advisory services they now missed. But as a retired farm advisor explained, this was a paternalistic kind of care, expecting obedience.

The social and economic health of the countryside, as well as the capacity for Spain to produce its own food, were considered vital to Franco's vision of national strength. To this end, the Servicio de Extensión Agraria (SEA), or Agricultural Extension Service, was established during the late 1950s as part of the United States' aid to Franco's regime. Spanish officials traveled to the United States and returned with an agenda to replicate the American model of farm advising. The program aspired to a Green Revolution for Spain, prioritizing economic efficiency through mechanization, intensification, and irrigation. In line with Franco's technocratic model, farmers were expected to be passive recipients of wisdom delivered by technical experts.[30] Yet, due to its decentralized model, the institution had significant autonomy. Its workers lived alongside farmers and they gradually adapted their activities to meet community needs, developing a suite of rural social services which evolved to recognize rural populations as the engines of their own development.[31] Care provided by the extension service was complex, and its capacity to troubleshoot emerging concerns for crop health as well as rural social wellbeing navigated a tension between a paternalistic mission and a more grassroots manifestation.

During the transition to democracy, between Franco's death in 1975 and the approval of the Spanish constitution in 1978, political power was decentralized from Madrid to 17 Autonomous Communities (*Comunidades Autónomas*), leaving the SEA in limbo. When Spain entered the European Union in 1986, SEA agents became responsible for administering benefits through the Common Agricultural Policy, transforming agricultural advisors into bureaucrats.

Mallorcan farmers recalled the SEA with nostalgic praise. Nearly everyone I spoke with remembered one enthusiastic farm advisor who had dedicated his career to troubleshooting almond trees, even though it was not his original assignment. Unlike the ministry's representatives, hidden away in offices shuffling stacks of paper, this man was out in the fields, running trials of new techniques, and answering their questions. Many felt that the money flowing in from the EU fundamentally changed public perceptions of farming from a livelihood to a paper chase. They wondered how the detection and response to *Xylella* might have gone differently if the SEA were still active.

Agricultural cooperatives were yet another institution many farmers felt had ultimately failed them, despite the best of intentions, and failed to provide the responsiveness which might prevent the proliferation of *Xylella*. In order for farmers to receive benefits from the EU, they were required to organize into agricultural cooperatives. These cooperatives were envisioned by policymakers to become the backbone of healthy, stable, agricultural communities by providing farmers with greater economic power through bulk purchases of inputs at lower prices and collective sales of their product at higher prices. Almonds require processing equipment to remove the hull and shell, and a cooperative would allow farmers to own the expensive processing equipment themselves, capturing a greater share of the sales value. Although farmers appreciated these gains in the short term, many felt that they had been tricked in the long term. The cooperatives were not run by farmers concerned with preserving the land, they said, but by businessmen seeking to satisfy their customers. The ability to sustain small-scale diversified farms was less important to them than ensuring buyers of almonds remained consistent customers. Farmers also reported that their long-term commitments to the land, rather than to commerce, could be perceived as a liability. The retired farm advisor said that in the early days he had proposed forming a cooperative composed only of farmers, but the administration rejected this proposal, saying that the farmers could not feasibly market their product.

The decline of institutional support through the SEA, and the lack of full alignment between farmers and cooperative employees left farmers with fewer people to consult when they encountered something strange in their almonds. When they did call on the ministry or their cooperative, a degree of distance, distraction or distrust seemed to limit the connection. Institutional networks of care for farmers are crucial for responding to disease, and, without them, both the social and biophysical health of the agroecosystem become less resilient to new stressors. The shift from farm advising to farm administrating revealed just how crucial relationships had been to the daily troubleshooting of living from the land. Yet nostalgia for the care provided by the SEA was complex, unable to be disentangled from the dishonor of Franco's legacy. Care is always relational, contextual, and political. Care can be conservative, sustaining existing structures. Care as an analytical lens thus resonates with the call to address multiple potentially competing agendas for what holistic health across species lines might entail. Recognizing the significance of care work does not presume that inherent benefits flow in all directions. Instead, it pushes us to articulate how to care, what is being maintained, and to what ends.

Little Reason to Invest

Among other structural factors, the property relations structuring landscape maintenance shaped responses to the *Xylella* outbreak. "The farms aren't actually mine," one farmer mentioned as an aside after our lengthy conversation discussing the rhythms of his days and years pasturing sheep among the almond groves. It was an afterthought to explain because the situation is so common. The small parcels of land which farmers in the early twentieth century could afford to buy were not big enough to sustain a family. Most full-time farmers were effectively sharecroppers, splitting the harvest fifty-fifty with various landlords under informal agreements. As profits from agriculture declined, those who continued working the land stayed afloat by cultivating a dozen or more parcels with various owners. Precarious land-tenure dynamics made it difficult to justify investments in new almond trees. Wheat gives an annual harvest so fertilizer costs are easily recuperated at the end of the season. Almond trees take five to eight years to begin producing under rainfed conditions, reaching their full productive capacity at around 15 years. Long-term arrangements with landowners were difficult to come by. With tourism rapidly raising property values, farmers never knew when they might not have their agreement renewed. Tree growth could not conform to the abrupt temporalities of tourism.

A manager at a small almond cooperative told me he had suspected a new disease in 2010, but he had difficulty distinguishing between drought, aging trees, and lack of care. People were not renewing their orchards but merely keeping older, weaker, trees from a previous generation, he explained, valuing them as antiques rather than living landscapes. Many urban-dwelling families inherited land from their elders with little knowledge of how to maintain the landscape.

According to almond growers and their advisors, this younger generation comes back just to harvest the nuts, without knowledge of how to care for and renew the orchard. As the graduate student of almond diversity explained, "There's no generational renewal. *Xylella* will make all this faster and more traumatic."

As in many parts of the world, the agrarian workforce in Mallorca at this time was aging. Some suspected almonds had suffered neglect because pruning trees is more physically demanding than other farm tasks. Others noted that farmers in their 70s and 80s were unlikely to plant trees because they might not live to see them bear fruit. The temporality of tree bodies and human bodies were intertwined. Many felt less invested in organizing to combat the *Xylella* epidemic because they saw no one who would take care of the land when they died. "We haven't put up much of a fuss because there's no social weight forcing us to do things otherwise ... after me there's no one else."

In 2018, the island's young farmers' association had just four members. The young farmer I met expressed a sense of obligation to care for the land, while acknowledging that such care takes more than just the work of farmers. "We understand that maintaining the environment is our responsibility, that it benefits everyone, and we want to do it. What happens is that, alone, alone it´s impossible." Others echoed the significance of almond trees as being essential to the Mallorcan landscape and valued for "more than just production," slipping fluidly in and out of the commodity form.[32] Landscape care, a far more expansive notion than agricultural output, was a task that demanded more than the actions of increasingly scarce farmers.

As feminist theorists point out, care reveals interdependency.[33] The health of the trees, as well as the social and economic wellbeing of the agricultural community, can never fully be addressed by isolating and eliminating a pathogen or its insect vector. The spread of *Xylella* represented a disease thriving in the wake of disconnections. Gaps in intergenerational knowledge and practice left trees more vulnerable. Farmers' motivation declined as continuity into the future became increasingly elusive, for their land-tenure arrangements and for their families. A younger group of farmers valuing agriculture for its ecological and cultural role found their ambitions frustrated by isolation. The devastation of the *Xylella* outbreak could not be separated from its conditions of possibility. The disease complex was able to spread, not only due to the combination of species and climate, but also due to the weakening of relationships experienced by the agricultural community, a relational breakdown encoded in the ecology of aging trees, overgrown orchards, and the fading presence of farmers.

Irrigated Hopes

In 2018, irrigation was emerging as a new form of almond tree care, an act which was once considered unthinkable for such a drought-tolerant species. Because *Xylella* affects the fluid artery of the tree, the xylem, its effects are much less severe when trees are irrigated. Intensive, modernist agricultural practices were promoted as a new regime of care, one which might secure a future for Mallorcan almond processors and vendors but not necessarily for most Mallorcan almond farmers. After I interviewed the man handling the *Xylella* case at the Ministry of Agriculture, he offered to take me on a tour of recently planted orchards. "I want to show you the best" he said, "not the stuff by the roadside." He took me to visit a new, irrigated, modern planting owned by a wealthy doctor and assured me the trees received abundant fertilizer, interpreted as a sign of care. At the next stop, we marveled at a large orchard owned by a hotelier. My guide felt responsible for performing progress, explaining that *Xylella* is not the real problem. For him, it was old trees and lack of care. As we drove along, he pointed out of the window to almond trees with lichen growing on the branches as evidence. Although lichens are not harmful to trees, he found them unsightly. Not enough pruning or treatment with copper fungicide, he explained. Farmers hadn't taken care of their trees.

The largest cooperative on the island was also planting an irrigated orchard. The beneficiaries would be the farmer-members, the cooperative manager explained, but it was not celebrated by all growers. "There is not enough water for everyone" several stated frankly, including those who had planted the new irrigated orchard. Mallorca has no permanent rivers, and threats of water shortage are constant. Tourism puts serious strain on water supplies, particularly as the government has incentivized higher-end resorts with golf courses and extensive landscaping.[34] Overexploitation of aquifers has produced saltwater intrusion, a terrifying prospect for an island depending on groundwater for 75–95 percent of its supply. And then, there is the omnipresence of climate change. Hydrologists predict increasingly limited water supplies. Higher winter temperatures and intense heat episodes are also suspected to boost the virulence of *Xylella*. Many believe the record temperatures of recent summers triggered *Xylella* to rapidly multiply and kill its almond host.

One farmer shared with me his caution around embracing irrigation as a long-term solution.

> Now, they are doing irrigation, but, in my opinion, they are going to end up without water, because it doesn't rain. Every year we surpass record summer temperatures. Every year we set records. It doesn't rain. Every day more people, more human pressure, more tourists, more pools, more needs and we are going to end up without water. I don't know if they are going to be able to irrigate, though they say they will use recycled water.

Another cooperative manager described hopes for a mixed system. Some intensive irrigated orchards could generate enough almonds to keep production facilities moving and allow farmers to revitalize rainfed almond landscapes. For those with access to water, capital, and land to secure their investment, irrigation presented a technological fix that could keep the almond industry afloat and possibly avert land abandonment in the near future. Strangely, many agronomists expected the polyculture of rainfed almond production to be saved (at least temporarily) by the monoculture of irrigated, fertilized, and pesticide-controlled plantations. For rainfed farmers and others anxious about Mallorca's water future, betting the future of almond cultivation on irrigation seemed careless, lacking in consideration of how it might be sustained for the long term.

Plant Disease and the Politics of Care

Plant epidemics, like that of *Xylella*, have been peripheral within One Health discourses, yet they are rich with insights. Plant disease underscores the significance of social and economic wellbeing across species lines. The resilience of an agroecosystem to the disturbance of a novel pathogen is determined not only by its biophysical attributes, but also by the structural conditions and maintenance relations which sustain it. Whereas the health of animals, including humans, tends to carry clear, ethical obligations to minimize suffering, plant disease reveals more explicit trade-offs between various social and economic possibilities.

As I have argued, thinking with care highlights the vital importance of maintenance relations in shaping disease response. Plant disease outbreaks often produce cycles of blame, resentment, and deepened community divides. This may stem in part from treating the phenomenon (1) as an interaction between separate biological and social actors—the pathogen is the enemy and the people respond well or poorly—or (2) as a unidirectional causal chain—a single event introduces a pathogen which wipes out a species.

If the pathogen were the clear enemy, then its eradication might be simpler to manage. But, like heat, awakening the bacterium within a tree, the pathogen's virulence is provoked by a broad assemblage of facilitating conditions. The vector of time in epidemiology illustrates disease spread as a series of discrete events. Yet, in the eruption of an epidemic, past wounds, present transformations, and future fantasies or fears permeate one another. To embrace a pathogen as emerging from within is to exchange the causal chain inherent to blame for collective responsibility. Whereas blame points to discrete individuals and actions, care indicates the continuous relationships which might be repaired or reimagined. These relationships are not incidental to the health of plants and the broader communities of which they are a part. They shape the possibilities for disease emergence, for effective control, and perhaps most importantly for confronting the political contours of an ecology under stress.

Care, by its subtle and sustaining nature, is often most perceptible in its absence. Farmers felt a lack of care by government and a tourism-oriented culture. Bureaucrats interpreted trees with minimal management as lacking care. When the prospect that almond trees might disappear from the landscape was raised, either gradually by disease or suddenly by bulldozers, the agroecological maintenance work—a form of care—which the farmers provided became clear. The trees not only produce almonds, they root more-than-human lives in place and across time.

Understanding disease as emerging within an existing whole seems at first glance antithetical to agricultural biosecurity tactics, which serve to shield unaffected areas from biological intruders. Yet, it can be complementary while also pushing responses to be more comprehensive as policymakers work towards a more expansive More-than-One Health approach. A politics of landscape care requires us to pay attention to the entanglements that allow an epidemic to exist and to respond to each relation accordingly. Eradication, while potentially worthy, is perpetually reactive and, while agroecological production practices, like crop diversification, are popularly understood to confer disease resilience, they are not enough. When more-than-human maintenance work is rendered visible and valued, it can be more intentionally directed towards desirable, if unpredictable, ends.

There is no saving Mallorca's almonds, at least not in the form they once took. What rural landscapes "living with" *Xylella* on the Iberian Peninsula might look like is an open question, potentially requiring dramatic reconfigurations of such rural life. The *Xylella* epidemic was always about more-than-almonds. A politics of landscape care is not a call for protecting, preserving, or proliferating any particular beings—plant, animal, or otherwise. It is a call to see the web of *relations* among plants, farmers, insects, aquifers, bacteria, governments, petrochemicals, scientists, soils, and matter of all kinds as the site of politics where the ongoingness of specific configurations of existence is negotiated. It is a call to question the focus on an individual actor—be it a bacterium or a bureaucracy—as uniquely blameworthy, for the solutions thus proposed are likely far too narrow. For One Health to make progress in advancing holistic wellbeing across species lines and beyond the mere absence of disease, it must be more than an additive exercise which applies existing biosecurity tactics to new arenas. Using care as an analytic lens for One Health helps to expose the slow processes at work in transforming communities—like tourism, land tenure, and climate change—as well as the sudden, discrete shock of a novel disease outbreak. In keeping with the political explicitness of More-than-One Health, care does not presuppose any normative qualities: forms of maintenance and relation-building will always create differentiated effects. Pathogenic conditions are not anomalous; they are endemic to late capitalist life. Confronting them requires not only swift, bold actions, but also the subtle and gradual maintenance work of understanding and transforming inherited relationships, broadening imagined constituencies, and acknowledging bonds of mutual responsibility—and, in doing so, weaving new social fabrics of health, wellbeing, and more-than-human care.

Acknowledgements

This chapter contains material from Emily Reisman, "Plants, Pathogens, and the Politics of Care: *Xylella fastidiosa* and the Intra-Active Breakdown of Mallorca's Almond Ecology," *Cultural Anthropology* 36, no. 3 (2021): 400-427. Funding for this research was provided by the Wenner-Gren Foundation.

Notes

1 "Joint Tripartite and UNEP Statement on Definition of 'One Health,'" UNEP, 2021, https://www.unep.org/news-and-stories/statements/joint-tripartite-and-unep-statement-definition-one-health.
2 Mark Pokras, "Understanding Lead Uptake and Effects Across Species Lines: A Conservation Medicine-Based Approach," in *Ingestion of Lead from Spent Ammunition: Implications for Wildlife and Humans*, eds. Richard T. Watson et al. (Boise: The Peregrine Fund, 2009).
3 "One Health in Action," Wildlife Conservation Society, November, 2020, https://c532f75abb9c1c021b8c-e46e473f8aadb72cf2a8ea564b4e6a76.ssl.cf5.rackcdn.com/2020/12/07/3y67zbyeuj_One_Health_in_Action_final.pdf.
4 Stephen Hinchliffe, "More than One World, More than One Health: Re-Configuring Interspecies Health," *Social Science & Medicine* 129 (2015): 28–35.
5 "Las Logicas del Cuidado," *Diagonal*, March 15, 2007, https://www.diagonalperiodico.net/humor/logicas-del-cuidado.html.
6 Marian Barnes, *Care in Everyday Life: An Ethic of Care in Practice* (Bristol: Policy Press, 2012).
7 Annemarie Mol, *The Logic of Care: Health and the Problem of Patient Choice* (London: Routledge, 2008), 84.
8 Aryn Martin, Natasha Myers, and Ana Viseu, "The Politics of Care in Technoscience," *Social Studies of Science* 45, no. 5 (2015): 1–17, 3.
9 María Puig de la Bellacasa, *Matters of Care: Speculative Ethics in More-than-Human Worlds* (Minneapolis: University of Minnesota Press, 2017).
10 Stephen Hinchliffe et al., *Pathological Lives: Disease, Space and Biopolitics* (Malden: Wiley Blackwell, 2017), xiv.
11 Julie Bergeron, Sylvain Paquette, and Philippe Poullaouec-Gonidec, "Uncovering Landscape Values and Micro-Geographies of Meanings with the Go-Along Method," *Landscape and Urban Planning* 122 (2014): 108–121.
12 S. Eben Kirksey and Stefan Helmreich, "The Emergence of Multispecies Ethnography," *Cultural Anthropology* 25, no. 4 (2010): 545–576; Laura A. Ogden, Billy Hall, and Kimiko Tanita, "Animals, Plants, People, and Things: A Review of Multispecies Ethnography," *Environment and Society* 4, no. 1 (2013), https://doi.org/10.3167/ares.2013.040102.
13 Sarah Whatmore, "Materialist Returns: Practising Cultural Geography in and for a More-than-Human World," *Cultural Geographies* 13, no. 4 (2006): 600–609; Paul Robbins and Brian Marks, "Assemblage Geographies," in *The SAGE Handbook of Social Geographies*, eds. Susan Smith et al. (Los Angeles: SAGE, 2010); Ruth Panelli, "More-than-Human Social Geographies: Posthuman and Other Possibilities," *Progress in Human Geography* 34, no. 1 (2010): 79–87; Bruce Braun, "Environmental Issues: Global Natures in the Space of Assemblage," *Progress in Human Geography* 30, no. 5 (2006): 644–654.
14 Kim TallBear, "Why Interspecies Thinking Needs Indigenous Standpoints," *Fieldsights*, November 18, 2011, 6; Zoe Todd, "Fish, Kin and Hope: Tending to Water Violations in *Amiskwaciwâskahikan* and Treaty Six Territory," *Afterall: A Journal of Art, Context and Enquiry* 43 (2017): 102–107; Juanita Sundberg, "Decolonizing Posthumanist Geographies," *Cultural Geographies* 21, no. 1 (2014): 33–47.
15 Karen-Beth G. Scholthof, "The Disease Triangle: Pathogens, the Environment and Society," *Nature Reviews Microbiology* 5, no. 2 (2007): 152–156.
16 Alexander Purcell, "Paradigms: Examples from the Bacterium *Xylella fastidiosa*," *Annual Review of Phytopathology* 51, no. 1 (2013): 339–356.
17 Paolo Baldi and Nicola La Porta, "Xylella Fastidiosa: Host Range and Advance in Molecular Identification Techniques," *Frontiers in Plant Science* 8 (2017): 1–22.
18 G. Ladizinsky, "On the Origin of Almond," *Genetic Resources and Crop Evolution* 43 (1999): 143–147.

19 Enric Tello et al., "From Feudal Colonization to Agrarian Capitalism in Mallorca: Peasant Endurance Under the Rise and Fall of Large Estates (1229–1900)," *Journal of Agrarian Change* 18, no. 3 (2018): 483–516.
20 Pere Ferrer Guasp, *Joan March: El Inicis d'un Imperi Financer, 1900–1924* (Palma de Mallorca: Edicions Cort, 2000), 2.
21 Ramon Molina de Dios, "De Pan de Pobres a Sofisticado Aditivo: Tecnología e Innovación en Torno a la Industria de la Algarroba: El Caso Balear (1930–2010)," *Revista de Historia Industrial*, no. 49 (2012): 147–179.
22 Antonia Morey Tous and Jaume Fornés Comas, "El Cultivo Tradicional Del Almendro En El Mediterráneo: Baleares En El Contexto Español (ca. 1770–2017)," *Historia Agraria* 84 (2021): 103–140.
23 Ivan Murray et al., "Biocultural Heritages in Mallorca: Explaining the Resilience of Peasant Landscapes Within a Mediterranean Tourist Hotspot, 1870–2016," *Sustainability* 11, no. 7 (2019): 1926.
24 Instituto Nacional de Estadística, "Cultivos Arbóreos y Arbustivos: Superficie, Producción y Valor, Año 1975," Anuario Estadistico (Madrid: Instituto Nacional de Estadística, 1977).
25 Esteban Bardolet, "Tourism in the Balearic Islands," *The Tourist Review* 35, no. 4 (1980): 18–21.
26 Balearic Islands Tourism Board, "Balearic Islands Regional Context Survey" (Interreg Europe, June 14, 2017).
27 Diego Olmo et al., "Landscape Epidemiology of *Xylella fastidiosa* in the Balearic Islands," *Agronomy* 11, no. 3 (2021): 473.
28 Diego Olmo et al., "Panorama Actual de *Xylella fastidios*a En Las Islas Baleares" (XIX Congreso de la Sociedad Española de Fitopatología, Toledo, Spain, October 10, 2018).
29 Natalia Gutkowski, "Bodies that Count: Administering Multispecies in Palestine/Israel's Borderlands," *Environment and Planning E: Nature and Space* 4, no. 1 (2020): 135–157.
30 Alba Díaz Geada et al., "Agricultural Extension Programmes in Post-War Europe: A Comparative Study of Two Extreme Cases: Spain and the Netherlands (1946–1973)" (9th European Social Science History Conference, Glasgow, 2012).
31 Cristóbal Gómez Benito and Emilio Luque Pulgar, "Modernización Agraria, Modernización Administrativa y Franquismo El Modelo Educativo y Administrativo Del Servicio de Extensión Agraria (1955–1986)," *Areas: Revista Internacional de Ciencias Sociales,* no. 26 (2007): 131–149.
32 Anna Lowenhaupt Tsing, *The Mushroom at the End of the World: On the Possibility of Life in Capitalist Ruins* (Princeton: Princeton University Press, 2015).
33 Joan C. Tronto and Berenice Fischer, "Toward a Feminist Theory of Caring," in *Circles of Care; Work and Identity in Women's Lives*, eds. Emily K. Abel and Margaret K. Nelson (Albany: SUNY Press, 1990), 40–62; Maria Puig de la Bellacasa, "Making Time for Soil: Technoscientific Futurity and the Pace of Care," *Social Studies of Science* 45, no. 5 (2015): 691–716; Martin, Myers, and Viseu, "The Politics of Care in Technoscience."
34 M. Kent, R. Newnham, and S. Essex, "Tourism and Sustainable Water Supply in Mallorca: A Geographical Analysis," *Applied Geography* 22, no. 4 (2002): 351–374.

WINTER 2019: THE NOVEL CORONAVIRUS INFECTS HUMANS. A DOMINANT THEORY HOLDS THAT THE VIRUS IS ZOONOTIC—THAT IS, IT MAY HAVE MIGRATED FROM ANOTHER SPECIES TO HUMANS.
THIS IS HAPPENING MORE OFTEN, AS WE DEPRIVE OTHER SPECIES OF THEIR HABITATS—OF THEIR RIGHT TO BREATHE.
THE VIRUS SWELLS QUICKLY AMONG OUR SPECIES. AT ITS HEIGHT, IT CHOKES THE LIFE OUT OF MILLIONS.
IN ITS WAKE, IT LEAVES MILLIONS MORE GASPING FOR BREATH. EVERYONE IS VULNERABLE, BUT NOT EQUALLY SO. BLACK AND BROWN PEOPLE, THE POOR, THE ELDERLY, AND THOSE WITH COMPROMISED IMMUNE SYSTEMS ARE HIT THE HARDEST.
ARE MORE WAVES COMING? HOW MANY?

7

WHAT CAN GRAPHIC MEDICINE CONTRIBUTE TO ONE HEALTH?

Susan Merrill Squier

Graphic Medicine—the movement dedicated to supporting the use of comics in health—grows out of narrative medicine, a movement to improve health care by training practitioners in the art of interpreting stories told by patients. But, unlike the literary texts central to narrative medicine, works of graphic medicine are comics. As such, they draw on the affordances of the comics medium.[1] These include the ability to present visual images on multiple scales, in different times and spaces, and to build narratives that are linear or nonlinear, single, or multiple, or all of them at once.

Although works of narrative medicine focus on the human illness story, or *pathography*, the focus of Graphic Medicine can extend beyond the narrowly human, presenting what I call a "porous pathography."[2] Works of Graphic Medicine may picture health at multiple scales, by offering space to the other illness stories occurring within and around the human story.[3] As comics, they engage in visual world-building, incorporating images that include bacteria, microbes, parasites, insects, companion or agricultural animals, fish, and even the planet. By telling stories at the human, animal, microbial, ecosystem, and planetary scales, works of Graphic Medicine can also elicit a wide range of reader engagement. Because comics are created through multidisciplinary collaborations between cartoonists, writers, scientists, and health care workers, works of Graphic Medicine incorporate different perspectives on health as well as the variety of approaches required by such alliances. For all these reasons, works of Graphic Medicine can provide a powerful means of amplifying the One Health message, as I will demonstrate in what follows.

FIGURE 7.1 "The Right to Breathe," by Maureen Burdock and Joanne Regulska. Used with permission.

DOI: 10.4324/9781003294085-10

First, however, let me clarify the definition of One Health I use. Four models are often advanced as interventions in health: One Health, EcoHealth, Planetary Health, and One Welfare. These models share the view that human beings, animals, and the ecosystem/environment occupy the same planetary spaces and are subject to many of the same threats to health: infections, pathogens, and environmental stressors. Yet, while they do share several assumptions, these programs differ in the scientific areas the programs incorporate and the fundamental values they espouse.

One Health as a field has been criticized for its human-centeredness and its relative conceptual narrowness, as compared with EcoHealth and Planetary Health.[4] Scott Gilbert and Salla Sariola claim that both the One Health paradigm and its genomic partner, Precision Public Health, lack an adequate understanding of the human–microbial relationship crucial for adequate health in the era of the Anthropocene. In their view, by ignoring the "symbiotic relationships that exist between bacterial species and between bacteria, viruses, and their eukaryotic hosts," One Health fails to take seriously "both the beneficial (indeed necessary) interactions of microbes with their [human] hosts as well as their pathogenic interactions."[5]

Although they frame One Health as both a focused collaboration of human and veterinary medicines and a broader umbrella-like embrace of interdisciplinary approaches, scientific fields, and key concepts, Henrik Lerner and Charlotte Berg have also argued that One Health is primarily focused on vertebrate health.[6] They charge that it draws its method and concepts from core areas in the biological sciences rather than the social and agricultural sciences. In contrast, EcoHealth incorporates an ecosystem model based on biodiversity, Planetary Health focuses primarily on global threats to health, and they argue a risk exists that we may be casting too wide a net if we create a successor model, such as One Welfare.

I come to One Health from the field of Graphic Medicine, and that is why I approach One Health not as a field but as a process: an interdisciplinary or multidisciplinary mode of drawing together insights relevant to health and welfare from proximate and even distant scientific, social scientific, and humanistic fields. This process-based understanding draws from Graphic Medicine, which in the last decade has evolved from a relatively narrow focus on human medicine to a broader mission dedicated to the improvement of health. Like narrative medicine, Graphic Medicine has retained use of the term "medicine" for reasons both historical and strategic, to maintain acceptance in the medical schools and clinics where it first appeared.

Yet, over the decade and more since its first formulation, Graphic Medicine has expanded its mission in sequential acts of re-framing. Moving beyond the clinic and its focus on pathology and illness, the field of Graphic Medicine incorporated the exploration of illness narratives in comics (and their relation to fiction and non-fiction illness narratives in the PathoGraphics project), attention to the lived experience of disability, non-neurotypicality, and mental illness (or mad pride), and the exploration of health, including Public Health. By 2019, Graphic

Medicine had "scaled up" its definition of health to include issues as wide-ranging as microbial infections, parasitic diseases, and climate change.[7] The most recent addition to the Graphic Medicine toolkit has been a rethinking of the formal fundamentals of comics as a medium that challenges its claim to universality. In *How Comics Travel: Publication, Translation, Radical Literacies* (2022), Katherine Kelp-Stebbins demonstrates that comics articulates anticolonial, feminist, and anti-racist meanings through a set of strategies she labels resistance, *détournement,* translation, reformatting, decolonial mapping, and flipping. Her argument redefines some of the central tenets of comic studies: the iconicity of the ligne claire, the universality of the iconic blank face, the cultural implications of comics' use of stereotype, and the global consistency of comics reading and reception. Rather than embracing the tradition established by comics scholar Scott McCloud that sees comics as articulating universal, transcultural truths, Kelp-Stebbins defines the comics medium as a GPS, or "graphic positioning system," which articulates the pressures of local realities through translational variations, formal differences, and divergent marketing practices.[8] Enacted by works of Graphic Medicine, such a redirection of the comics medium enables a more precise attention to the local and situational realities of health and health disparities.

Stephen Hinchliffe is surely correct in arguing in the Foreword to this volume that the concept "One Health" can seem a tempting lure or a decoy, promising simplicity, uniformity, and consensus. Yet when understood as a process, One Health aligns with the decolonial and anti-universalist project of Graphic Medicine which frames health in embodied particularity.

As the comics medium bridges disciplinary divisions between the life sciences, human sciences, and veterinary medicine, it also engages the reader viscerally. Comics uses posture and gesture to express and elicit feeling, and thus we experience in our bodies what the comic page shows, whether we view it as activating our mirror neurons or creating a connection between body and comic text through the drawn line. This connection to the body is best exemplified in an image from Will Eisner's classic volume, *Comics and Sequential Art*.[9] Entitled a "MicroDictionary of Gestures," this full-page image itemizes part of the visual lexicon a cartoonist draws on when creating a comic narrative. In seven tiers of human images, with five to seven human beings arrayed in each tier, it documents the variety of stances by which human beings express what Eisner calls "internal feelings." Labeled anger, fear, joy, surprise, deviousness, threat, and power, these rows of simply sketched postures reveal the great variation in the ways we human beings act out our emotions. To take one example, the feeling of fear has eight different illustrations, from an arms-outstretched gesture of warding off, to a troubled skulking, or even creeping, to running away.

Eisner's comics-based strategy has its parallel beyond the realm of human medicine. Consider a poster I encountered on the wall in my veterinarian's office, entitled "The Spectrum of Fear, Anxiety, and Stress."[10] The poster combines words and images to explain a dog's emotional or physical state. Down the left side, a series of texts describe indications that a dog is stressed.

Changing from red to yellow and finally to green, these notations indicate the "severe signs," then "moderate signs," and finally "subtle signs" of a dog's state of health. Next to these words, down the right side of the poster, is a series of expressive dog postures beginning with dogs snarling, skulking head down, turning tail and running, then going on to dogs cowering, flagging their tails, licking their lips, and avoiding eye contact. At the bottom of the poster, text on the left is juxtaposed to three images on the right to show what a dog looks like when it is "alert/excited/anxious," "perked/interested/anxious," or simply "relaxed."

As these examples show, both human medicine and veterinary medicine assert the values of attending to nonverbal signs such as posture, facial expression, and gesture to diagnose disease or support health. To my knowledge, the gestural connection between human and animal emotions has not been discussed explicitly in One Health literature, although analysis of the expression of emotion in human beings has been claimed as part of the evolutionary toolkit from Charles Darwin to Paul Ekman. But the visual depiction of gesture and posture in comics gives the medium the ability to express a wide range of psycho-physiological states, including the experiences of illness, disability, medical treatment, caregiving, or simply wellbeing. Their combination of visual and verbal expression of embodied experiences may explain why works of Graphic Medicine are increasingly being used to communicate scientific findings by groups ranging from government agencies, such as the Center for Disease Control, to professional organizations of veterinarians and physicians, research laboratories, and think tanks.

Works of Graphic Medicine are more than merely utilitarian, more than simple transparent vehicles for facts. Like other modes of visual, verbal, and aural expression, they are also rhetorical tools, reflecting and shaped by the social and institutional context within which they are created and deployed. They transmit layered and complex emotional and social meanings that can be interpreted, or *read*, to uncover the audience they address and the explicit and implicit meanings they convey. I take all of this into consideration in what follows, as I look closely at how several works of Graphic Medicine portray the entanglement of illness and health as an interspecies relationship. I further consider how such works offer a new set of representational and methodological strategies to broaden the reach of One Health.

Though the field of One Health originates in a focus on the transmission of microbial diseases, it has more recently expanded from its focus on microbe-borne illness to include the intertwined forces affecting multispecies health. These comics can be categorized by the three areas they focus on: illnesses caused by germs that pass between humans and animals (traditionally called *zoonoses*); feelings of joy or wellbeing arising from the connection between humans and animals (more recently termed *zooeyia*); and awareness of the mingled positive and negative impact of the microbes that connect humans and animals (which I call *zooambivalence*).

Zoonoses

The Center for Disease Control defines zoonotic diseases or zoonoses as "diseases caused by germs that spread between animals and people."[11] The most recent example of a zoonotic disease is, of course, the global SARS-CoV2 pandemic.[12] *The Covid Chronicles,* a collection of comics responding to the pandemic, includes works of Graphic Medicine that assess the impact of COVID-19 across the human and animal landscape. "The Right to Breathe," by Maureen Burdock, with Joanna Regulska, connects the zoonotic origins of the coronavirus pandemic to the injurious effects of racism and poverty. The comic explicitly tracks two linked devastations: deforestation and wild animal displacement, and the structural racism, consequent ill-health, and rising curve of infection created by those possessing "unchecked might and wealth." Although the focus is on human health, the reach of this comic extends to nonhuman creatures and the environment in which they live. On its second page, the comic depicts how bats, whose native habitat is shrinking under human pressure, are entangled with the people hit the hardest by COVID-19. In the top panel, the foreground reveals bats hanging from a cave ceiling, while behind them cars and vans fill a crowded street, and exhaust fumes spin into the air. To the right, a worried young woman seems to be taking in both scenes as if trying to make sense of their juxtaposition. The middle panel moves into fantasy, picturing the virus as a terrifying green monster, skin erupting with COVID spheres, which threatens the terrified, shrinking citizens to its lower right. On the bottom panel at the left, a young woman very like the one in the top panel is in bed wearing a respirator, her brow furrowed in grief or pain. To the right stand three mask-wearing individuals as if embodying the truth of the text that erupts between the second and third panels: "Black and Brown people, the Poor, the Elderly, and those with Compromised Immune Systems are hit the hardest," as readers can observe in Figure 7.1.

On the next page, a dense image expresses the reality that, just as human helpers can be found in unexpected places, so too threats to human health come in many shapes, sizes, species, and structural conditions. On the top right, a masked woman works to "flatten the curve" by sewing masks for health care workers, undaunted by the monstrous virus that towers over her. Between the top and bottom panels, a text banner contrasts those fighting against the rising curve of the pandemic to those profiting from "the curve of corruption—of unchecked might and wealth." In the bottom panel, the order Parasitiformes (mites and ticks) morphs into a human parasite, a corpulent Trump look-alike in a sweatshirt branded with the dollar sign with a coronavirus crown and scepter. "Like ticks, those inflated with corruption depend on the sweat and the blood of others"—as one can see in Figure 7.2.

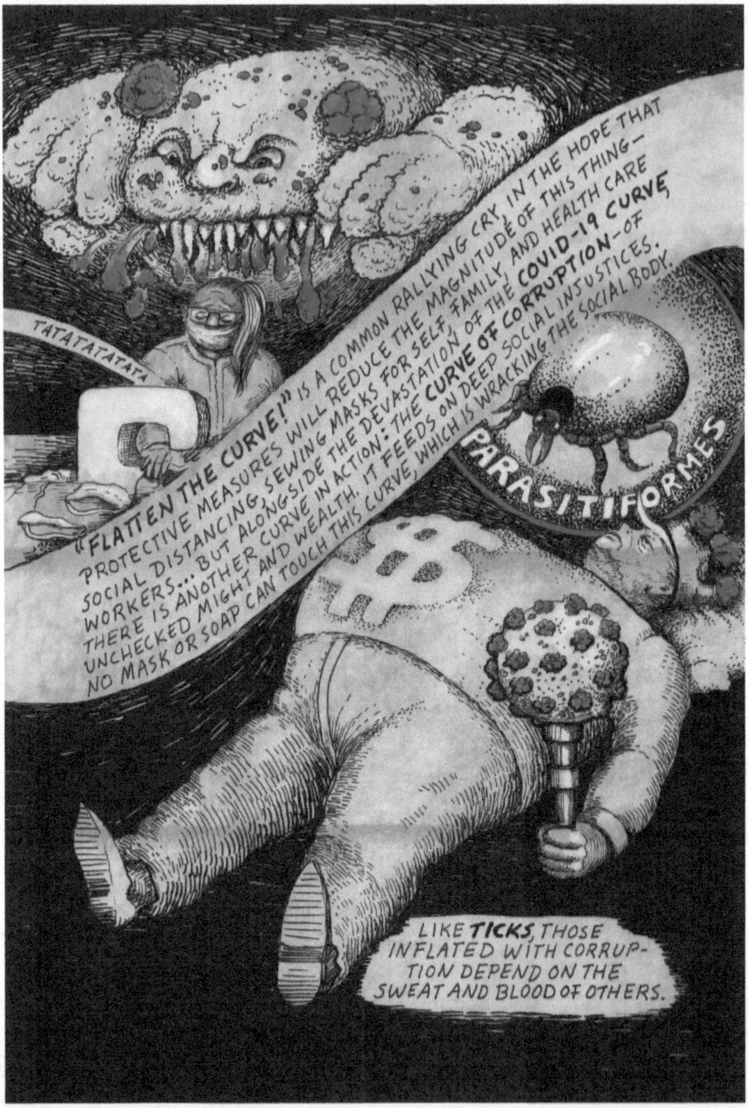

FIGURE 7.2 "The Right to Breathe," by Maureen Burdock and Joanne Regulska. Used with permission.

Using a zoonotic illness narrative as its pivot point, *Elmer*, by Ernest Alanguilan, tells the story of a global eruption of H5N1, or avian flu. Told from the perspective of a rooster who is part of the first generation of chickens to gain the same consciousness as human beings, the comic turns a powerful reverse lens on movements for racial equity and social justice, connecting them to debates over animal rights and personhood.[13] The transspecies connections made by the comic stretch to toxic racial and ethnic relations between human beings as well as to harsh cruelty between humans and animals. Once they have become aware of the horrors of industrial agriculture, the chickens unite. They rebel against their abusive keepers and create a multispecies world government in which chickens live as equals with humans. But this brief idyll ends with the eruption of avian flu and the pandemic-linked suspicions and hostilities it brings with it. A mash-up of industrial agriculture, contagion, and debates about animal personhood, *Elmer*'s plot crosses species lines to explore processes of stigmatization, exploitation, and abuse. Along the way, *Elmer* reframes industrial agriculture's primary security measure against zoonotic disease—the cull—as genocide.[14] This reverses the lens on human and animal health, potentially tipping us into a One Health perspective.

Both comics express a posthumanities perspective on multispecies health informed by what Joyelle McSweeney calls the necropastoral, "a political-aesthetic zone in which the fact of mankind's depredations cannot be separated from an experience of 'nature' which is poisoned, mutated, aberrant, spectacular, full of ill-effects and affects." As McSweeney describes it,

> The Necropastoral is a non-rational zone, anachronistic, it often looks backwards and does not subscribe to Cartesian coordinates or Enlightenment notions of rationality and linearity, cause and effect. It does not subscribe to humanism but is interested in non-human modalities, like those of bugs, viruses, weeds and mold.[15]

Taking the reverse perspective enabled by the necropastoral, these comics understand the metaphoric potential of zoonosis, exposing the negative effects that have already spread from humans to the other life forms around us.

Zooeiya

"Zooeyia, the positive benefits to human health from interacting with animals," is a concept proposed by Canadian veterinarians Kate Hodgson and Marcia Darling.[16] This concept builds on the World Health Organization's definition of "health as a state of complete physical, mental and social wellbeing and not merely the absence of disease or infirmity."[17] While many works of Graphic Medicine address the issue of zoonoses, there are also works that explore zooeyia.

A comic published in the *Annals of Internal Medicine* in 2015, "Mom's Flock," by Sharon Rosenzweig, tells the story of some "renegade hens" who gave a new lease on life to the cartoonist's mother, an older woman bedridden with Alzheimer's disease.[18] When her family realizes that "Mom" is losing the will to live and becoming isolated and depressed, they get her some backyard chickens. The poultry cure works, and, within six months of getting her new hens, Mom is out of her bed, navigating the backyard in her walker, and intently observing and enjoying her new avian friends. Mom's sharp observations on the chicken society reveal her newly attained physical and mental health, but they do more than that. They encourage us to see the connections between human and animal behavior, human wellbeing, and animal health.

"Head Lice and My Worst Boyfriend," by the MacArthur award-winning cartoonist Lynda Barry, also addresses the way that other species can shape a person's health.[19] This "autobiofictionalography" details how the mental and emotional health of the protagonist is threatened first by her classmates, who bully her, then by maternal abuse, and finally by the contempt of a verbally abusive boyfriend. Yet, when she visits relatives in the Philippines, she falls in love with "the kid everyone called the Professor," because "he was the smartest kid I ever met and he liked bugs." Learning from him to admire the adaptive genius of parasites, the young Barry accomplishes her own adaptation. She casts off the traumatic maternal abuse of her earlier years as well as the abuse of her current boyfriend, finding in each a similar, self-centered character. By learning to love head lice, she has come to accept all the variations of human difference, including her own. "Although it sounds far-fetched to think different colors of people could have different colors of headlice, it turns out to be true. In certain evolutionary ways, lice are geniuses."[20] Lice are geniuses, Barry suggests, because they appreciate that the benign interspecies influence of head lice brings about zooeyiatic healing.

We have seen how cartoonist Maureen Burdock uses the order Parasitiformes to turn zooonosis on its head, revealing the virulent aspect of human relations with each other and with nonhuman beings, and thus experiencing zooeiya. Like the way Lynda Barry adopts the head louse as her autobiographical avatar, cartoonist Miriam Katin chooses bedbugs as her surrogates in *Letting it Go*, a comic exploring how the experience of fleeing the Nazis as a child has colored her visit to Germany as an adult. In a conversation between briefcase-bearing bedbugs, Katin ironically affirms her lingering connection to Berlin. As one bedbug says to the others, "Her blood will be all over the city. That will call her back."[21] As Burdock explains in an essay exploring how female cartoonists turn to nonhuman beings to enact their embodied histories, "These transspecies unions are awkward and uncomfortable; but drawing them helps us to draw ourselves out of painful situations, out of intolerable isolation and fixed identities. We have the ability to transform what is bugging us."[22] Indeed, through an exploration of zooeiya, these cartoonists attain a One Health perspective.

Zooambivalence

In a 2020 article in *Microorganisms*, Salla Sariola and Scott F. Gilbert challenge us to recognize "the ambivalence of microbes in the Anthropocene."[23] They argue that, since the discovery of antibiotics and the rise of public health, the human relationship with microbes has changed in three crucial ways: the "Western world" has replaced "unplanned and unplannable nature" with a sterile, domesticated, and industrially managed world; microbes themselves have changed drastically, incorporating antibiotic resistance; and medicine and public health have just begun to understand how microbes and human bodies collaborate in "mutualistic symbiosis."[24]

Although I may be more optimistic than they are about the potential for One Health, I agree that the perspective that they advance, which we might call "zooambivalence," is a crucial one. I will conclude by looking closely at two comics that articulate the realities of zooambivalence.

The Wellcome Center for Integrative Parasitology, in Glasgow, focuses on the parasite, maintaining the tradition of veterinary epidemiologist Calvin W. Schwabe.[25] Attention to parasites and their ecologies is central to the WCIP's most recent comic, *Helminths: The Secret World of Parasitic Worms* (2020), written by Jamie Hall and Edward Ross, and illustrated by Edward Ross.[26] This comic takes a systems-based approach to disease reduction, telling the story of Namazzi, a Ugandan girl who is infected by the schistosome larvae when filling her water jug in Lake Victoria. The comic tracks the network of people studying the ecological and environmental conditions around Lake Victoria, "biologists ... economists, social scientists, and even actors." Hoping to understand the factors giving rise to repeated infections with schistosomiasis, they must work together to address drug resistance, inadequate village sanitation, and poverty. Although it focuses explicitly on human health, through its system-based approach, this porous pathography reveals the environmental and ecological health of the region as well.

Zooambivalence is also central to *The Antibiotic Tales,* a comic co-created by Eisner-award-winning cartoonist Sonny Liew and Dr. Hsu Li Yang, head of the Infectious Diseases Program at Saw Swee Hock School of Public Health at the National University of Singapore.[27] *The Antibiotic Tales* dramatizes the issue of antimicrobial resistance in a narrative that braids two tales: the realistic story of a family accompanying their grandmother to the doctor and a comic telling the story of a "worst-case post-apocalyptic world" set in 2119. The comic within this comic, *The Post-Antibiotic Apocalypse*, is set in the speculative future when the rise of antibiotic resistance and the effects of climate change have led to "a collapse of civilization."[28]

From its very cover illustration, this comic relies on visual strategies to prepare the reader to face the fact that microbes can be either beneficial or highly harmful. The contrast between the two vertical half-panels that comprise the cover illustration engenders a sense of unease. The brightly colored left-hand panel pictures a schoolboy wearing teal shorts and a bright teal backpack standing next to a colorful food market display of eggs, vegetables, and fruit. He is reading a comic whose title, "The Antibiotics," echoes the title that stretches across both left and right vertical panels in the cover illustration: *The Antibiotic Tales.*

An art historian or literature scholar will easily recognize the visual structure the left-hand panel draws on. It's a *mise en abyme*, an image within an image or a story within a story.[29] This visual structure signals that there are multiple levels of interpretation available to the reader/viewer. Yet, while the *mise en abyme* produces a sense of infinite recursion, self-referentiality, or even chaos, in this case

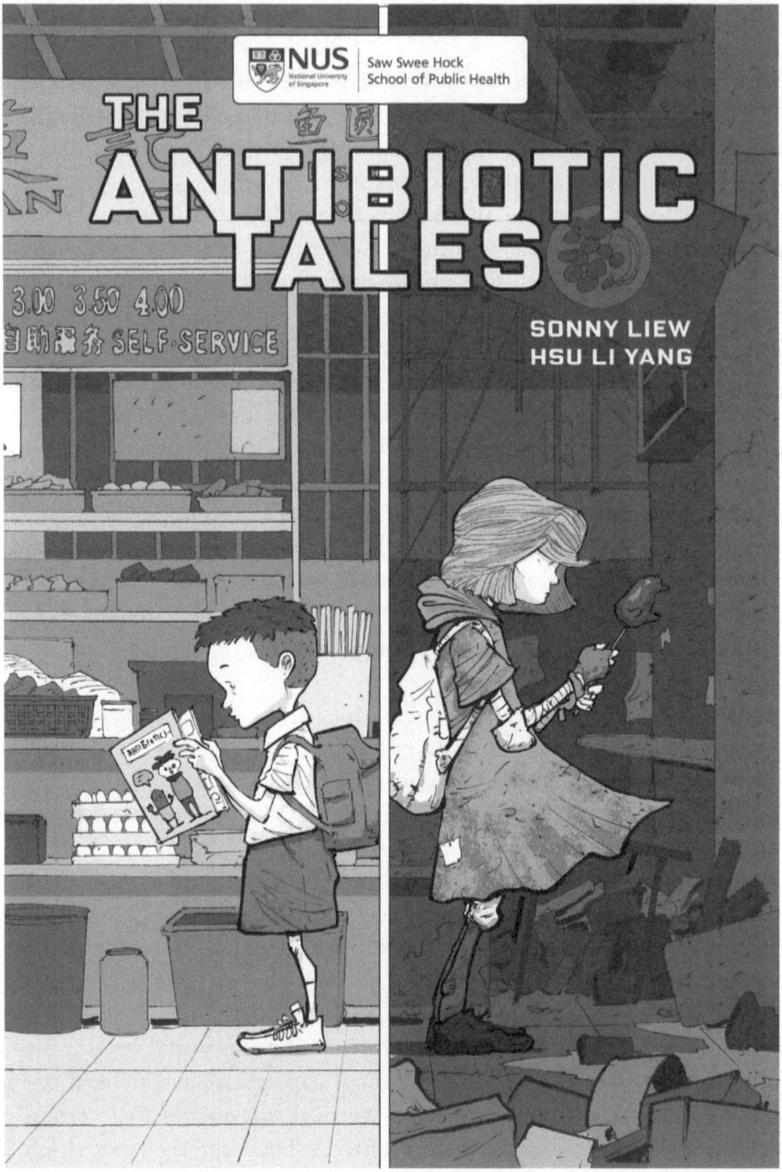

FIGURE 7.3 Cover, *The Antibiotic Tales*. Liew and Yang. Used with permission.

the image is only a partial *mise en abyme*, with important effects on the reader. Rather than showing the cover of a comic picturing the cover of a comic on which a boy gazes at the cover of a comic showing a boy gazing at the cover of a comic (you get the idea), *The Antibiotic Tales* interrupts its use of the *mise en abyme*. The cover of the comic the schoolboy is holding pictures not another child but two anthropomorphic pill capsules, one yellow and the other blue-and-red. This image interrupts the expected endless reiteration and asks us to shift our focus to the antibiotics themselves. How do they interrupt the ongoing human story?

We get the beginnings of an answer when our eyes cross the narrow white gutter separating the left and right-hand panels. Unlike the tidy schoolboy at the food market in the left-hand panel, in the right-hand panel we discover a young girl wearing a worn burlap backpack and a patched red coat. Her blue hair the only bright note in the image, she is standing in a murky gray, rubble-strewn setting, possibly the wreck of the food market from the panel on the left. Rather than a colorful comic, she holds in her hands an unidentifiable small mammal on a skewer.

The Antibiotic Tales uses its cover image to ask which tale we will tell about antibiotics in the future. Microbes connect our bodies, our fellow species, our environments, and our ecosystems, sometimes for good and sometimes for ill. In modern medicine, antibiotics have been part of the broad group of antimicrobial agents that play a central role in combating the diseases spread by bacteria. But recently, the valuable role played by antibiotics has been at risk because of the rise of antimicrobial resistance (AMR). Liew and Yang describe antimicrobial resistance as "probably the greatest infectious disease public health threat today." They decided to create a comic that would carry that narrative to readers of all ages. Although they considered both time-travel narratives and a story told by talking bacteria, they eventually decided to use a "comic-within-a-comic" approach.[30] The final comic they created thus juxtaposes two plots: a present-day story of a routine doctor's visit and a futuristic story of a family's evening dinner in 2119, "post-antibiotic apocalypse." In the first story, a little boy accompanies his mother and grandmother on a train trip to the doctor. He brings with him a comic he was given at school, which he explains is "about antibiotics": "That if we are not *careful* using them, it could lead to the *end* of modern medicine …!" "What do they mean by being 'not careful'" the mother asks. "Hmm. mainly by taking antibiotics when we don't need them. And something about farms, too."

In the clinic, the mother shares her son's comic with the doctor. "My son got this comic about antibiotics at school," she tells him, and asks "is what it says about an *apocalyptic* future true?" The doctor assesses the comic plot as "far-fetched … the idea that medicine and civilization will collapse completely." Yet he agrees the comic is "*not wrong* about how the careless overuse of antibiotics is helping bacteria to develop resistance … nor about their amazing ability to transfer those resistance genes to other bacteria." Nevertheless, the grandmother still asks the doctor to prescribe antibiotics for her. Only when the grandmother returns home and decides to read the comic for herself does the genre pivot to speculative dystopian fiction.

"*Post-Antibiotic Apocalypse* Part II: Animal Farms," is the science fiction tale that prompts the grandmother to refuse antibiotics. This comic-within-a-comic is set in the worst-case future, when microbial evolution has led to antibiotic-resistant bacteria, the collapse of civilization, and a serious food shortage. The title panel is a cutaway shot of a burned-out building, within which ten people huddle around a fire on which a small animal cooks on a spit. The tale features a campfire conversation between a grandfather, a mother, her blue-haired granddaughter, and a scraggly band of wanderers.

Three panels down the right-hand side offer a small vignette to situate the reader in this new future. We see the granddaughter—whom we recognize from the right-hand panel on the cover of *The Antibiotic Tales*—about to bite into a morsel on a stick which we also recognize from the cover. Her toothless grandfather asks her, "How's the rat meat, Oren?" and she answers, "It's delicious, Grandpa!"

How did we get from the present to *this* future? Sitting at the campfire, the grandfather reminisces about the brightly lit stores of his youth: "We had supermarkets filled with *all* kinds of meats." His adult daughter joins the conversation, explaining the complex chain of events that has led to the current food scarcity. Her explanation, imaged as a flow chart behind her as she speaks, links human beings and animals through their food, their waste, their shared medications, and the environment. Her contribution to this dystopian fireside conversation drives home the possibility of a future in which people no longer have access to clean and wholesome food but instead must scrounge for their meals.

Reading the tale, the tale-within-a-tale, and considering the concept linking both tales, the reader must come to grips with the causes of AMR. These include the medical overuse of antibiotics to treat conditions that are viral rather than bacterial and the agricultural use of antibiotics to help chickens lay more eggs, piglets grow up faster and bigger, and cows stay healthy in crowded conditions. Taken together, these constitute what Laura Kahn has called the politics of antimicrobial resistance: the clashing sites of expertise, combative professional hierarchies, and national and international regulations—and the failures to regulate that led to the call for One Health.[31]

Let me step back a bit and talk about the strategies this comic uses to communicate a One Health message. *The Antibiotic Tales* targets multiple readerships, including secondary school and university students and a general readership, as well as readers with a more extensive understanding of both cartooning and science. These multiple readerships are evident in their comic-within-a-comic structure, their incorporation of the doctor's meta-commentary on the uses of comics in patient education, and their visual representations of the complex schematics of horizontal bacterial gene transfer and antimicrobial resistance in agriculture.

But the comic does even more to address those multiple readerships. It also includes a preface and explanatory notes. The preface explains the complexity of AMR, and the notes expand on the elements of the historical background to *The Antibiotic Tales*: the discovery and synthesis of antibiotics; the distinction between antibiotics and the larger set of antimicrobials (or antimicrobial agents) to which they belong; the tragic story of Constable Albert Alexander who was first treated by penicillin for an infection contracted when he was scratched by a rose thorn and died when the supply of penicillin ran out; the role of antibiotics in the practice of medicine; and finally their misuse in farming as growth promoters, which has given rise to antibiotic resistance in livestock, the environment,

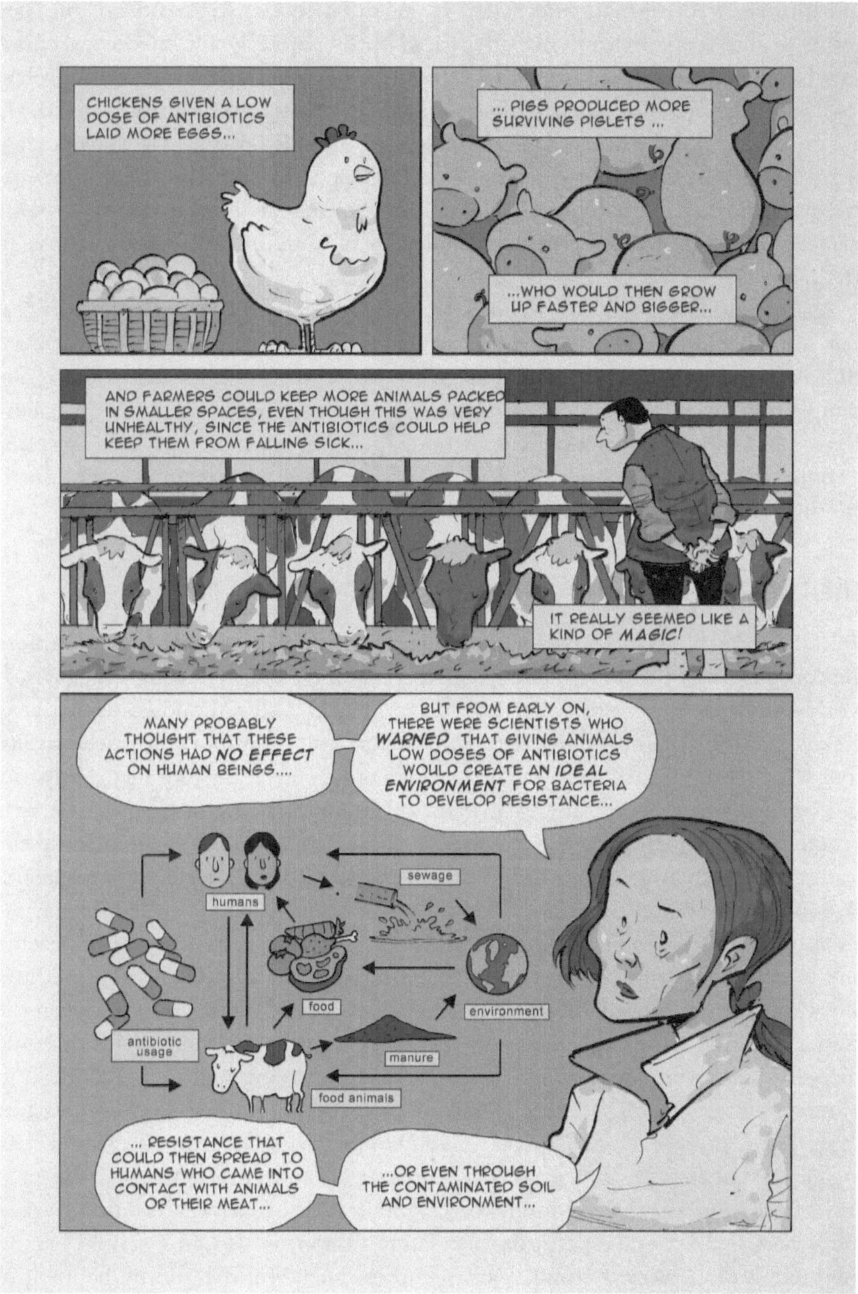

FIGURE 7.4 *The Antibiotic Tales*. Liew and Yang. Used with permission.

and ultimately in human infections. Concluding sections expand on four key points in the comic by drawing on articles in the scholarly and medical archive: how bacteria develop resistance to antibiotics; the global impact of antimicrobial resistance; the meaning of the color in mucus in relation to indications of bacterial infection; and the unclear relation between organic farming and the decrease in antibacterial-resistant bacteria in food. The appendix stipulates, "Unlike what the grandmother in the comic suggests, the evidence is not clear that 'organic farming' as a whole results in the reduction of antibiotic-resistant bacteria in food."[32]

As if responding to the call of Sariola and Gilbert for increased awareness that both beneficial and pathogenic interactions take place between bacteria, microbes, animals, and human beings, the focus of *The Antibiotic Tales* is the complexity of our symbiotic relations with the microbes that surround, permeate, and indeed enable, as well as endanger, our human lives. The result is a comic that expresses and encourages us to experience zooambiguity. But is showing and telling in a comic enough?

The Uses of Graphic Medicine as Practice

By telling and showing, comics can educate and involve lay readers to explore the complexities of zoonosis, zooeyia, and zooambivalence. Through their visual world-building, comics can also move us to identify and respond emotionally to life as it unfolds at multiple scales. This is particularly important in the health context, where comics have been documented to increase physician empathy and patient comprehension.[33] The field of Graphic Medicine also encourages people to start making their own comics. This is in addition to advocating the reading and teaching of comics about medicine, illness, disability, and health care in colleges and universities, medical schools, and public libraries. The argument for this focus on practice is multilayered. Learning to make a comic teaches one to appreciate the demands of cartooning and the affordances the medium offers to express embodied experiences. Because of the connection between the drawn line, the hand, and the body of the cartoonist, the very act of drawing energizes emotional connection with the material being pictured. Recall Will Eisner's comic "*MicroDictionary of Gestures*" or the Fear-Free Pets comics-like chart "*The Spectrum of Fear, Anxiety, and Distress*" that I mentioned earlier. Both images reveal that gesture unlocks identification and emotions. And the process of creating a universe down to the smallest detail, a very familiar part of the writing of fiction, especially speculative fiction, is also an essential part of the process of creating a comic. But in comics, such world-building happens in multiple ways. Not only does the narrative text of a comic build its world, but the visual space within which the sequential comic images unfold can contain multiplicities: foreground and background, protagonist(s) and minor characters, flora and fauna, scenery, and landscape, atmospheres, and colors, and onomatopoeic sounds. At every scale, the images of a comic not only tell a story, but also beckon us imaginatively to embody it.

We can see this process at work if we consider "*Life Cycle Comics*," a simple exercise by Marek Bennett, the Vermont cartoonist and teacher at the Center for Cartoon Studies. A frequent guest teacher in elementary schools, Bennett has created a lesson that coaxes a budding cartoonist through the creation of a mini-comic by supplying prompts in parentheses on succeeding blank pages. As he describes it, this is "a 1-sheet mini-comic template for biology (or heck, maybe mythology) projects."[34]

The prompts are simple. First, choose a subject:

> Pick your favorite animal/plant /fungus /protist /microbe ... "Then, on the title page and each succeeding page, draw: 1) (intro image) and (author) 2) "(how life begins) egg? seed? Etc." 3) "(early stage) hatch? Sprout? Larva? etc. ... "; 4–5) the progression from young to adult in "2 single pages or 1 double-page spread!"); 6) "(Interactions w/ecosystem) or (surprise facts!)"; 7)" (End of life) and/or (Ideal situation)," and finally, 8)" (credits) (thanks) (sources) (note to readers) etc.

What Bennett does in this simple exercise is encourage the novice cartoonist to use the life course as a plot, and to consciously choose a protagonist for his comic tale from the whole range of living beings. Decentering the human being and instead affirming the existence of many species whose voices and lives are all potential protagonists, Bennett's cartooning exercise enlists the life cycle as something a comic can explore, something common to every living species, be they "animal/plant/fungus/protist/microbe." Be they even, as the concluding ellipsis wonderfully hints at but refrains from specifying, human, the default mode for many works of Graphic Medicine.

Conclusion: Graphic Medicine and One Health

I began by asking what Graphic Medicine can bring to One Health. Surveying comics that investigate the three modes of interspecies relations affecting health—zoonosis, zooeiya, and zooambivalence—I found a multispecies entanglement that demonstrates just how much each transdisciplinary approach draws from, and can enhance, the other. Originating in hospital-based clinical experience, Graphic Medicine has been challenged to expand its focus beyond individual human *medicine* to human *health,* environmental health, and the ultimate health of the planet, including the threat posed by climate change. One Health has also been critiqued for a human-centered bias, and for understating anthropogenic damages to the environment and the planet. So, for example, Sariola and Gilbert assert that "the three components of the One Health model [people, animals, and the environment] are not equal, and the framework is still used to prioritize protection of humans from zoonotic diseases."[35] Yet Graphic Medicine increasingly presents a porous pathography, building worlds beyond the hospital and the clinic that incorporate the scaled health and illness experiences of the many species beyond (and within) human beings. So, too, One

Health has the potential to teach us how to move beyond our human-centered focus, by incorporating comics in its communication and practices.

I conclude with Marek Bennett's "*Life Cycle Comics*" because, in its focus on the practice of creating comics, its attention to biology, and its invitation to choose from a wide range of protagonists, it provides an inspiring three-part model for "thinking with" both Graphic Medicine and One Health. Coaching its budding cartoonist to engage in an improvisational "Yes, and ...," Bennett's template for creating a "*Life Cycle Comic*" models a way to imagine life from the perspective of endless other living beings. As a porous pathography, it draws on and expands the reach of Graphic Medicine while simultaneously revealing the pragmatic and visionary methods now available to One Health.

Notes

1 Sathyaraj Venkatesan and Sweetha Saji, "Rhetorics of the Visual: Graphic Medicine, Comics and its Affordances," *Rupkatha Journal on Interdisciplinary Studies in Humanities* VIII, no. 3 (2016): 0975–2935.
2 Susan Squier, "Scaling Graphic Medicine: The Porous Pathography, a New Kind of Illness Narrative," in *PathoGraphics: Narrative, Aesthetics, Contention, Community*, eds. Susan Merrill Squier and Irmela Marei Krüger-Fürhoff (University Park: Penn State University Press, 2020), 205–226, 206.
3 The comics I address in this chapter include all these elements.
4 They even wonder whether microbes might be important for good human mental health: "What if, in addition to protection against allergies and asthma, bacteria were protecting us against mental health conditions such as schizophrenia, bipolar disease, and autism?" Salla Sariola and Scott F. Gilbert, "Toward a Symbiotic Perspective on Public Health: Recognizing the Ambivalence of Microbes in the Anthropocene," *Microorganisms* 8 (2020): 746 and Henrik Lerner and Charlotte Berg, "A Comparison of Three Holistic Approaches to Health: One Health, EcoHealth, and Planetary Health," *Frontiers in Veterinary Science* 4 (2017): 163. See also in Irus Braverman's introduction to this volume.
5 Ibid., 7.
6 Lerner and Berg, "A Comparison of Three Holistic Approaches."
7 Susan Squier, "'Scaling Graphic Medicine.'"
8 Katherine Kelp-Stebbins, *How Comics Travel: Publications, Translations, Radical Literacies*. (Columbus OH: Ohio State University Press, 2022), 1.
9 Will Eisner, *Comics and Sequential Art* (Tamarac: Poorhouse Press, 1985), 102.
10 "FAS Spectrum and Pain Algorithm," Fear Free Pets, 2022, https://fearfreepets.com/fas-spectrum/.
11 "Zoonotic Diseases," Centers for Disease Control and Prevention, July 1, 2021, https://www.cdc.gov/onehealth/basics/zoonotic-diseases.html.
12 During this pandemic, Graphic Mundi, a new US publishing imprint, has dedicated itself to publishing "graphic novels [that] ... 'scale-up' the concept of health," taking into account the intersecting worlds not only of humans but also of microbes, plants, and nonhuman animals to highlight our myriad connections." Kendra Boileau and Rich Johnson, eds., *The Covid Chronicles* (University Park: Graphic Mundi, 2021), x. Full disclosure: this statement is drawn from Susan Squier, "Comics and Graphic Medicine as a Third Space for the Health Humanities," in *Routledge Companion to the Health Humanities*, eds. Paul Crawford et al. (New York: Routledge, 2020), 61–62.

13 In that sense, this comic carries out the same strategy that Frédéric Keck observed among the diasporic Chinese: using a focus on a bird-borne virus to speak about issues relating to human migration and sociopolitical alienation.
14 Alanguilan chooses this controversial term because he is talking about industrial agriculture. In a dialog with Elizabeth Roudinesco about violence against animals, Jacques Derrida recalls making the same word choice: "When I spoke on this question … at the law school of a Jewish university, I used this word *genocide* to designate the operation consisting, in certain cases, in gathering together hundreds of thousands of beasts every day, sending them to the slaughterhouse, and killing them *en masse* after having fattened them with hormones. This earned me an indignant reply. One person said that he did not accept my use of the word genocide: 'We know what genocide is.' Let's withdraw the word then. But you see very well what I'm talking about." Earlier, Derrida explains that he hesitates to use the word *genocide* "only in order not to abuse the inevitable associations," although "in fact, the word would not be so appropriate." Jacques Derrida and Elizabeth Roudinesco, *For What Tomorrow …: A Dialogue*, trans. Jeff Fort (Stanford: Stanford University Press, 2004), 73.
15 Joyelle McSweeney, "What is the Necropastoral?," in *Just Us: An American Conversation*, ed. Claudia Rankine (Minneapolis: Graywolf Press, 2020), 87–88.
16 Kate Hodgson and Marcia Darling, "Zooeyia: An Essential Component of 'One Health,'" *The Canadian Veterinary Journal* 52, no. 2 (2011): 189–191, 189.
17 Ibid. They explain that they coined the term by combining "the Greek root words for animal (zoion) and health (Hygeia was the ancient Greek goddess of health, the same source as 'hygiene.') Zooeyia is the positive inverse of zoonosis (from the same 'zoion' and 'nosos' or disease.") Ibid.
18 Sharon Rosenzweig, "Mom's Flock," *Annals of Internal Medicine* (2015), https://www.acpjournals.org/doi/10.7326/G14-0003.
19 Lynda Barry, *One! Hundred! Demons!* (Seattle: Sasquatch Books, 2002).
20 Ibid., 18.
21 Miriam Katin, *Letting It Go* (Toronto: Drawn & Quarterly, 2013), 152.
22 Maureen Burdock, "Shapeshifters: Metamorphosing Transgenerational Trauma through Comics," in *Contested Selves: Life Writing and German Culture,* eds. Katja Herges and Elisabeth Krimmer (Rochester: Camden House, 2021), 229–247.
23 Sariola and Gilbert, "Toward a Symbiotic Perspective on Public Health," 746.
24 Ibid.
25 Rachel Mason Dentinger, "The Parasitological Pursuit: Crossing Species and Disciplinary Boundaries with Calvin W. Schwabe and the *Echinococcus* Tapeworm, 1956–1975," in *One Health and its Histories: Animals and the Shaping of Modern Medicine*, eds. Abigail Woods et al. (Basingstoke: Palgrave, 2018), 161–191, 162.
26 Jamie Hall and Edward Ross, *Helminths: The Secret World of Parasitic Worms* (Glasgow: Wellcome Centre for Integrative Parasitology, 2020).
27 Sonny Liew and Hsu Li Yang, *The Antibiotic Tales* (Singapore: Saw Swee Hock School of Public Health, 2019). All references in what follows are to this work, which is unpaginated.
28 Ibid.
29 Comics scholars also call this the Droste Effect, which, in its full form, can be seen on the website "Graphic Medicine," Graphic Medicine, 2022, http://graphicmedicine.org.
30 Kami Navarro, "Tackling Antimicrobial Resistance through Comic Books," *Asian Scientist*, January 21, 2002.
31 Laura Kahn, *One Health and the Politics of Antimicrobial Resistance* (Baltimore: Johns Hopkins University Press, 2016). See also Kahn, this volume.
32 Liew and Yang, "Notes to *The Antibiotic Tales*."

33 Patricia F. Anderson, Elise Wescom, and Ruth C. Carlos, "Difficult Doctors, Difficult Patients: Building Empathy," *Journal of the American College of Radiology* 13, no. 12 (2016): 1590–1598. See also Anna Brand et al., "Medical Graphic Narratives to Improve Patient Comprehension and Periprocedural Anxiety Before Coronary Angiography and Percutaneous Coronary Intervention: A Randomized Trial," *Annals of Internal Medicine* 170, no. 8 (2019): 579–581.

34 Marek Bennett, "*Life Cycle Comics*" (1-Sheet Template), personal communication via email from his Patreon account, October 19, 2022.

35 Sariola and Gilbert, "Toward a Symbiotic Perspective on Public Health."

PART III

Othering One Health

Toward Multibeing Justice

8455

8

THE ONE HEALTH INITIATIVE AND A DEEPER ENGAGEMENT WITH ANIMAL HEALTH AND WELLBEING

Moving Away From Animal Agriculture

Maneesha Deckha

Introduction

The vision statement of the One Health Initiative, which is a subset of its overall mission statement, states that One Health "is dedicated to improving the lives of all species—human and animal—through the integration of human medicine, veterinary medicine, and environmental science."[1] A little lower down in its mission statement, the One Health Initiative declares that "[r]ecognizing that human health (including mental health via the human-animal bond phenomenon), animal health, and ecosystem health are inextricably linked, One Health seeks to promote, improve, and defend the health and well-being of all species."[2] Conventional medicine is premised on a sharp reason–body and human–animal divide and a masculinist individuated understanding of health.[3] By contrast, the integrative conceptualization of health envisioned by the One Health Initiative (the "OHI") is boundary breaking. However, insofar as the OHI's vision and mission statements ostensibly resist an anthropocentric worldview, the OHI has demurred from taking a public position against globally pervasive animal-use industries that pose continuing harm to humans, animals, and the environment.

Specifically, the OHI has remained silent regarding the ethics surrounding one such prominent anthropogenic use of animals: the consumption of animals for their flesh, milk, or eggs. The OHI is presumably aware of the socially dominant and extractive nature of the animal-based food systems given the scale of the industries involved. Agricultural exploitation consumes the lives of tens of billions of terrestrial animals every year, with chickens comprising about 90

FIGURE 8.1 Calf in a veal industry hutch, a by-product of the dairy industry, photographed in 2018. The separation of calves from their mothers is one source of the intense suffering animals experience in both the veal and the dairy agricultural industries. Courtesy of Pexels. Used with permission.

DOI: 10.4324/9781003294085-12

percent of this number globally.[4] Commercial fishing and fish farming consumes the lives of at least a trillion fish and other aquatic animals per year.[5] Dietary consumption of animals also exacts devastating harm on humans and the environment,[6] and miserable captive living conditions ending in fatal outcomes for animals.[7] Whether as a byproduct of farming conditions or by genetic design, many of these animals must tolerate infection, disease, and deformity.[8] As just one instance, domesticated chickens can be rendered blind by the toxic levels of ammonia generated by their own urine in inadequately ventilated feeding operations.[9]

What should a commitment to the OHI, and the multispecies mission that it aspires to embrace, mean in the context of such massive exploitation of animal lives? If the intent of the OHI is to effectively counter anthropocentric values, it would seem it must adopt an ethical animal-free agenda in opposition to animal-use industries that maim and kill animals or, as this volume argues, a "more-than-one health" approach. This seems particularly so when such industries pose serious risks to human health and the environment and thus directly challenge all three prongs of the OHI's integrated vision. This chapter argues that the OHI should move from passive acceptance of commercial animal agriculture to adopt a clear position statement, policy, or fact sheet that acknowledges that commercial animal agriculture is incompatible with its integrated vision[10] and to call for an accelerated phasing out of this industry.[11]

The argument is divided into three parts. The first part reveals the current non-response and thus passive acceptance from the OHI and WHO, its principal constitutive member, to commercial animal agriculture (hereafter, "animal agriculture"), and situates it within larger, transnational policy discourse on animal agriculture. This part reviews some of the interspecies harms that flow from animal agriculture to the environment in the course of uncovering the passive acceptance by the OHI and WHO of the industry despite its environmentally catastrophic impact. The second part deepens the discussion of the interspecies harms of animal agriculture by returning to the OHI's formative concern about zoonotic diseases to illustrate another major way that animal agriculture is incompatible with the OHI's integrated vision.[12] The third part briefly considers and replies to anticipated objections.

Anthropocentric Policy Silences as a General Backdrop to the OHI's Non-Response

Scholarly criticism of animal-based food systems has grown into a recognizable movement. More and more scientists are exposing the animal agricultural and aquacultural sectors' outsized role in climate change, biodiversity loss, pollution,[13] zoonotic disease risk,[14] and chronic disease risk in humans.[15] With a growing cohort of the scientific community calling for a steep reduction in animal farming, and advocating instead for a societal transition to healthy plant-based diets, governments and sustainability-focused organizations should take heed.[16]

This has yet to occur. Although fish farming attracts attention from policy actors as a questionable practice,[17] commercial fishing and the animal agricultural industry are typically excluded from calls for systemic change at the level of government or international policy.[18] There are very few mainstream institutional voices (prominent media, think tank, non-profit sector, or other civil society groups) promoting plant-based diets.[19]

Affluent countries with the influence to lead global dietary policy have been notably slow to embrace a transition to plant-based diets. Wealthy countries have the highest per capita consumption rates of animal-derived foods, and their endorsement of plant-based diets is necessary for fundamental change.[20] Human psychology and cultural and economic factors help explain why eating animals is widely viewed as "natural," "normal," and "necessary."[21] While public discourse promoting plant-based diets has grown modestly, vegan and animal rights media remain primarily responsible for advancing this message.[22] When mainstream media does provide reporting on plant-based diets, coverage rarely includes the multiple ethical and climate benefits of a global dietary shift.[23]

Transnational Organizational Voices Breaking the Silence

There are some important exceptions to the silence of our civic institutions. Two prominent international authorities have recently endorsed the adoption of plant-based diets (understood in this section to be primarily but not exclusively plant-based) as a necessary or important step for counteracting global environmental devastation. The United Nations' Intergovernmental Panel on Climate Change (IPCC), which was awarded the Nobel Peace Prize in 2007, and the EAT-*Lancet* Commission have both emerged as vocal advocates of vegetable-based diets.[24] Previous IPCC reports initially attracted criticism for failing to promote fully plant-based eating (i.e., veganism) or seriously underestimating the "true carbon cost of our meat and dairy consumption" due to a flawed methodology for measuring animal agriculture's carbon footprint.[25] However, the IPCC's 2019 Special Report on Climate Change and Land has received scholarly approval for their recommendations to reduce animal-based consumption and for advocating that governments employ incentives to encourage this transition.[26]

A strong plant-based endorsement is the core message of the EAT-*Lancet* Commission Report.[27] This report prescribes a healthy human "reference diet" that "largely consists of vegetables, fruits, whole grains, legumes, nuts, and unsaturated oils, includes a low to moderate amount of seafood and poultry, and includes no or a low quantity of red meat, processed meat, added sugar, refined grains, and starchy vegetables."[28] The report further asserts that a plant-based diet is feasible for supporting a forecasted worldwide human population of 10 billion by 2050, but cautions that "even small increases in consumption of red meat or dairy foods would make this goal difficult or impossible to achieve."[29]

The Commission asserts that the reference diet is compatible with all human cultural cuisines (which is not to say it is identical to these cuisines or will be simple to implement).[30] Lastly, this diet would meet "scientific targets for the safe operating space of food systems [according to] six key Earth system processes."[31]

To achieve global adoption of the reference diet by its stated goal of 2050, the EAT-*Lancet* report calls for a bold revamping of present food systems. Given the resistance displayed by affluent nations to the adoption of plant-based diets thus far, it is no surprise that the EAT-*Lancet* report's recommendation to reduce or eliminate animal-derived food consumption has triggered vocal criticism. Objections have emerged from the animal agricultural industry, allied trade publications, and traditional media,[32] as well as from scholarly sources, whose rebuttals to the report were co-published by *The Lancet* in response to the EAT-*Lancet* report. For instance, a letter to the editor published in the *Correspondence* stated that irregularities in documentation, methodology, data collection, and modeling led the authors to question the report's conclusions.[33] However, the letter's authors also self-reported having multiple industry ties, including grant funding from the National Cattlemen's Beef Association. The authors of the EAT-*Lancet* report provided a refutation of each of the industry-supported critics' points.[34]

WHO and OHI—No Bold Statements

Prior to the EAT-*Lancet* report, the WHO had not promoted wholesale or even predominantly plant-based diets even for high-income countries, other than cautioning against excessive consumption of animal products known to be carcinogenic.[35] As of this writing, which occurred during the SARS-CoV2 pandemic, the WHO has still not endorsed the EAT-*Lancet* report. The extent of the WHO's response on the topic (discussed below) is a 2021 publication of a fact sheet by its European Office describing the health benefits of plant-based diets.[36]

In the lead-up to the United Nations Climate Change Conference in Glasgow in November 2021, the WHO published *The WHO COP26 Special Report on Climate Change and Health* (COP26 Report), which listed ten "priority actions" for governments concerning food programs and policy.[37] Action Item 8 is "to promote health, sustainable, and resilient food systems."[38] The discussion section of this Action Item refers to "unhealthy foods" without singling out animal-based foods as a member of this category.[39] "Action Points" to concretize the Action Item are provided, but none explicitly single out animal agriculture as a desirable target for intervention, nor are plant-based diets distinguished as being healthier. The COP26 Report highlights the ability of plant-based diets to lower global methane levels, but does not extend itself further. On the whole, the Action Points ask governments to "nourish our future," "remove harmful agricultural subsidies," "support a just agricultural transition," and "mainstream biodiversity for nutrition and health." With such equivocally worded objectives, the

WHO seems intent on not taking a stand against animal agriculture despite the disproportionate harm it poses to climate change and health.[40]

The COP26 Report's most pointed statement comes just before it lays out the Action Points for Action Item #8. It states: "Sustainable food production requires a shift ... This includes the vital area of farming, which could be transformed by ending subsidies harmful to health or the environment, subsidising nature-positive production, and by high-income nations consuming more plant-based diets and cutting food waste."[41] Is this an obtuse criticism of animal agriculture? One can hardly tell. Failure by omission in identifying animal farming or animal agriculture as a sizable contributor to adverse climate change is still failure.[42] It may be inferred that the WHO has opted for avoidance and obfuscation in order to avoid criticism, in a manner characteristic of international policy discourse on the negative externalities of animal agriculture.[43]

A term search for "plant-based diets" on the WHO Institutional Repository for Information Sharing website yields a WHO Healthy Diet Factsheet from 2021 that praises plant-based diets.[44] Entitled "Plant-Based Diets and Their Impact on Health, Sustainability, and the Environment," this factsheet atypically concludes that "considerable evidence supports shifting populations towards healthful plant-based diets that reduce or eliminate intake of animal products and maximize favourable 'One Health' impacts on human, animal and environmental health."[45] Again, however, the promotion of a reduction in or elimination of animal-based consumption does not translate into a proposal to phase out animal agricultural industries (although other industry and agricultural interventions are topic considerations of the WHO). This approach is consistent with the most recent WHO "Healthy Diet" Fact Sheet from 2020, which emphasizes that healthy diets "include" a variety of plant-based foods without explicitly endorsing purely or predominantly plant-based diets.[46] Similar to the COP26 Report, the factsheet calls on governments to incentivize healthy diets through systems change, including "increasing incentives for producers and retailers to grow, use and sell fresh fruit and vegetables," without any coupled recommendations for phasing out even intensive animal production.[47]

The WHO, with a mandate to promote human health around the world, is clearly anthropocentric in purpose. It has highlighted climate change as a threat to human health without connecting the dots of causality to animal agriculture or commenting on animal health. We should expect more, however, from health organizations such as the OHI that seek to steer society away from a human-first approach to global health, and instead, foster an interdependent view of human health as integrated with environmental and animal health. The more recent Planetary Health Initiative (PHI), which espouses a similar integrated vision to the OHI,[48] has also declined to champion plant-based diets as necessary for human health, animal health, or the environment. Both the OHI and PHI continue to favor anthropocentric values, despite their professed commitment to restoring environmental baselines to pre-Anthropocene levels.[49] The time is long past due for these initiatives to cease ignoring the "cow in the room," and to rise to their articulated integrative visions of global health.

Zoonotic Disease Risk also Escaping Comment

As noted earlier, animal agriculture generates multiple negative externalities for humans, but also for animals and the environment.[50] Space precludes further detailing all of these harms. However, it is worth focusing at some length on what is perhaps the most obvious and timely reason that the OHI must condemn animal agriculture and call for a global transition from animal-based agriculture to plant-based models: the causal links between animal agriculture and zoonotic diseases. The OHI has recognized the links between human health and animal health in terms of the risk of infectious diseases in humans emanating from animals. Indeed, better containment of the animal vectors of zoonotic diseases was the catalyst for the formation of the OHI.[51] Yet, as discussed above, the OHI has not taken a stand against animal agriculture, which is a primary catalyst of zoonotic diseases, such as COVID-19 and SARS, and the Ebola and Nipah viruses.[52]

We can observe this catalyzing effect in three ways. A first way to see the connection between animal agriculture and zoonotic diseases is to observe the encroachment of the former upon the habitat of wild animals. Animal agriculture and particularly the practice of ranching occasion such incursions through the deforestation necessary to raise cows on the current industrial scale,[53] a connection even the WHO has now highlighted in its manifesto discussed above.[54] There is thus corresponding injury to wildlife health and ecosystems caused by intensive farming operations,[55] but also increased risk of zoonotic disease since industrial agricultural incursion into habitat then leads to intensified human–wildlife contact.[56] Animal agriculture also drives habitat loss and intensified human–wildlife contact because intensive farming pushes small-scale farmers out to the hinterlands where they and the animals they farm are likely to have more interface with wild animals.[57] Enhanced opportunities for interface with wild animals also occurs through the wildlife trade and the farming and captivity of wild-caught animals for medicinal markets, but also for live animal food ("bushmeat").[58]

A second way to see intensive farming's production and "alarming" acceleration of infectious zoonotic diseases is to consider that the cause of zoonotic diseases (which comprise roughly 75 percent of all new diseases) can emanate from non-viral elements such as "bacteria, fungi, and parasites from a variety of animal sources."[59] These animal-sourced pathogens are produced through "grossly aberrant crowding of animals for human purposes" as in agricultural operations and other live animal markets which severely stress animals, making them more prone to disease.[60] Animal agriculture breeds and raises genetically monolithic, health-compromised animals, who are then more susceptible to infections and, once sick, to spreading infections to their health-compromised counterparts due to their mutual intensive confinement and extreme overcrowding.[61]

Consider the case of chickens. Chickens are the most intensively farmed animal on the planet. In 2019, it is estimated that 50 billion chickens were consumed worldwide,[62] and that there are roughly "23 billion on the planet at any one time."[63] As David Wiebers and Verly Feigin note, "transmission of (avian influenza) from chickens to humans was almost nonexistent 25 years ago; now serious outbreaks are occurring regularly—more in the past 15 years than in the entire 20th century," a phenomenon they attribute to the rise of concentrated animal feeding operations the world over.[64] Others have also highlighted these repeated links between intensive animal farming and the spread of infectious diseases.[65] Scholars further note that the transport of animals central to agricultural operations—across countries, oceans, and continents—increases the risk of zoonotic disease as well.[66]

A third way to see the causal link between animal agriculture and zoonotic diseases is to trace the effects of climate change. Consider the following causal chain: animal agriculture is a leading cause of climate change; climate change is causing mass migrations of humans and animals; climate-related migration is causing more and more humans to invade the traditional territory of wild animals, thereby creating more opportunities for zoonotic transfer.[67] This causal chain is a reality. It was highlighted in the Executive Summary of the recent Report, *Escaping the Era of Pandemics* ("the Report"), by the Intergovernmental Science-Policy Platform on Biodiversity and Ecosystem Services. The Report states, "[t]he underlying causes of pandemics are the same global environmental changes that drive biodiversity loss and climate change. These include land-use change, *agricultural expansion and intensification*, and wildlife trade and consumption."[68] The Report goes on to note that pandemics will recur unless the anthropogenic activities which cause them are stopped. One of the major policy changes the Report calls for is the institution of "levies on meat consumption, livestock production, or other forms of high-pandemic-risk consumption."[69]

These causal connections and the media spotlight that the global SARS-CoV-2 pandemic has aimed at the elevated zoonotic risk of industrial animal farming have escaped WHO comment. The WHO has recently acknowledged that wildlife farming is the most likely spillover pathway for the pandemic.[70] In May 2020, the WHO released a "Manifesto for a Healthy Recovery from COVID-19" with corresponding "Actionables." Like the COP26 Report, the manifesto encourages "healthy, sustainable food systems," without issuing any explicit indictment of animal agriculture. To the contrary, the manifesto endorses "sustainable fisheries."[71] The heightened role of the "clearing of land to rear livestock" in producing greenhouse gases is mentioned, followed by "land use change is the single biggest environmental driver of new disease outbreaks. There is a need for a rapid transition to healthy, nutritious, and sustainable diets."[72] No indictment of animal agriculture follows. In contrast, a subsequent call to "stop using taxpayers money to fund pollution" delivers a clear condemnation against continued subsidies for the fossil fuel industry.[73] I have uncovered no WHO or OHI position statement, *explicitly* identifying animal agriculture as a risk factor for zoonotic pandemics.

Is "Humane" Farming Better?

What about non-intensive animal agriculture or so-called "humane" or "ethical" farming? What if such farming could be shown to avoid aggravating the risk of zoonotic diseases in terms of avoiding intensive farming's scale of wildlife habitat destruction and confinement of sickly animals and pathogen-laced operations? Even if this were the case, we would have to confront the fact that non-intensive farming is even less environmentally sustainable due to the extra land that would be required to give farmed animals proper freedom of movement.[74] Further, there are convincing reasons to be skeptical about the connotations of "humane" or "ethical," notably since painful branding, body modification, and slaughter, as well as denial of kinship bonds, are still routine industry practices in non-intensive farming.[75] Even if such industry practices were to change, even the "gentlest" animal agriculture we can imagine would seek to exploit animals' reproductive capacities and/or kill animals for consumption. The particular gendered nature of violence inflicted on farmed animals through breeding, laying, and milking continues on in less intensive contexts.[76] Such endpoints are not easily reconciled with OHI's mission to "promote, improve, and defend the health and wellbeing of all species" as there is no defensible benefit for animal species, however well cared for, that are regarded simply as "food" for humans and repeatedly breedable and killable.[77]

That being said, it is not necessary to settle the debate about the ethical viability of non-intensive farming from an animal rights or more-than-one health perspective. The argument presented here calls for the OHI to take a stand against commercial animal agriculture, which overwhelmingly takes the form of intensive farming, and is an interspecies phenomenon which is far less defensible as protective of the health and wellbeing of animals for all the reasons referenced already.[78]

Other Reasons to Favor Animal Agriculture?

Proponents of intensive animal agriculture see things differently and often argue that such practices are necessary to ensure human food security.[79] Other objections include that intensive plant-based agriculture also harms animals and marginalized human workers,[80] and that such diets are culturally imperialist or anti-heritage cuisines.[81] There are arguments to counter such objections.[82] For those particularly concerned that a plant-based diet reflects a "Western" or "Global North" perspective, or is otherwise imperialist or elitist, this is a complicated framing that I and others have unpacked elsewhere and have ultimately argued is misguided.[83] If anything, the globalization of Western models of animal agriculture is a leading example of capitalist, colonial, and gendered extractive exploitation of both animals and humans.[84] It is worth recalling that the present argument is for the OHI to call for a phasing out of commercial animal agriculture as well as provide the necessary financial supports to bridge the transition for those humans whose lives or livelihood would be placed at risk. Such a transition can attend to context and be responsive to global disparities.[85] Furthermore, such a transition is expected to ameliorate the global food insecurity caused by current animal-based food systems.[86]

Assuming otherwise and preferring a neutral position in these debates because of "culture," "tradition," or any other human priority compromises the OHI's mission. The OHI espouses the view that health is an interdependent variable such that medical and allied health professionals must grapple with the multispecies realities of our globalized world. An integrated vision such as the OHI's cannot remain indifferent to industries that rely on animal instrumentalization, objectification, and fragmentation. It is not enough to promote collaborations between veterinarians and physicians or seek to break down disciplinary boundaries in the pursuit of best-ensuring human, animal, and Planetary Health outcomes. It is also insufficient to claim that the integrated vision is only concerned with animals at the population level and not with individual animal health or wellbeing.

After all, it is the status quo of instrumentalizing animals for humans or corporate purposes that is a symptom of the greater disease of anthropocentrism and human exceptionalism that OHI is attempting to combat. When animals are denied individuality or further fragmented into body parts for human consumption, it is not possible to regard them as multispecies partners. Such consumption requires their active material subordination and reinforces a larger discursive apparatus that classifies animals as property, constricting the global solutions that seem reasonable and achievable.[87] Indeed, the sharp species-based compartmentalization of health-based professional designations that OHI is seeking to integrate did not emerge from a vacuum. The demarcations between health professionals for humans and a single health profession for all other animal species only make sense from within a human exceptionalist mindset. The mindset infuses the professional ethos of physicians as well as veterinarians.[88] If the OHI is to surmount an entrenched species divide and move towards an integrated and entwined more-than-one-health model, the belief that humans are superior to all other animals must disintegrate.

Conclusion

Even if present-day uses of animals did not contribute to the accelerated rate of global climate change, pandemics, and the incidence of zoonotic diseases in humans, the treatment animals endure through being put to human use or affected by anthropogenic activities still harms animals and the environment on an enormous and devastating scale, particularly in animal-based agriculture, aquaculture, and fishing. Any health-focused organization like the OHI that "seeks to promote, improve, and defend the health and well-being of all species" needs to interrogate the anthropogenic practices that compromise the health and wellbeing of animals. A more-than-one-health vision is in order to contest the human exceptionalism that the OHI is attempting to uproot.

Even if the departure point of this analysis is wrong, and the OHI is ultimately purely motivated by anthropocentric reasons for the integrated vision it takes, this human-focused motivation should still compel the OHI to consider an organizational stance against industries that also compromise the health of humans and the environment. A central and important first step in this process is for the OHI to promote the phasing out of animal agriculture, with the introduction of related supports for those entering new livelihoods to avoid food insecurity or financial precarity during a global transition to plant-based food systems. As long as animals are viewed as consumable resources rather than living beings deserving of their own rights or similarly robust protections, the mission of the OHI and similar initiatives remains compromised and the attainment of their objectives elusive.[89]

Notes

1 "Mission Statement," One Health Initiative, accessed January 3, 2022, http://dev.onehealthinitiative.com/mission-statement/.

2 Ibid.

3 Rosine Kelz, "Genome Editing Animals and the Promise of Control in a (Post-) Anthropocentric World," *Body and Society* 26, no. 1 (2019): 3–25, 4, 13; Anjana Menon, "Understanding Colonial Masculinity and Native Bodies: Rereading the Discourse of Homeopathy as a Feminist Form of Medicine," *Rupkatha Journal on Interdisciplinary Studies in Humanities* 12, no. 5 (2020): 1–5, 1–3; Kelly Struthers Montford and Chloë Taylor, "Beyond Edibility: Towards a Nonspeciesist, Decolonial Food Ontology," in *Colonialism and Animality: Anti-Colonial Perspectives in Critical Animal Studies*, eds. Kelly Struthers Montford and Chloë Taylor (Abingdon: Routledge, 2020), 129–156, 144–145.

4 For the United States, see Stephanie Marek Muller, "Zombification, Social Death, and the Slaughterhouse: U.S. Industrial Practices of Livestock Slaughter," *American Studies* 57, no. 3 (2018): 81–101, 82; David Arthur Cleveland et al., "How Many Chickens Does It Take to Make an Egg? Animal Welfare and Environmental Benefits of Replacing Eggs with Plant Foods at the University of California, and Beyond," *Agriculture and Human Values* 38, no. 1 (2021): 157–174, 158. For global figures, see Khurram Muaz et al., "Antibiotic Residues in Chicken Meat: Global Prevalence, Threats, and Decontamination Strategies: A Review," *Journal of Food Protection* 81, no. 4 (2018): 619–627, 619.

5 Troy Vettese, Becca Franks, and Jennifer Jacquet, "The Great Fish Pain Debate: What Happens When Scientists Get Hooked on a Question that Could Be Argued Forever?," *Issues in Science and Technology* 36, no. 4 (2020): 49–53, 53.

6 Anka Trajkovska Petkoska and Anita Trajkovska-Broach, "Mediterranean Diet: A Nutrient-Packed Diet and a Healthy Lifestyle for a Sustainable World," *Journal of the Science of Food and Agriculture* 101, no. 7 (2021): 2627–2633, 2627–2628; Adriano Profeta et al., "Preferences of German Consumers for Meat Products Blended with Plant-Based Proteins," *Sustainability* 13, no. 2 (2021): 1–17, 1–2. See also Chris Walzer and Laura Kahn in this volume.

7 Dinesh Wadiwel, *The War Against Animals* (Lieden: Brill/Rodopi, 2015), 5–6.

8 Erika Cudworth, "Breeding and Rearing Farmed Animals," in *The Palgrave International Handbook of Animal Abuse Studies*, eds. Jennifer Maher, Harriet Pierpoint, and Piers Beirne (London: Palgrave Macmillan, 2017), 159–177.
9 Muller, "Zombification, Social Death, and the Slaughterhouse," 92.
10 For reasons of space, I limit my focus to animal agriculture and to its commercial practices. This does not mean that other industries or non-commercial or subsistence killing of animals through hunting, fishing or agriculture should remain unquestioned. For more on this question regarding subsistence livelihoods and the ethics of killing animals, see Montford and Taylor, "Beyond Edibility."
11 Public supports will be necessary to ensure adequate standards for human lives and livelihoods and retraining as these practices phase out and systems transform.
12 Melanie Rock, Chris Degeling, and Cindy Adams, "From More-than-Human Solidarity to Multi-Species Biographical Value: Insights from a Veterinary School about Ethical Dilemmas in One Health Promotion," *Sociology of Health & Illness* 42, no. 4 (2020): 789–808, 791–792.
13 Joseph Poore and Thomas Nemecek, "Reducing Food's Environmental Impacts Through Producers and Consumers," *Science* 360, no. 6392 (2018): 987–992, 987.
14 David Wiebers and Valery Feigin, "What the COVID-19 Crisis Is Telling Humanity," *Neuroepidemiology* 54, no. 4 (2020): 283–286, 284.
15 Walter Willett et al., "Food in the Anthropocene: The EAT–*Lancet* Commission on Healthy Diets from Sustainable Food Systems," *The Lancet* 393, no. 10170 (2019): 447–492, 459; "The Eat-Lancet Commission on Food, Planet, Health," EAT, accessed May 20, 2022; Aysha Akhtar et al., "Health Professionals' Roles in Animal Agriculture, Climate Change, and Human Health," *American Journal of Preventive Medicine* 36, no. 2 (2009): 182–187, 183.
16 Akhtar et al., "Health Professionals' Roles," 183–184; Profeta et al., "Preferences of German Consumers," 1, 4; Laura Fernandez, "Using Images of Farmed Animals in Environmental Advocacy: An Antispeciesist, Strategic Visual Communication Proposal," *American Behavioral Scientist* 63, no. 8 (2019): 1137–1155, 1138; Poore and Nemecek, "Reducing Food's Environmental Impacts," 4–5; Federico Regaldo, "Who Is Going to Pay for Causing Pandemics?," *Global Jurist* 21, no. 1 (2021): 73–100, 87.
17 Louis-Etienne Pigeon and Lyne Létourneau, "The Leading Canadian NGOs' Discourse on Fish Farming: From Ecocentric Intuitions to Biocentric Solutions," *Journal of Agricultural and Environmental Ethics* 27, no. 5 (2014): 767–785, 768–769; Nesar Ahmed, Shirley Thompson, and Marion Glaser, "Global Aquaculture Productivity, Environmental Sustainability, and Climate Change Adaptability," *Environmental Management* 63, no. 2 (2019): 159–172, 160–165.
18 Paula Arcari, "Normalised, Human-Centric Discourses of Meat and Animals in Climate Change, Sustainability and Food Security Literature," *Agriculture and Human Values* 34, no. 1 (2017): 69–86, 70; George Monbiot, "Seaspiracy Shows Why We Must Treat Fish Not as Seafood, but as Wildlife," *The Guardian*, April 7, 2021; Charlotte Blattner, *Protecting Animals Within and Across Borders: Extraterritorial Jurisdiction and the Challenges of Globalization* (New York: Oxford University Press, 2019), 9–10. Documentaries covering both silences are *Seaspiracy* and *Cowspiracy*. More recently, the first "concrete" steps by the new U.S. Biden administration to target methane emissions, for example, did not target animal agriculture, but natural gas and oil extraction, and refrigeration and air-conditioning units. See Lisa Friedman, "Washington Moves to Cut Greenhouse Gases," *New York Times*, May 5, 2021.
19 Fernandez, "Images of Farmed Animals," 1138.
20 Willett et al., "Food in the Anthropocene," 455–456; Trajkovska Petkoska and Trajkovska-Broach, "Mediterranean Diet," 2627; Profeta et al., "Preferences of German Consumers," 1, 3.

21 Melanie Joy, *Why We Love Dogs, Eat Pigs, and Wear Cows: An Introduction to Carnism* (San Fransisco: Conari Press, 2009), 96–97; Cudworth, "Breeding and Rearing Farmed Animals," 172.
22 Matthew Adams, "Communicating Vegan Utopias: The Counterfactual Construction of Human-Animal Futures," *Environmental Communication* 16, no. 1 (2022): 125–138, 126–127, 130, 132.
23 Marina G. B. Aleixo et al., "Using Twitter® as Source of Information for Dietary Market Research: A Study on Veganism and Plant-Based Diets," *International Journal of Food Science & Technology* 56, no. 1 (2021): 61–68, 66; Richard Buttny and Etsuko Kinefuchi, "Vegans' Problem Stories: Negotiating Vegan Identity in Dealing with Omnivores," *Discourse and Society* 31, no. 6 (2020): 565–583, 566, 579–580; Adams, "Communicating Vegan Utopias," 132.
24 See, for example, "Global Warming of 1.5°C," The Intergovernmental Panel on Climate Change, accessed May 20, 2022, https://www.ipcc.ch/sr15/; Willett et al., "Food in the Anthropocene."
25 Arcari, "Normalised, Human-Centric Discourses," 70, 82; George Monbiot, "The IPCC Land and Climate Report Fails Miserably – It Vastly Underestimates the True Carbon Cost of Our Meat and Dairy Consumption, Which Is Astonishing. We Can't Keep Eating Like This," *The Guardian*, August 8, 2019; Paula Arcari, Fiona Probyn-Rapsey, and Haley Singer, "Where Species Don't Meet: Invisibilized Animals, Urban Nature and City Limits," *Environment and Planning E: Nature and Space* 4, no. 3 (2021): 940–965, 948.
26 Quirin Schiermeier, "Eat Less Meat: UN Climate-Change Report Calls for Change to Human Diet," *Nature* 572, no. 7769 (2019): 291–292, 291. The IPCC Reports continue to instigate heated public discussion on social media with some animal advocates criticizing the failure of the IPCC to promote fully plant-based diets, i.e., veganism. Mary Sanford et al., "Controversy Around Climate Change Reports: A Case Study of Twitter Responses to the 2019 IPCC Report on Land," *Climatic Change* 167 (2021): 1–25, 13–14.
27 Willett et al., "Food in the Anthropocene," 447–449.
28 Ibid., 447.
29 Ibid.
30 Ibid.
31 Ibid.
32 Andrew Anthony, "How Diet Became the Latest Front in the Culture Wars," *The Observer*, March 17, 2019; Stephen B. Smith, "EAT-*Lancet* Commission Report: Does it Reflect Reality?," *Beef*, April 25, 2019; Philip Clarke, "New Anti-Meat Report Lambasted by AHDB," *Farmers Weekly*, January 17, 2019.
33 Francisco J. Zagmutt, Jane G. Pouzou, and Solenne Costard, "The EAT–*Lancet* Commission: A Flawed Approach?," *The Lancet* 394, no. 10204 (2019): 1140–1141.
34 Walter Willett, Johan Rockström, and Brent Loken, "The EAT–Lancet Commission: A Flawed Approach? – Authors' Reply," *The Lancet* 394, no. 10204 (2019): 1141–1142.
35 Allison Aubrey, "World Health Organization Report Links Red, Processed Meats to Cancer," *National Public Radio*, October 26, 2015.
36 "Plant-Based Diets and Their Impact on Health, Sustainability and the Environment: A Review of the Evidence," WHO European Office for the Prevention and Control of Non-communicable Diseases, 2021, Licence: CC BY-NC-SA 3.0 IGO.
37 "COP26 Special Report on Climate Change and Health: The Health Argument for Climate Action," World Health Organization, 2021, 11. Licence: CC BY-NC-SA 3.0 IGO.
38 Ibid., 46.
39 Ibid., 47.
40 Ibid.
41 Ibid.
42 Arcari, "Normalised, Human-Centric Discourses," 71, 77.

43 Ibid.
44 WHO European Office for the Prevention and Control of Non-communicable Diseases, "Plant-Based Diets"; "Sustainable Procurement of Health-Care Commodities," World Health Organization, 2022, https://www.who.int/teams/environment-climate-change-and-health/air-quality-and-health/sectoral-interventions/health-care-activities/strategies.
45 WHO European Office for the Prevention and Control of Non-communicable Diseases, "Plant-Based Diets," 6 (emphasis added).
46 "Healthy Diet Fact Sheet," World Health Organization, April 29, 2020, https://www.who.int/news-room/fact-sheets/detail/healthy-diet.
47 Ibid.
48 David et al., "Pandemics in the Age of the Anthropocene," 1142.
49 Ibid., 1150; Vural Özdemir, "'One Nature': A New Vocabulary and Frame for Governance Innovation in Post-COVID-19 Planetary Health," *OMICS* 24, no. 11 (2020): 645–648, 645.
50 Leo Horrigan, Robert S. Lawrence, and Polly Walker, "How Sustainable Agriculture Can Address the Environmental and Human Health Harms of Industrial Agriculture," *Environmental Health Perspectives* 110, no. 5 (2002): 445–456, 445.
51 David et al., "Pandemics in the Age of the Anthropocene," 1143.
52 Wiebers and Feigin, "COVID-19 Crisis," 284; "IPBES Pandemics Report: Escaping the Era of Pandemics," Workshop Report on Biodiversity and Pandemics of the Intergovernmental Platform on Biodiversity and Ecosystem Services, Intergovernmental Platform on Biodiversity and Ecosystem Services, October 29, 2020, 5–6, 14, 50, https://ipbes.net/sites/default/files/2020-12/IPBES%20Workshop%20on%20Biodiversity%20and%20Pandemics%20Report_0.pdf, 5–6, 14, 50; Harman S. Sandhu et al., "Pandemic Prevention and Unsustainable Animal-Based Consumption," *Bulletin of the World Health Organization* 99, no. 8 (2021): 603–605, 603.
53 Gala Argent, "The Human-Animal Studies Report," Animals & Society Institute, October 29, 2020, https://www.animalsandsociety.org/human-animal-studies/the-october-2020-human-animal-studies-report-investigates-the-nexus-between-climate-change-species-migrations-and-pandemic-emergence/; Aysha Akhtar, "Nonhuman Animals, Public Health, and Ethics: A First Step, but...," *Journal of Applied Animal Welfare Science* 20, no. 1 (2017): 106–107, 107.
54 "Actionables for a Healthy Recovery from COVID-19," World Health Organization, May 26, 2020.
55 Lee Skerratt, "Wildlife Health Systems," *Animal Sentience* 30, no. 18 (2020): 1–5, 1; Ben Jones, "Eating Meat and Not Vaccinating: In Defense of the Analogy," *Bioethics* 35, no. 2 (2021): 135–142, 136.
56 IPBES, "Escaping the Era of Pandemics," 43–45, 50–51.
57 Ibid., 19.
58 Gabriele Volpato et al., "Baby Pangolins on My Plate: Possible Lessons to Learn from the COVID-19 Pandemic," *Journal of Ethnobiology and Ethnomedicine* 16, no. 1 (2020): 1–12, 5; IPBES, "Escaping the Era of Pandemics," 25–29; Bhaskara L. Reddy and Milton H. Saier, Jr., "The Causal Relationship Between Eating Animals and Viral Epidemics," *Microbial Physiology* 30, nos. 1–6 (2020): 2–8, 2.
59 Wiebers and Feigin, "COVID-19 Crisis," 284.
60 Ibid., 284.
61 Akhtar et al., "Health Professionals' Roles," 183.
62 Alex Thornton, "This Is How Many Animals We Eat Each Year," *World Economic Forum,* February 8, 2019.
63 Adams, "Communicating Vegan Utopias."
64 Wiebers and Feigin, "COVID-19 Crisis," 284.
65 Jones, "Eating Meat and Not Vaccinating," 139.
66 Wiebers and Feigin, "COVID-19 Crisis," 284–285.
67 Horrigan et al., "Sustainable Agriculture," 448.

68 IPBES, "Escaping the Era of Pandemics," 2 (emphasis added).
69 Ibid., 5.
70 Michaeleen Doucleff, "WHO Points to Wildlife Farms in Southern China as Likely Source of Pandemic," NPR, March 15, 2021.
71 "Actionables for a Healthy Recovery from COVID-19," World Health Organization, July 23, 2020.
72 Ibid.
73 Ibid.
74 Monbiot, "IPCC Land and Climate Report"; Poore and Nemecek, "Reducing Food's Environmental Impacts," 3–4; Andrew McGregor and Donna Houston, "Cattle in the Anthropocene: Four Propositions," *Transactions of the Institute of British Geographers* 43, no. 1 (2018): 3–16, 10.
75 Borkfelt et al., "Closer to Nature?," 1055–1056.
76 Cudworth, "Breeding and Rearing Farmed Animals," 166–168; Jessica Eisen, "Milked: Nature, Necessity, and American Law," *Berkeley Journal of Gender, Law & Justice* 34, no. 1 (2019): 71–115, 92–95.
77 One Health Initiative, "Mission Statement."
78 Cudworth, "Breeding and Rearing Farmed Animals," 160.
79 Arcari, "Normalised, Human-Centric Discourses," 73, 76, 79.
80 See the arguments raised in Mike Archer, "Ordering the Vegetarian Meal? There's More Blood on Your Hands," *The Conversation,* December 15, 2011; and Leigh Huggins, "What We're Getting Wrong About Ethical Eating," *Greatest,* January 7, 2020.
81 Anthony, "How Diet Became the Latest Front in the Culture Wars."
82 McGregor and Houston, "Cattle in the Anthropocene," 4; Monbiot, "IPCC Land and Climate Report"; Cudworth, "Breeding and Rearing Farmed Animals," 169–171; Eisen, "Milked," 75–83.
83 Montford and Taylor, "Beyond Edibility," 129–130; Deckha, "Veganism, Dairy, and Decolonization," 4–6; Maneesha Deckha, "Postcolonial," in *Critical Terms for Animal Studies*, ed. Lori Gruen (Chicago: University of Chicago Press, 2018), 160–174, 162–173.
84 Cudworth, "Breeding and Rearing Farmed Animals," 161–162, 168.
85 Sandhu et al., "Pandemic Prevention and Unsustainable Animal-Based Consumption," 604.
86 Willett et al., "Food in the Anthropocene," 480–481, 485.
87 Arcari, "Normalised, Human-Centric Discourses," 82.
88 Jakob Zinsstag et al., "Mainstreaming One Health," *Ecohealth* 9, no. 2 (2012): 107–110, 107.
89 The Planetary Health vision launched by the U.S.-based Rockefeller Foundation and the British medical journal *The Lancet* comes to mind as one such model. See Pierre-Marie David, Nicolas Le Dévédec, and Anouck Alary, "Pandemics in the Age of the Anthropocene: Is 'Planetary Health' the Answer?," *Global Public Health* 16, nos. 8–9 (2021): 1141–1154, 1145–1146.

राम
ॐ
राम

9

CAN CAMARADERIE HELP US DO BETTER THAN COMPASSION AND LOVE FOR NONHUMAN HEALTH?

Some Musings on One Health Inspired by the Case of Rabies in India

Deborah Nadal

Introduction

The COVID-19 pandemic has kept people's minds busy on many fronts, from physical to mental health, from the passing away of loved ones to the loss of jobs, from the increase of existing inequalities to the creation of new social fractures. For some, existential thoughts have also surfaced, questioning the role, the place, and the very being of human and nonhuman animals (hereafter, animals) in relation to SARS-CoV-2 and much beyond it. Are bats to be feared? Are pangolins a new concern? Are minks to be precautionarily "dealt with"? Are cats and dogs to be kept at a distance?[1] And regarding ourselves as humans: considering the features of the current Anthropocene era, would it not be better for everybody if the human race became extinct? Look at the photos (original or fake, it does not matter) of wild animals enjoying quiet city streets during the first lockdown—does nature not flourish as soon as we step back?

For those involved in One Health, all these questions essentially merge into one. How do we make this approach, which is by definition inclusive and cohesive, work and be as effective as possible when many people find it hard to accept that human health intimately depends on the health of animals and that of the environment? This question is not a new one. Finding and stressing previously unseen connections at the human–animal–environmental interface, and safeguarding and improving health for all the involved parties has been the goal of the experts of One Health at least since this vision of health received its official name, in 2003.[2] Yet the COVID-19 pandemic, and the year 2021 in

FIGURE 9.1 A free-roaming dog rests on the steps of a traditional building in Jodhpur, Rajasthan, Western India. This dog was vaccinated against rabies to protect himself and the people and animals around him from this lethal disease. Photo by author, July 2012.

DOI: 10.4324/9781003294085-13

particular, marked a turning point for One Health. At the G7 and G20 meetings held in 2021, this agenda turned into a public health priority policy across the world.[3]

Without underestimating this key achievement, my experience with rabies in India and an eye-opening paper by Chris Degeling et al. curb my enthusiasm.[4] Even though One Health is progressing quickly toward being recognized as a global policy, it is—paradoxically—still making slow progress in having its core guiding values and practical principles properly defined. Chris Degeling et al. ascribe the ultimate effectiveness of One Health policies to their alignment with, or the modification of, public values. They urge policymakers to prepare for emerging diseases (i.e., infectious diseases that have newly appeared in a population or have existed but are rapidly increasing or expanding) with a set of principles that integrate ethics into decision-making. By doing so, people will be prepared for the kinds of choices that will have to be made when the disease emerges. By that time, at least in democratic countries, conflictual beliefs about "what is valuable [and why], what is to be protected [and why] and, ultimately, what is dispensable [and why]" will have already been addressed and negotiated, and some consensus over what is in the public interest will have eventually been reached.[5] On a local, regional, and national level, each community will have come up with their own set of values, possibly stable but obviously not immutable. Although some common ground across different sets is desirable, at least to coordinate action when One Health issues are of global reach, principles, beliefs, and ideals are specific to local histories, religion- and culture-based worldviews, political and economic dynamics, and much more.[6]

I support the recommendation of Chris Degeling et al. and I argue that it should apply equally to endemic diseases (i.e., diseases, like rabies, that are constantly present in a geographic area). The interest in One Health—especially that of the general public and policymakers—has been recently fueled by global concerns relating to emerging diseases with pandemic potential. Nevertheless, early advocates of One Health became aware of the importance of this approach when working on endemic diseases, like rabies, that unfortunately are still hiding in plain sight because they affect socially and economically marginalized multispecies communities.[7]

In this chapter, I propose interspecies camaraderie as an addition to the list of principles that we, as humans presented with the urgent responsibility of deciding what to do with human and more-than-human health, can use to guide our One Health actions. I came up with the concept of interspecies camaraderie when writing my book on the coexistence of humans, free-roaming animals (specifically, dogs, cows, and rhesus macaques), and the rabies virus in the public spaces of Delhi and Jaipur, in India.[8] This multispecies ethnography project involved interviewing a variegated range of humans who were, each in their own way, engaged with animals; shadowing animals around the streets and observing their relationship with others; following debates on human–animal cohabitation in media; and volunteering at three animal shelters.

During my work on dogs, I struggled with two concepts: compassion (i.e., a sympathetic pity and concern for the sufferings or misfortunes of others) and love (i.e., a strong feeling of affection and attachment toward someone). Intuitively appealing—love and compassion are usually considered two positive feelings that help make the world a better place—during my research, they actually proved thorny both for myself and for the dog–human assemblage I observed. As someone who always considered herself an animal lover, I found myself at an impasse when, during conversations with other animal lovers, I felt "stuck" in this role and limited by it. Then, when I started data analysis, I again noticed that the ubiquitous and seemingly innocuous presence of compassion and love in dog-related narratives was somehow limiting my ability to provide new, alternative, practical input. Interspecies camaraderie helped me overcome this dual impasse.

In this chapter, I will firstly introduce the aspects of rabies and the human–dog relationship that are relevant to this discussion. Then, I will explain why I find compassion and love tricky, rather impractical, and potentially counterproductive within the context of One Health. I do not consider these feelings to be challenging *per se*, but they risk becoming so if we choose them as values or tenets to guide our One Health actions. Finally, I will describe what I mean by interspecies camaraderie and how it differentiates from compassion and love. Through the example of rabies, I will demonstrate how camaraderie can contribute, as a guiding tenet, to One Health in general, and to the current debate on care and multispecies justice in particular.[9] This chapter also provides a chance to start discussing the still unexplored issue of emotions in One Health.

Rabies

Rabies is the oldest zoonotic disease (i.e., a disease that infects both animals and humans) known to humankind. Despite being vaccine-preventable since 1885, it still kills an estimated 59,000 people per year, mainly in Asia and Africa, and an uncalculated number of animals.[10] After the onset of its brutal symptoms, the disease is incurable and lethal in almost 100 percent of the cases. Dogs, through their bites and scratches, are the main carriers of rabies to humans (in 99 percent of the cases) and to pet, farm, and wild animals.[11]

Zoonotic diseases have historically been central to One Health[12] and dog-mediated rabies is often considered a model disease for this approach,[13] because the most cost-effective and sustainable way to reduce human and animal rabies cases is by vaccinating dogs. Post-exposure prophylaxis in humans can complement mass dog vaccination, but it often remains inaccessible for the poorest and most remote communities. In any case, it does not apply to animal victims. Hence, the current, ambitious goal of the World Health Organization, the World Organisation for Animal Health, and the Food and Agriculture Organization is to vaccinate 70 percent of a susceptible dog population in order to eliminate dog-mediated human rabies deaths.[14] When and where mass dog vaccination is not implemented, rabies is usually addressed by indiscriminately killing free-roaming dogs.[15] Yet this strategy is often rejected by communities for ethical reasons and, from a scientific perspective, it is not only ineffective—because rabies transmission does not depend on dog population density[16]—but also counter-productive. In fact, dog owners often hide or move their (unvaccinated) dogs to other areas or acquire new (unvaccinated) dogs to replace those who have been killed, and this movement of dogs contributes to the spread of rabies. Moreover, without clear identification of vaccinated dogs, culling of free-roaming dogs can include vaccinated dogs, reducing population immunity.[17]

The year 2021 was important for One Health and rabies in India as well. Whereas animal rabies has long been notifiable, human rabies was made notifiable only in fall 2021, when the country also announced the imminent implementation of the National Action Plan for Rabies Elimination.[18] This plan has been formally presented as a One Health program, where mass dog vaccination is at the center of a three-pronged strategy that also includes post-exposure prophylaxis for bitten individuals and environment-based interventions such as improved waste management. Interestingly, the plan mentions "free-roaming dogs" (or "stray dogs" in the common language, in India as well as worldwide), but also "street dogs." The latter is an expression that, in India, has a specific connotation since the Animal Birth Control (Dogs) Rules (hereafter, ABC Rules) coined it in 2001.[19] Unlike in many Western countries where stray dogs are considered illegal dwellers in public spaces, India's street dogs have been granted by the ABC Rules the right to live on the street. As the street is acknowledged to be their home, they are not homeless and, from a legal perspective, the presence or absence of an owner, or simply a caretaker, is irrelevant.

India's rabies control strategy was built on the legal concept of street dogs and to their rights as such. The main focus of the ABC Rules, as their name suggests, is dog population control, with anti-rabies vaccination (ARV) being an addition. Practically, dogs must be humanely caught by properly trained dog-catchers; safely transported to an ABC-ARV center; housed; identified through a coded collar or a V-shaped notch in the left ear; spayed or neutered; given adequate post-surgical care; inoculated against rabies; and taken back to the exact location from which they were collected. When I carried out my fieldwork in 2012, 2013, and 2015, the ABC-ARV program was being executed by local municipalities and animal welfare organizations in several Indian cities, including Delhi and Jaipur.

This program was often discussed in local and national newspapers and the general feeling was of high—and, according to most of my informants, growing—social disagreement. The bones of contention, so to speak, were numerous, and there were also several different perspectives on each of them. A similar scenario has also been observed by Kritika Srinivasan et al. in the city of Chennai, India.[20] First of all, rabies is generally not considered a priority concern either by (most) policymakers and lawmakers, or by (most) animal welfare organizations, or by society at large.[21] Many other dog-related issues attract people's interest and fuel debate, such as dogs barking at night, chasing motorbikes, and attacking people, as well as dogs suffering from starvation, being injured by cars, and falling victim to intentional acts of violence.

Then, in practical terms, there are very specific attitudes toward the ABC-ARV program. Dog-catchers, who have the crucial role of catching and releasing street dogs, describe neighborhoods where people are eager to help them during the dog-catching and dog-releasing phases. But they also happen to work in neighborhoods where people obstruct them in the catching phase, because they do not want dogs to be taken away, mindful of past incidents where animals ended up mass-poisoned with strychnine. Finally, there are neighborhoods where protests arise when dogs are brought back after sterilization and vaccination. For their part, the animal welfare organizations that work in the ABC-ARV program often lament that neither the government nor the public seem to appreciate their work; the former by under-funding this program and the latter by showing little interest in the cause of street dogs. Meanwhile, accusations of cruelty and carelessness are thrown at government-owned ABC-ARV centers and rumors about unethical conduct circulate around some animal welfare organizations.

Furthermore, petitions and campaigns are organized by citizens, but for two very different reasons. On the one hand, there are people who advocate for the physical elimination of street dogs, seen as a mounting menace, both for their number and for their behavior.[22] On the other hand, there are those who would like to do more (e.g., mass food distributions) for animals considered the helpless, voiceless, and innocent victims of people's callousness, anthropocentrism, and disregard for nature.[23] Finally, a third approach is voiced mainly by scientists, who are skeptical about the whole ABC-ARV idea truly being the best that can be done for free-roaming dogs (both for their welfare and their rights) and their relationship with humans.[24] Their skepticism is usually due to two different reasons. First, to be effective in the long-term and all over urban and rural India (which has an estimated population of 60 million dogs[25]), the ABC-ARV program requires massive funding, infrastructure, and human resources. Hence, its overall feasibility is being questioned by scientists[26] (and, quite paradoxically, even the very animal welfare organizations that are implementing it[27]). Second, ABC-ARV advocates maintain that this program, and its overall purpose of ensuring street dogs a happy and healthy life, can be easily supported by all those people who provide food, water, or shelter to street dogs, for example through the abovementioned mass food distributions. Those skeptical toward people's real and effective interest in street dogs point out the fact, also noticed by Kritika Srinivasan,[28] that the acts of care toward street dogs from the public (e.g., food, water, or shelter provision) are mainly unstructured and irregular, and hence unlikely to effectively sustain the health of dogs over time.

Additionally, the religious background behind the ABC-ARV program plays a role in the same. In Hinduism and Islam, the two main religions of India, dogs hold an ambivalent position that exemplifies the notorious, transcultural liminality of dogs as living on the boundary of the wild and the domestic, the tame and the feral.[29] In these two religions, free-roaming dogs are often associated with physical and symbolic dirt and impurity, disease (which they both spread to humans and also symbolically take away), and death (which they sense beforehand, announce, or accompany people to and beyond, in their journey through the afterworld).[30]

The scenario is evidently complex. No matter how well designed the brand-new, One Health-inspired National Action Plan for Rabies Elimination is, and how well implemented it will be, I argue that its realization and success depend first and foremost on people's attitude toward dogs and their health. On the streets of India, dogs are such a ubiquitous presence that the city's human dwellers have an intimate, everyday relationship with them. More often than not, this relationship is experienced, described, judged, and eventually managed around two tenets: compassion and love.

Compassion

The term "compassion" originates from the Latin noun "compassion," from the verb "compăti," "to feel sorry for." Compassion is a cardinal principle in many South Asian religions and abundant literature exists on its application toward other living beings.[31] Gilles Tarabout describes the use of this notion in actual legal cases on various animal-related topics discussed in Indian law courts, including human coexistence with street dogs.[32] Most of the decisions in these cases are based on the court's interpretation of Article 51A(g) of the Constitution of India: "It shall be the duty of every citizen of India to protect and improve the natural environment including forests, lakes, rivers and wild life and to have compassion for living creatures." Compassion is not mentioned elsewhere in the Constitution and is also not defined in it. The courts thus provided a working definition for this term.

In the case of Animal Welfare Board of India vs. A. Nagaraja and others from 2014, the Indian Supreme Court defined compassion as "concern for suffering, sympathy, kindliness."[33] According to the High Court of Himachal Pradesh, a northern state in India, "the core of religion, based upon spiritual values, which the Vedas, Upanishads and Puranas [ancient Hindu scriptures] were said to reveal to mankind, seem to be 'love others, serve others, help ever, hurt never.'"[34] In a judgement of the High Court of Gujarat, "Compassion … means 'a strong feeling of sympathy for those who are suffering and a desire to help them.' … Compassion is suggestive of sentiments, a soft feeling, emotions arising out of sympathy, pity and kindness."[35] Compassion hence applies to the "'weak and the meek' … and those subjected to cruel treatments."[36] Compassion for animals is elicited not only because of intentional human brutality and consequent animal suffering,

> but also because they [animals] are helpless and voiceless. … "It is further pleaded that human being has [*sic*] to show compassion to all animals

including stray dogs who are unable to protect themselves [and] have to be protected by the Society and Courts."[37]

Another application of compassion, this time as the avoidance of unnecessary pain, is animal farming and animal research (see Maneesha Deckha's chapter, this volume).

"Karuna" and "krupa," the Hindi equivalents for compassion, appear rather commonly in the names of Indian animal welfare organizations. The logo of the Animal Welfare Board of India includes the phrase "Compassion towards all living creatures." According to the first president of the Animal Welfare Board of India, "by instituting these [animal protection] reforms, India is going to set a unique example to the whole world. No other country in the world could so easily have laws which put mercy above the fruits of cruelty and have wide support from the people."[38] Nowadays, several countries show a constitutional concern for the interests of animals, but the Indian Constitution is the only one in the world that directly references compassion in relation to animals.[39]

Constitutional texts are only one of the many to inscribe compassion toward animals. Compassionate conservation (i.e., a discipline that combines conservation and animal welfare, guided by the funding principles of Do no harm, Individuals matter, Inclusivity, Peaceful coexistence)[40] is another field where this approach is practiced and prescribed. The debate between compassionate conservationists and conservation biologists is useful here.[41] Very briefly, and taking the case of invasive species as an example, compassionate conservationists argue that it is morally wrong to kill an animal in the name of conservation, whereas conservation biologists argue that it is morally wrong to allow (native) species to go extinct, especially if this is caused by human actions such as the movement of animals to new places. This debate provides interesting inputs for the argument I am making in this chapter, that compassion and love are fundamental to relationships among living beings, but, if we look at them as potential guiding values for One Health, they present several limitations. Many of these can be noticed when looking at the coexistence of people, street dogs, and other animals in the streets of India.

First, compassion is triggered by suffering. Although most One Health issues become such because of their negative impact on health and wellbeing, not all of them directly involve an amount of suffering "big" or "touching" enough to activate an intense feeling of sympathy and desire to help. Even though rabies has devastating symptoms and invariably leads to death, its acute phase—the only one when rabies can be "seen," but no longer prevented—usually lasts a week or so. Compared to other diseases that afflict (human or animal) victims for prolonged periods of time, rabies is more transient and does not allow significant help.

Next, compassion mainly aims at alleviating suffering, not at preventing it or sustaining health. According to the World Health Organization, health "is a state of complete physical, mental, and social wellbeing and not merely the absence of disease or infirmity."[42] In the past two years, critics of the language of war that was used during the pandemic pointed to the need for shifting the focus from merely fighting against an "invisible enemy" to developing new forms of solidarity and coexistence.[43] Even though One Health has been conceived to reduce the shared burden of disease, vulnerability, and risk, its full potential is arguably expressed in the prevention or early mitigation of health threats and the creation of a healthier world. The suffering caused by terminal rabies can be reduced only through intense palliative care (in humans, only in well-equipped hospitals) or humane euthanasia (in animals[44]). In January 2022, an animal welfare organization filed a Public Interest Litigation questioning a judgement passed by the Supreme Court of India in 2018.[45] The judgement had allowed passive euthanasia in cases of particular human health conditions, but not for rabid (human) patients. This petition sparked the debate on the right to die with dignity. Meanwhile, this right is denied also to clinically confirmed rabid dogs who, according to the ABC Rules, cannot be euthanized.

Moreover, compassion mainly applies to sentient beings who, by definition, can perceive positive and negative feelings. This is a highly limiting factor for a One Health approach that aims at including not only animal species whose sentience has not yet been ascertained, but plants as well (see Emily Reisman's chapter in this book).[46] Additionally, only a small proportion of the world's countries have formally acknowledged animal sentience.[47] Thus, if action to protect animal health depended on compassion, and compassion only involved sentient animals, sharp differences and even inequalities would be present across countries. For a disease like rabies, where mass dog vaccination programs should ideally be cross-border, this would be deleterious.

Additionally, compassion makes the sufferer dependent on the feelings of somebody else, arguably increasing their sense of vulnerability and deprivation of their right to health as individuals. This seriously undermines the goal of One Health of reducing inequality and injustice, among humans and toward other living beings.[48] Free-roaming dogs in the Global South should be granted the same right to rabies vaccination that pet dogs in Western countries are regularly granted, even more so considering their higher exposure to rabies.

Finally, for many, compassion is linked to religion. Whereas spiritual health is an important dimension of health, and thus of One Health, this may create a prejudice toward atheists or people who follow a religion in which compassion is not described as a cardinal tenet. Additionally, religious beliefs have an impact on animals. Free-roaming dogs are usually stigmatized in orthodox Hinduism[49] and this can easily result in a disregard for their health and their life. Across rabies-endemic countries, there is evidence of religion being an important, yet understudied, determinant of dog-keeping practices.[50] In India, this situation becomes paradoxical if we consider that despised free-roaming dogs are the main source of rabies infection for cows, animals that are—at least, in theory—highly valued in Hinduism.

Love

Whenever my four-year-old daughter asks me what my job is, I tell her that I study to help Indian dogs. "Why?" she asks. "Because I love them" has always worked well for us, as she dotes on our dog and I have always found it very easy to connect with animals. Yet during my fieldwork, I felt increasingly uncomfortable whenever my interlocutors asked me whether I am a dog lover. In what follows, I describe the role played by love (and hate) within the Indian debate on street dogs, dog population management, and rabies control.

This debate is most visible in traditional and social media, where the division of the public between "dog lovers" and "dog haters" is evident, and so the risk is higher for this polarization of emotions to undermine the chances of coming up with workable solutions. Interestingly, each side accuses the other of the same thing: blindness, fanaticism, and irrationality. Let us look at some examples.

Those who fear street dogs and consider them a threat—the dog haters—"see dog lovers as naïve in expressing blind trust in these animals, as if humans, having chosen them as their 'best friends' would automatically eliminate any risk of interspecies conflict."[51] This kind of caricature of dog lovers is well captured in the opening sentence of a newspaper article: "Her love for strays proved costly."[52] The article describes what happened to Sanjana, a young girl who encountered an unknown street dog next to her house. Used to interacting with the dogs living in her neighborhood, Sanjana approached the new dog, who bit her severely on her face and hand. According to the journalist, the dog had recently been chased out of his neighborhood and attacked by local people and other dogs. For these reasons, "The dog was not in a [*sic*] mood to take Sanjana's gesture kindly."

Some critics of the "street dog idea" accuse its supporters of being heartless elitists, who care more about dogs than they do about people. And not just people in general, but vulnerable people such as "those at the periphery of our own species"[53] (such as slum dwellers, pastoral communities, or rural poor—and their children), who are the most common victims of rabies because they are more exposed to dog bites and struggle to access post-exposure prophylaxis. Street dog lovers are criticized for being armchair activists "who dispense their opinions from a privileged position ... With their misplaced priorities, sentimentality, ideological pretensions, antidemocratic insolence, and social and economic advantages, opponents charge, these dog-loving no-kill armchair activists put human lives at risk."[54] The comments left on the website of *The Times of India* to a street dog-related article make this point clear: "You people live in well-off areas and travel in cars and won't allow these dogs to enter inside your streets, at least it [*sic*] roams in our streets" and "There are more dog lovers in usa than india [*sic*] but such a casual approach is never made. [T]hey should be in the homes of the lovers not in the street eating garbage and biting innocent people."[55] Interestingly, the term "innocent" is also used, by street dog advocates, to describe these animals.[56]

At the other end, dog lovers and animal activists passionately present their counterarguments. ABC-ARV advocates claim that the urban poor, such as slum and street dwellers, are actually those who love street dogs the most and who derive the main benefits from having them around, as dogs protect them at night, guard their belongings, and provide them with company and affection. According to a well-known animal activist in Delhi, "if in India some humanity is left, for sure it is among the poor."[57] Yet because of their illiteracy and vulnerability, the poor are somehow looked at and portrayed by the media and the "anti-stray dog lobby" as people who have no choice, informed opinion, or right to assert their love for street dogs.[58]

As happens among those who ask for an alternative to the ABC-ARV strategy, dog lovers also refer to their counterparts as an obstacle. On the website of a well-known animal welfare organization, the Colony Animal Caretaker Card (i.e., an ID card issued by the Animal Welfare Board of India to street dog feeders) is described as "useful if people complain to the police or neighbors and animal haters prove to be a nuisance." Surprisingly, factions exist even within the dog-lover community and are based on the differing amount of genuine love for dogs. A young, enthusiastic animal welfare activist I met at a puppy adoption stall did not like the idea that one might think of oneself as an animal lover by merely possessing the Colony Animal Caretaker Card. According to him, being a street dog lover has become, in some circles, fashionable to the extent that love has lost its authenticity.

Like compassion, love is a powerful and "world-moving" feeling but, as a guiding principle of One Health, it is a problematic one. The case of the complex human–street dog relationship highlights some key points, several of which are also addressed in social psychology research.

First, love is selective. Research into what Herzog's book title eloquently synthetizes as "Some We Love, Some We Hate, Some We Eat: Why It's so Hard to Think Straight About Animals," has shown that we tend to sort animals into different categories for a combination of reasons based on human psychology, cultural norms (especially culinary culture), direct experience, social pressure, and media exposure.[59] Provided that One Health's inclusivity can grow gradually, its full potential is expected to coincide with the overcoming of speciesism.[60] In the context of my study, loving animals does not necessarily mean loving dogs, and loving dogs does not necessarily mean loving street dogs. The latter circumstance is part of the Indian debate around street dogs in two ways. On the one hand, animal welfare organizations are increasingly targeting dog lovers and prospective dog owners to convince them to adopt a local street dog instead of buying a foreign breed dog.[61] On the other hand, anti-ABC-ARV dog lovers claim that loving dogs is incompatible with loving street dogs. This is not because street dogs are not worthy of love but, quite the opposite, because creating the legal category of "street dogs" causes them additional problems (e.g., exposure to road accidents, the prohibition of euthanizing them when terminally ill with rabies, etc.) and short circuits in logic. One case in point is street dog adoption. According to ABC

Rules, removing street dogs from the street is illegal (except for sterilization and vaccination) but, at the same time, their adoption is encouraged. We can agree on the fact that there are advantages and disadvantages to the physical and mental health of both pet and free-roaming dogs, and that dogs do not necessarily do better under human care.[62] Yet doubts remain as to why dog adoption is advocated by ABC-ARV supporters if the street is the best home for free-roaming dogs.[63]

Furthermore, love changes, across time and space. Both at the individual and community level, affection for and interest in animals may evolve. The current interest in bees and the boom of beekeeping were probably unfathomable a mere 20 years ago.[64] Across cultures, humans feel attached to different animal species. Even though dogs are found across virtually all human societies, they do not enjoy the same status everywhere.[65] This influences dog-keeping practices and people's care for dogs' health. In India, the States of Kerala[66] and Nagaland[67] are often accused (for different reasons: tourist beautification in Kerala and dog meat consumption in Nagaland) of hatred toward street dogs and cruel actions such as mass dog culling. One Health, for its far-reaching objectives—in time and space—is likely to be favored by a principle that is less volatile than love.

Finally, love is instinctive toward the innocent. Babies, whether human or animal, generally evoke greater empathy than adults because of their innocence and defenselessness.[68] Although this is understandable and praiseworthy, if we look at the flipside of this, some concerns arise: can animals be guilty? Guilty of what? And if they were, would they deserve less from us? Victim blaming has unfortunately long occurred in the context of human health (e.g., cancer[69], AIDS[70], and now COVID-19[71]). Extending it to animals, for example to rabid dogs, appears unreasonable and, we can argue, unethical. With rabies, historical research shows that the dogs of the poor, and thus the poor themselves, have generally been held responsible for the spread of the disease.[72] In today's India, and arguably in all rabies-endemic countries, this is problematic for One Health for two reasons. First, it gives people the false idea that pet dogs are not at risk of contracting rabies. Second, it worsens the social fracture between the rich and the poor, and increases social blaming toward the latter.

Interspecies Camaraderie

Camaraderie is very different from compassion and love. The word comes from the French "*camarade*," comrade, which in turn originates from Latin "*camara*," chamber. Camarade (or, more commonly, comrade) is who we share our room with (for example, in a boarding school dormitory, or where soldiers sleep, etc.).[73] In socialism and communism, the word "comrade" has been adopted by fellow members to address each other. Yet its first and most general meaning is a friend, mate, companion, buddy, and partner with whom we spend a lot of time or share an experience (which is usually, but not necessarily, difficult or dangerous). Camaraderie pertains to the realm of social relations and can be understood as the inclination to associate with peers and to develop a shared spirit of closeness, understanding, respect, trust, support, goodwill, loyalty, complicity, and solidarity. It is rooted in a community of interests and creates and fosters a sense of bond, belonging, self-confidence, and self-efficacy toward the achievement of a common goal. According to the Dictionary of Psychology of the American Psychological Association, "Camaraderie is an important component of the morale, unit cohesion, and *esprit de corps* required in forming and sustaining unit dynamics. It can also serve as a buffer in protecting members of a unit."[74] Its antonyms—forlornness and loneliness—indicate a feeling of sadness and hopelessness caused by social isolation.[75]

Some context-specific examples will add important nuances to the term. Among youths, camaraderie is usually instinctive, warm, convivial, and lighthearted. Among fellow soldiers, pride and patriotism facilitate its strengthening. In business, specific exercises exist for the development of workplace camaraderie among colleagues.[76] In amateur team sports, camaraderie increases cooperation and thus the chances of winning, but it also helps the creation of an environment in which people enjoy each other's company and feel happy while going through this experience.

To sum up, camaraderie has several characteristics that make it helpful as a One Health tenet to guide how we, as humans, can do good for other living beings. First, it is a social relation in itself—not an emotion that can translate into meaningful action for others *only if* it is present in the "giving" subject. Camaraderie builds on a feeling of togetherness and the sharing of time, experiences, interests, and goals. It is positive and fulfilling, stimulates individual and collective improvement, implies social equality, and aims at resolving abuses of power in a system-thinking way,[77] not allowing conflicts to divide the unity. Camaraderie certainly requires commitment, but it is not exhausting, unlike, for example, compassion fatigue (i.e., the feeling, experienced at the physical, emotional, and psychological level, that we want to keep on helping others but we are too overwhelmed by the continuous exposure to their suffering).[78] Moreover, camaraderie can be nurtured and, as a collective ideal, it can grow gradually, unlike mutually exclusive dualisms such as love/hate. Finally, camaraderie is rooted in other values and can accommodate feelings such as compassion and love.

If applied as a principle to inform the One Health approach toward dog-mediated (human and animal) rabies, interspecies camaraderie can resolve several of the current tensions. People and dogs will become comrades with the dual goal of avoiding rabies infection and, in the process, building and enjoying a less conflictual relationship. Mutual care, a tenet of camaraderie—and of One Health[79]—will ensure that *both* dogs and people are protected against rabies through vaccination. Mass dog vaccination and sterilization have been described as convenient—hence questionable—biopolitical moves to intervene on dogs (as a "disposable population") rather than on humans (as "privileged individuals," for example by investing in post-exposure prophylaxis). As Kritika Srinivasan puts it:

> In other words, it is *simpler* to intervene on dogs than [to] work with people. These judgments about efficiency are ultimately rooted in the fact that issues relating to consent and harm are easily elided when it comes to nonhuman animals, i.e., they are rooted in anthropocentrism.[80]

I do not share this view, according to which mass dog vaccination and sterilization make dogs tools in the hand of anthropocentrism, for two reasons. First, mass dog vaccination and sterilization are not easy—logistically, financially, ethically. Second, even if dogs are usually described as carriers of rabies to animals and humans, they are, in fact, the first and most numerous victims of this disease. Thus, to rephrase the above quote, I argue that "it is *more just* to intervene on dogs than [to] work with people." Vaccination is beneficial to dogs in the first place, both as individuals and as a population (through the benefit of herd immunity toward unvaccinated dogs). Additionally, camaraderie-inspired vaccination automatically excludes dog culling as an option,[81] both for its unethicality and its lack of logic. Moreover, vaccinated dogs protect not only humans from rabies, but also all the other animal species (i.e., wildlife), which would be difficult to protect via direct vaccination.[82] Finally, human–dog camaraderie does not exclude compassion and love; on the contrary, it welcomes those who share these feelings toward dogs to advocate for a camaraderie-inspired One Health approach to rabies.

FIGURE 9.2 A free-roaming dog naps in the shade of the umbrella that a shopkeeper has just opened for him on a hot summer day in Jaipur, Rajasthan, Western India. Photo by author, June 2012.

Once complete, the virtuous cycle of interspecies camaraderie maximizes care, where "to care is to become subject to another, to recognize an obligation to look after another,"[83] and ensures multispecies justice. The reduction of social injustice—visible in the disproportionate burden of disease on marginalized people, their dogs, and their livestock—is part of the discourse on rabies elimination, but, so far, it has been limited to humans.[84] Specifically, dog vaccination is considered more just than post-exposure prophylaxis, because the latter is often hard to access for geographically and economically marginalized people. Interspecies camaraderie, for the reasons described in the above paragraph, creates a tangible possibility for the much-needed achievement of equity in One Health—not only for humans, but for dogs and other animals too.

Conclusion

One Health is intuitively appealing, yet little attention has been paid to what the core values are that practically guide how we make decisions for human, animal, and environmental health.[85] Provided that a set of tenets—not just one—are to be preferred, I propose interspecies camaraderie as a guiding principle for One Health. Taking rabies, a model One Health disease, as an example, this chapter discussed interspecies camaraderie in comparison to the feelings of compassion and love. I argue that, if we use compassion and love to guide One Health action, they are more problematic and unpractical than they look. Interspecies camaraderie incorporates these two emotions but, at the same time, goes beyond them. Across the spectrum of One Health issues, interspecies camaraderie can help us ground our actions on a demanding but rewarding spirit of togetherness that ultimately results in just care for others.

Notes

1 Sue VandeWoude, Angela Bosco-Lauth, and Christie Mayo, "Deer, Mink and Hyenas Have Caught COVID-19—Animal Virologists Explain How to Find the Coronavirus in Animals and Why Humans Need to Worry," *The Conversation*, February 17, 2022, https://theconversation.com/deer-mink-and-hyenas-have-caught-covid-19-animal-virologists-explain-how-to-find-the-coronavirus-in-animals-and-why-humans-need-to-worry-176666.
2 Paul Gibbs, "The Evolution of One Health: A Decade of Progress and Challenges for the Future," *Veterinary Record* 174, no. 4 (2014): 85–91.
3 Italian G20 Presidency et al., "Call to Action on 'Building One Health Resilience,'" G20 Health, 2021, https://www.salute.gov.it/imgs/C_17_pagineAree_5459_10_file.pdf.
4 Chris Degeling, Angus Dawson, and Gwendolyn L. Gilbert, "The Ethics of One Health," in *One Planet, One Health*, ed. Merrilyn Walton (Sydney: Sydney University Press, 2019), 65–84.
5 Ibid., 67.
6 Andrew A. Cunningham, Ian Scoones, and James L. N. Wood, "One Health for a Changing World: New Perspectives from Africa," *Philosophical Transactions of the Royal Society B* 372 (2017): 20160162. The criticism of the One-ism in One Health

is also discussed in this volume's Foreword by Stephen Hinchliffe (also in Stephen Hinchliffe, "More than One World, More than One Health: Re-Configuring Interspecies Health," *Social Science & Medicine* 129 (2015): 28–35), and this volume's Introduction by Irus Braverman.

7 Jo E. B. Halliday et al., "Endemic Zoonoses in the Tropics: A Public Health Problem Hiding in Plain Sight," *The Veterinary Record* 176, no. 9 (2015): 220–225.

8 Deborah Nadal, *Rabies in the Streets: Interspecies Camaraderie in Urban India* (University Park: Penn State University Press, 2020).

9 Melanie J. Rock, Dawn Rault, and Chris Degeling, "Dog-Bites, Rabies and One Health: Towards Improved Coordination in Research, Policy and Practice," *Social Science and Medicine* 187 (2017): 126–133; Danielle Celermajer et al., "Multispecies Justice: Theories, Challenges, and a Research Agenda for Environmental Politics," *Environmental Politics* 30, nos. 1–2 (2021): 119–140; Sophie Chao, Karin Bolender, and Eben Kirksey, *The Promise of Multispecies Justice* (Durham: Duke University Press, 2022); Jennifer A. Mather, "Ethics and Care: For Animals, Not Just Mammals," *Animals* 9, no. 12 (2019): 1018; Maria Puig de la Bellacasa, *Matters of Care: Speculative Ethics in More than Human Worlds* (Minneapolis: University of Minnesota Press, 2017).

10 Katie Hampson et al., "Estimating the Global Burden of Endemic Canine Rabies," *PLoS Neglected Tropical Diseases* 9, no. 5 (2015): e0003709.

11 WHO, *Expert Consultation on Rabies: Third Report* (Geneva: WHO, 2018).

12 Marcello O. Sato et al., *Zoonotic Diseases and One Health* (Basel: MDPI, 2020).

13 Sarah Cleaveland et al., "Rabies Control and Elimination: A Test Case for One Health," *Veterinary Record* 175, no. 8 (2014): 188–193; Kurt C. Vercauteren et al., "Rabies in North America: A Model of the One Health Approach," *USDA National Wildlife Research Center* 1202 (2012): 56–63.

14 WHO, FAO, and OIE, *Zero by 30: The Global Strategic Plan to End Human Deaths from Dog-Mediated Rabies by 2030* (Geneva: WHO, 2019).

15 Sreejith Radhakrishnan et al., "Rabies as a Public Health Concern in India—A Historical Perspective," *Tropical Medicine and Infectious Disease* 5, no. 4 (2020): 162.

16 Michelle K. Morters et al., "Evidence-Based Control of Canine Rabies: A Critical Review of Population Density Reduction," *Journal of Animal Ecology* 82, no. 1 (2013): 6–14.

17 Cleaveland et al., "Rabies Control," 192.

18 "National Action Plan for Dog-Mediated Rabies Elimination from India by 2030," Ministry of Health & Family Welfare Government of India, 2021, http://www.awbi.in/awbi-pdf/NationalActiopPlan.pdf.

19 "Central Government Act: The Animal Birth Control (Dogs) Rules, 2001," Indian Kanoon, accessed May 17, 2022, https://indiankanoon.org/doc/131470747/.

20 Kritika Srinivasan et al., "Reorienting Rabies Research and Practice: Lessons from India," *Palgrave Communications* 5 (2019): 152.

21 Deborah Nadal et al., "Perspectives of India's Animal Welfare Organization Members on Dog-Mediated Rabies Control and Dog Population Management Strategies," forthcoming; Srinivasan et al., "Reorienting," 8.

22 People for the Elimination of Stray Troubles v. State of Goa, (2003) 4 BomCR 588.

23 Srimoyee Chowdhury, "Man Tries to Kick Stray Dog but Ends Up Falling Hard in Viral Video. Perfect Karma, Says Internet," *India Today*, February 20, 2022, https://www.indiatoday.in/trending-news/story/man-tries-to-kick-stray-dog-but-ends-up-falling-hard-in-viral-video-perfect-karma-says-internet-1915530-2022-02-20.

24 Shireen Bhalla and Abi T. Vanak, "Killing with Compassion: Why Feeding Dogs in Public Places Must Stop!" *Down To Earth*, July 2, 2020, https://www.downtoearth.org.in/blog/wildlife-biodiversity/killing-with-compassion-why-feeding-dogs-in-public-places-must-stop--72092.

25 Matthew E. Gompper, "The Dog–Human–Wildlife Interface: Assessing the Scope of the Problem," in *Free-Ranging Dogs and Wildlife Conservation*, ed. Matthew E. Gompper (Oxford: Oxford University Press, 2013), 9–54.

26 Aniruddha Belsare and Abi T. Vanak, "Modelling the Challenges of Managing Free-Ranging Dog Populations," *Scientific Reports* 10, no. 1 (2020): 18874.
27 Nadal et al., "Perspectives."
28 Krithika Srinivasan, "Remaking More-than-Human Society: Thought Experiments on Street Dogs as 'Nature,'" *Transactions of the Institute of British Geographers* 44, no. 2 (2019): 376–391.
29 Philip Howell, "Between Wild and Domestic, Animal and Human, Life and Death: The Problem of the Stray in the Victorian City," in *Animal History in the Modern City: Exploring Liminality*, eds. Clemens Wischermann, Aline Steinbrecher, and Philip Howell (London: Bloomsbury Academic, 2018), 145–160.
30 Wendy Doniger O'Flaherty, *The Origins of Evil in Hindu Mythology* (Berkeley: University of California Press, 1976); Sarra Tlili, "The Canine Companion of the Cave: The Place of the Dog in Qur'ānic Taxonomy," *Journal of Islamic and Muslim Studies* 3, no. 2 (2018): 43–60.
31 Lisa Kemmerer and Anthony J. Nocella, *Call to Compassion: Reflections on Animal Advocacy* (New York: Lantern Books, 2011).
32 Gilles Tarabout, "Compassion for Living Creatures in Indian Law Courts," *Religions* 10, no. 6 (2019): 1–21.
33 Animal Welfare Board of India v. A. Nagaraja, (2014) 7 SCC 547.
34 Taraboud, "Compassion," 4.
35 Ibid., 9.
36 Ibid.
37 Ibid.
38 Ibid., 18.
39 The Egyptian Constitution recommends "the kind treatment of animals." Jessica Eisen, "Animals in the Constitutional State," *International Journal of Constitutional Law* 15, no. 4 (2017): 909–954.
40 Mark Bekoff, ed., *Ignoring Nature No More: The Case for Compassionate Conservation* (Chicago: University of Chicago Press, 2013).
41 Andrea S. Griffin et al., "Compassionate Conservation Clashes with Conservation Biology: Should Empathy, Compassion, and Deontological Moral Principles Drive Conservation Practice?" *Frontiers in Psychology* 11 (2020): 1–9.
42 "Constitution," World Health Organization, 2022, https://www.who.int/about/governance/constitution.
43 Francesca Brencio, "Mind Your Words: Language and War Metaphors in the COVID-19 Pandemic," *Psicopatologia Fenomenológica Contemporânea* 9, no. 2 (2020): 58–73.
44 It is to be noted that euthanizing a terminally rabid dog to reduce their suffering is different, ethics-wise, from killing a biting dog to detect the possible presence of the rabies virus in its brain, as illustrated in Susan McHugh's chapter.
45 Dhananjay Mahapatra, "Can Rabies Patients Opt for Euthanasia?" *Times of India*, January 9, 2020, https://timesofindia.indiatimes.com/india/can-rabies-patients-opt-for-euthanasia/articleshow/73164110.cms.
46 Didier Andrivon, Josselin Montarry, and Sylvain Fournet, "Plant Health in a One Health World: Missing Links and Hidden Treasures," *Plant Pathology* 71, no. 1 (2022): 23–29.
47 "Database Legislation," Global Animal Law, 2022, https://www.globalanimallaw.org/database/national/index.html.
48 Benjamin Capps, "One Health Ethics," *Bioethics* 36, no. 4 (2022): 348–355.
49 David G. White, *Myths of the Dog-Man* (Chicago: University of Chicago Press, 1991), 71.
50 Maria Digna Winda Widyastuti et al., "On Dogs, People, and a Rabies Epidemic: Results from a Sociocultural Study in Bali, Indonesia," *Infectious Diseases of Poverty* 4, no. 1 (2015): 1–18.
51 Nadal, *Rabies*, 91.

52 "Stray Dog Attacks 6-Year-Old Girl in Bangalore," *Times of India*, January 26, 2013, https://timesofindia.indiatimes.com/city/bengaluru/stray-dog-attacks-6-year-old-girl-in-bangalore/articleshow/18191056.cms.
53 Bjørn Ralf Kristensen, "Welcome to the Viraloscene: Transcorporeality and Peripheral Justice in an Age of Pandemics," *Medium* (blog), May 19, 2020, https://medium.com/@bjornkristensen/viralocene-66a954260487.
54 Nadal, *Rabies*, 115.
55 Abhinav Garg, "NDMC Told to Halt Drive Against Mongrels in Lodhi Garden," *Times of India*, August 18, 2013, https://timesofindia.indiatimes.com/city/delhi/ndmc-told-to-halt-drive-against-mongrels-in-lodhi-garden/articleshow/16857732.cms.
56 Also, in Srinivasan et al., "Reorienting," 382.
57 Nadal, *Rabies*, 118.
58 Anuradha Ramanujan, "Violent Encounters: 'Stray' Dogs in Indian Cities," in *Cosmopolitan Animals*, eds. Kaori Nagai et al. (Basingstoke: Palgrave Macmillan, 2015), 216–232.
59 Victoria C. Krings, Kristof Dhont, and Alina Salmen, "The Moral Divide Between High- and Low-Status Animals: The Role of Human Supremacy Beliefs," *Anthrozoös* 34, no. 6 (2021): 787–802.
60 Marie Pelé et al., "Perceptions of Human–Animal Relationships and Their Impacts on Animal Ethics, Law and Research," *Frontiers in Psychology* 11 (2020): 1–4.
61 Sriya Narayanan, "Ten Years of 'Adopt, Don't Shop,'" *The Hindu*, December 11, 2018, https://www.thehindu.com/society/ten-years-of-adopt-dont-shop/article25718318.ece.
62 Krithika Srinivasan, "The Welfare Episteme: Street Dog Biopolitics in the Anthropocene," in *Animals in the Anthropocene. Critical Perspectives on Non-Human Futures*, ed. Human Animal Research Network Editorial Collective (Sydney: Sydney University Press, 2015), 201–220.
63 Meghna Uniyal, "Who Let the Dogs Out?," *Down to Earth*, June 4, 2019, https://www.downtoearth.org.in/blog/india/who-let-the-dogs-out--64908.
64 Stephan Lorenz and Kerstin Stark, "Saving the Honeybees in Berlin? A Case Study of the Urban Beekeeping Boom," *Environmental Sociology* 1, no. 2 (2015): 116–126.
65 Sophia Menache, "Dogs and Human Beings: A Story of Friendship," *Society and Animals* 6, no. 1 (1998): 67–86.
66 Arjun Raghunath, "Kerala HC Intervenes After Frequent Cases of Animal Cruelty," *Deccan Herald*, July 1, 2020, https://www.deccanherald.com/national/south/kerala-hc-intervenes-after-frequent-cases-of-animal-cruelty-1003782.html.
67 Ishan Kukreti, "A Sharp Bite: Dog Meat Ban Evokes Some Sharp Reactions in Nagaland," *Down to Earth*, August 18, 2020, https://www.downtoearth.org.in/news/food/a-sharp-bite-dog-meat-ban-evokes-some-sharp-reactions-in-nagaland-72865.
68 Jack Levin, Arnold Arluke, and Leslie Irvine, "Are People More Disturbed by Dog or Human Suffering? Influence of Victim's Species and Age," *Society and Animals* 25, no. 1 (2017): 1–16.
69 Susan Sontag, *Illness as Metaphor* (New York: Farrar, Strauss, & Giroux, 1978).
70 Susan Sontag, *AIDS and Its Metaphors* (New York: Farrar, Strauss, & Giroux, 1989).
71 Annalisa Pelizza, "Blame Is in the Eye of the Beholder: Beyond an Ethics of Hubris and Shame in the Time of COVID-19," *Harvard Kennedy School (HKS) Misinformation Review* 1, no. 3 (2020), https://misinforeview.hks.harvard.edu/article/blame-is-in-the-eye-of-the-beholder-beyond-an-ethics-of-hubris-and-shame-in-the-time-of-covid-19/.
72 Kathleen Kete, "La Rage and the Bourgeoisie: The Cultural Context of Rabies in the French Nineteenth Century," *Representations* 22 (1988): 89–107.
73 "Camaraderie," Merriam-Webster.com Dictionary, 2022, https://www.merriam-webster.com/dictionary/camaraderie.

74 "Camaraderie," APA Dictionary of Psychology, 2022, https://dictionary.apa.org/camaraderie.
75 "Camaraderie," Merriam-Webster.com Thesaurus, 2022, https://www.merriam-webster.com/thesaurus/camaraderie.
76 Christine M. Riordan, "We All Need Friends at Work," *Harvard Business Review*, July 3, 2013, https://hbr.org/2013/07/we-all-need-friends-at-work.
77 David H. Peters, "The Application of Systems Thinking in Health: Why Use Systems Thinking?" *Health Research Policy and Systems* 12 (2014): 1–6.
78 Holly Monaghan et al., "Compassion Fatigue in People Who Care for Animals: An Investigation of Risk and Protective Factors," *Traumatology* (2020): 1–9.
79 Melanie J. Rock et al., "Animal-Human Connections, 'One Health,' and the Syndemic Approach to Prevention," *Social Science and Medicine* 68, no. 6 (2009): 991–995.
80 Srinivasan et al., "Reorienting," 9 (my emphasis).
81 Chris Degeling, Zohar Lederman, and Melanie Rock, "Culling and the Common Good: Re-Evaluating Harms and Benefits Under the One Health Paradigm," *Public Health Ethics* 9, no. 3 (2016): 244–254.
82 Sarah Cleaveland et al., "Canine Vaccination—Providing Broader Benefits for Disease Control," *Veterinary Microbiology* 117, no. 1 (2006): 43–50.
83 Thom Van Dooren, "Care," *Environmental Humanities* 5, no. 1 (2014): 291–294.
84 Diorbhail Wentworth et al., "A Social Justice Perspective on Access to Human Rabies Vaccines," *Vaccine* 37, no. 1 (2019): 3–5.
85 Degeling et al., "The Ethics"; Joost van Herten, Bernice Bovenkerk, and Marcel Verweij, "One Health as a Moral Dilemma: Towards a Socially Responsible Zoonotic Disease Control," *Zoonoses and Public Health* 66, no. 1 (2019): 26–34; Jane Johnson and Chris Degeling, "Does One Health Require a Novel Ethical Framework?," *Journal of Medical Ethics* 45, no. 4 (2019): 239–243; Joe Copper Jack et al., "Traditional Knowledge Underlies One Health," *Science* 369, no. 6511 (2020): 1576.

10

ANTHRODEPENDENCY, ZOONOSES, AND RELATIONAL SPILLOVER

Bjørn Ralf Kristensen

Introduction

In the wake of what is now being referred to as the "Anthropause"—the lockdowns and changes in routine human behavior meant to curb the spread of COVID-19—we have had the opportunity to witness widespread connections between human, animal, and environmental health, as well as a revelation of previously invisibilized relationships and networks of dependency.[1] Social media posts circulated early in the pandemic often touted how beneficial the removal of humans from their regular day to day activities was for many animals. However, for many anthrodependent species—those whose lives are wrapped up in varying degrees of dependency on humans—the opposite was often true, and such cases received much less attention.

In this chapter, I consider two distinct examples of anthrodependency in wild animals tied to zoonotic disease to exemplify cases where conventional rigid principles in relation to such animals fall short. I propose that John Dewey's approach to moral deliberation brings necessary adaptive flexibility to complex situations such as these, which defy habitual understandings of wild animals. I call for an approach to moral deliberation grounded in sympathy and reflection that is adaptable to particular situations and their unique complexities. The case studies of anthrodependency that I consider exemplify the need for rejecting a harsh division between what is perceived as non-moral and moral. A broader relational understanding of zoonotic disease emphasizes an increased sensitivity

FIGURE 10.1 Rock hyraxes (*Procavia capensis*) in Jerusalem exemplify the moral complexity involved in seemingly trivial decisions such as the discarding of piles of rock near the city, an urban development practice which imitates the wild habitat of this species, invites them in, and leads to the development of dependencies on humans and the spillover of zoonotic disease. Photo by Charles J. Sharp, Erongo, Namibia, 2003. CC BY-SA 4.0.

DOI: 10.4324/9781003294085-14

to the multitude ways in which humans are already involved in the lives of more-than-human others.

One crucial strength of One Health approaches is a recognition that human health is tied to animal and environmental health. And yet, despite claims of interconnection, such One Health approaches tend to operate unidirectionally in the sense that concern for animals is generally viewed instrumentally toward the wellbeing of humans. As I argue—and as the following case studies exemplify—humans often exist as the necessary background conditions under which many wild animals thrive. Under many circumstances, these conditions have led to deep dependencies. The direct and indirect implications of zoonotic disease often lead to realizations of these connections—what I refer to as "relational spillovers." Given this, the current chapter makes the case for a more sympathetic and reflective approach to zoonotic disease mitigation policy that is mindful—not just with respect to the impact that animal health and wellbeing has on humans but also to more-than-human dependencies developed through anthropogenic conditions which often remain unrealized until direct or indirect consequences of zoonotic spillover events occur.

Anthrodependency, Dualism, and Responsibility

Commensalism is a term used to denote a relationship in which one organism associates with another to bring about a particular benefit, usually without harming or benefiting the other organism. These types of relationships exist throughout the more-than-human world, and they are also widespread in the context of relationships that animals have with humans. Ardern Hulme-Beaman and colleagues introduced a more specific term to denote the particular kind of relationship in which an organism is commensal in the context of anthropogenic environments: "anthrodependent."[2] Although I will be using this term throughout this chapter, I want to preface its use with the observation that my understanding of anthrodependency goes beyond its original definition which largely focuses on invasive species and generalists—namely, species adaptive to a wide variety of circumstances with flexible dietary needs. Indeed, part of the purpose of this chapter is to argue that our notion of which animals depend on humans is deficient, and that this has been made evident through the revelation of many previously unexamined dependencies connected to zoonotic disease. Such a realization underlies the motivation for a more reflective and sympathetic approach, not simply with regard to those dependencies of which we are aware, but also toward less well-known dependencies. I move forward by considering two case studies: a seabird colony in the Baltic Sea that was disrupted due to lack of tourist presence during COVID-19 protocols, and the influx into Jerusalem, Israel of rock hyraxes carrying a zoonotic skin disease due to urban building practices.

Considerations of animals in the context of zoonotic disease often focus on the animals themselves as carriers of disease. Certainly, early discourse surrounding COVID-19 highlighted the zoonotic spillover from animals to humans stemming from a wet market environment.[3] Such cases draw attention to the wider implications of everyday interactions between humans and other animals, and the dependencies that often manifest outside of our awareness through such interactions. Part of the move in this regard must be to see beyond animals as the *causes* of disease themselves, but rather as tied to a wider relational understanding of the conditions (both intentional and unintentional) that become embodied through them—thus, my call for an increased sensitivity to the ways in which our lives become implicated in each other. I explore such an example in the rock hyrax case study.

I would also like to draw attention to the indirect implications of behavior change as a result of zoonotic disease. Most examples of anthrodependency in the age of coronavirus are of this variety. One example includes street dogs, cows, and rhesus macaques adapted to a life of dependence on regular human activity in urban India which were under threat of starvation in the absence of humans during lockdowns.[4] Another is the rise in poaching of many species of animals in parts of sub-Saharan Africa and Latin America due to protective systems in place for these species that depend on economic inputs from tourism, which experienced unexpected declines due to crises such as the COVID-19 pandemic.[5]

The dualism between humans and animals (particularly humans and wild animals) is also pervasive in conservation discourses. It is often assumed that wild animals would be better off free from human intrusion. Such a perspective has been inscribed into legal documents which influence policy decisions, such as the Wilderness Act of 1964 in the United States, which creates a strict division between human habitats and "the earth and its community of life," which should remain "untrammeled by man."[6] This legacy of human/nature dualism neglects to recognize our unavoidable dependence on the more-than-human world, while simultaneously divorcing us from significant responsibility toward it.[7]

The dualistic nature–human approach is also pervasive in animal ethics discourse, whereby ethicists often prioritize negative (non-interference) approaches to other animals rather than positive approaches that focus on improving the wellbeing of animals.[8]As Tom Regan writes with regard to wild animals, they "Do not need help from us in the struggle for survival."[9] Likening human intervention in the lives of wild animals to paternalism, Lori Gruen proposes that interference is inappropriate "[i]n the case of individuals who are capable of exercising their freedom to live their lives in their own ways."[10] However, others argue that we are obligated to intervene to alleviate the suffering of animals in the wild. Martha Nussbaum suggests along these lines that humans should even interfere with predator–prey relationships if doing so would bring about less harm than not interfering.[11] Jeff McMahan agrees that "Suffering is intrinsically bad for those who experience it and there seems always to be a reason, though not necessarily a decisive one, to prevent it—a reason that applies to any moral agent who is capable of preventing it."[12]

A pragmatic approach would recognize that neither end of the interference/non-interference spectrum will be universally applicable. Instead, we must strive to respect the particularity of each situation. McMahan himself admits this much with his statement that there is a reason for interference in such cases of overt animal suffering, yet not necessarily a decisive way to respond. Importantly, historical entanglements will give way to the possibility of responsibility and recognition in response to the role humans play in the lives of wild animals.[13] Yet, such entanglements often exist outside of our awareness with regard to the dependencies that such animals have on humans. The murre bird colony at Stora Karlsö, Sweden is one such situation. I will turn now to discussing this case, through which anthrodependency was revealed by the absence of tourism during COVID-19.

The Seabird Colony at Stora Karlsö

The common murre (common guillemot, *Uria aalge*) is a species of seabird that has widespread distribution yet has vulnerable breeding colonies in some areas. Stora Karlsö, a Swedish island in the Baltic Sea, has a large colony of around 60,000 murres which has been protected since the late nineteenth century.[14] It is also a popular tourist site for this reason. In early 2020 when the COVID-19 pandemic first hit, the island saw a 92 percent decline in human presence due to measures put in place to curb the virus.[15] Many people might assume such circumstances would benefit the populations of wild animals living in the colony, yet this was not the case for all.

Researchers working at Stora Karlsö discovered, much to their shock, that 2020 ended up being the worst breeding season on record for the murre breeding colony.[16] A variety of circumstances brought about this result. White-tailed sea eagles had previously been frightened away from coming close to the colony by the presence of tourists. Whereas these eagles do sometimes prey on the murres and their eggs, most of the impact from their presence due to the absence of tourists is connected to the fear they caused for the murres when flying over the colonies, which caused the murres to leave their nests unattended.[17] This disruption caused stress which resulted in a delay in egg laying, and, when the murres left their nests which already contained eggs, it also allowed herring gulls and hooded crows to prey on these eggs.[18] Both factors contributed to drops in reproduction. In responding to these circumstances, researchers working at Stora Karlsö argued that we should recognize that the regular presence of humans at Stora Karlsö is part of the ecological processes of the colony. Indeed, they even framed the tourists at Stora Karlsö as "seabird guardians."[19] Rather than understanding regular human presence as necessarily problematic, this example clearly reveals a previously unknown dependency that the murre colony had on humans.

This example—the revelation of which was made possible through measures meant to curb COVID-19—pushes against the pervasive dualism between humans and other animals present in conservation discourse. Interference is certainly widespread in conservation—consider, for example, captive breeding of threatened species or protective enclosures for wildlife preserves. Yet, there is a pervasive interpretation that a life largely independent from regular human interference is taken as an objective good, and thus we miss the particularity of each situation. This point will be crucial in engaging with Dewey's ethics in the context of anthrodependency. In the next section, I look at a case of anthrodependency in which a population of animals is the direct carrier of disease: rock hyraxes in Jerusalem.

Urban Rock Hyraxes in Jerusalem

The rock hyrax (*Procavia capensis*) is a small mammal that lives throughout the Middle East and sub-Saharan Africa, typically choosing to live on cliff edges or rock piles which provide shelter from harsh desert temperatures and weather, as well as from predators.[20] They have long populated the Judean Desert in what is now Israel and the Occupied Palestinian Territories.[21] In the last 30 years, they have begun taking up residence in urban Jerusalem.[22] Rock hyraxes are attracted to city life because of prevalent food sources. In fact, they are particularly attracted to more religious, Hasidic neighborhoods due to religious beliefs around placing food waste outside rather than disposing of it in the trash.[23] Rock hyrax populations also intersect with poorer areas of the city which do not seem to have the same access to sanitation services as more affluent neighborhoods.[24] In addition, poorer parts of the city often have accumulated waste and temporary building structures, the latter providing shelter for hyraxes that mimics their traditional wild habitats.[25]

One of the primary reasons that rock hyraxes have taken up residence in urban Jerusalem is a common urban construction process which results in discarded rock piles being left relatively close to the city in the wake of finished constructions. As Noam Ben-Moshe and Takuya Iwamura state:

> Urban building practices and heavy machinery, used for the excavation and laying of new roadways, have pushed discarded piles of rocks down the mountains and into the valleys below. This[sic] debris have built up over time, and have become urban simulations of the hyrax habitats enveloping new neighborhoods. We found that these artificial rock piles are great shelters for the hyraxes and they come with the added benefits of providing access to human waste and rich foraging grounds in the adjacent city parks.[26]

Clearly, the presence of rock hyraxes in Jerusalem is due largely due to people not recognizing the more-than-human impact of decisions carried out with human goals in mind. Whereas the seabirds of Stora Karlsö did not have an immediate negative impact on human populations, rock hyraxes in Jerusalem certainly do. Rock hyraxes are frequently reservoir hosts for a variety of sandfly species that carries the zoonotic disease leishmaniasis.[27]With respect to humans, this disease can cause skin lesions, scarring, various disabilities, and death.[28] Since the circumstances that attracted the rock hyraxes to certain parts of Jerusalem often parallel the distribution of socioeconomic status and religious practices, it also reflects the distribution of leishmaniasis in the city.[29] Furthermore, the status of rock hyraxes as potential vectors of the disease has had a profound negative impact on the rock hyraxes as well. To curb the spread of leishmaniasis, Israel's Ministry of Environmental Protection now allows rock hyraxes, which were once a protected species, to be culled within the city.[30]

Relational Spillover: Toward a Sympathetic Contextual Approach

A zoonotic spillover is when a zoonotic disease crosses over from animals to humans. I maintain that common language around zoonoses, such as the World Health Organization's statement that zoonotic disease is primarily "[A]n infectious disease that has jumped from a non-human animal to humans"[31] obscures deeper, relational spillovers between humans and the more-than-human world.[32] Stacy Alaimo cautions against an overemphasis on attempts to pin down some root originary cause of COVID-19, as origin stories may serve to minimize the importance of considering relationality.[33] Focusing merely on the physical properties of the disease also misses both broad and contextual power relations exhibited through the ongoing propagation of zoonoses.

Outlooks on public health that recognize the interconnections of human, animal, and environmental health—such as One Health initiatives—take a primarily unidirectionaland ethically anthropocentric approach. This is to say that zoonoses are understood predominantly through what Val Plumwood refers to as an *incorporative* and *instrumental* lens: animals, the lives of whom are tied to zoonotic disease, are typically defined in relation to a presumed center: humans.[34] The proposed international legal treaty entitled "Animal Protection for Public Health, Animal Welfare, and the Environment" from the Lawyers for the Convention on Animal Protection seeks to remedy some of this unidirectional approach by presenting a call for animal welfare provisions (among other considerations) to be tied to zoonotic disease mitigation measures.[35]An open letter commentary in the inaugural issue of *CABI One Health*, signed by numerous leading academics and researchers, also makes explicit the need for a move beyond simply an instrumental understanding of animals in relation to One Health—rather, they propose that relevant entities should "better reflect the value of improving animal health and welfare not only for the sake of humans but also for the sake of the animals themselves."[36] These examples illustrate that there is hopeful movement pushing against the overwhelmingly unidirectional approach of zoonotic disease mitigation, although this conversation currently remains at the periphery.

Beyond the welfare concerns brought on by considerations of animals who are viewed merely instrumentally in relation to zoonotic disease, I also want to highlight that, by defining animals only in relation to humans, we run the risk of "backgrounding" humanity as the condition necessary 0for many animals to thrive.[37] By looking at the situation through a unidirectional lens, we miss the ways in which we are implicated in particular ways to more-than-human lives. I suggest that the two case studies I discuss in this chapter exemplify what I call a "relational spillover," illuminating how deeply embedded humans are in the lives of other animals. More than simply the passage of disease into human bodies, such encounters result in the spillover of wider understandings of the degree to which humans and other animals are entangled in relationships and processes, leading to the actual disease spillover events

John Dewey proposes that the very way in which we understand ourselves and the world is through a landscape of habits which ground our conception of reality.[38] Most of these habits operate beneath our conscious awareness. Many are also rigidified, focused merely on what Dewey refers to as "occasions for imitation," as with much of the legacy of Western moral philosophy, which emphasizes appeals to detached and objective reason, or mechanical principles.[39] So too, as we see, it is often perceived that the place of humans apart from wild animals is necessarily beneficial.

The habits through which we interpret the world and the rigid principles we often use to address situations are disrupted in the cases I examined previously. Certainly, for much of the world, the COVID-19 pandemic has operated as a mass disruption of habits, many of which had externalized impacts which previously were easier to disguise. In the wake of these circumstances, we might ask how to move forward in a world that does not align with the neatly arranged perception of categories which we have previously perceived it through. I want to consider these circumstances of relational spillover through looking at some of Dewey's thoughts on moral deliberation. Specifically, I propose that sympathetic reflection is essential in broadening our understanding of what even counts as moral. Indeed, the case studies I have highlighted above are prime examples of Dewey's statement that "there is no gulf dividing non-moral knowledge from that which is truly moral."[40] This is akin to Peter Singer's "the expanding circle," by which our ethical project involves necessarily seeking to understand the scope to which we are implicated in the lives of others.[41] By contrast, I argue that the expanding circle does not simply involve including literal *others* within our moral community, but also a heightened awareness of *how* they are already included. As such, one key implication is that the application of so-called objective principles dictating the place of humans in the lives of wild animals must be interrogated toward a more flexible and adaptive approach grounded in the needs of each particular situation.

Dewey maintains that "Sympathy is the animating mold of moral judgment."[42] One essential tenet of Dewey's ethical theory is that moral situations are complex problems which require attention being paid to the unique circumstances of the given moment. This requires humans to engage with their imaginative and sympathetic capacities in search of new information and new perspectives beyond one's own. This also involves using the tools we have at hand, which certainly include past experiences, dominant perspectives, and moral principles. However, Dewey rejects placing too much faith in detached and universally applicable rules. He repeatedly refers to this as "the philosophical fallacy," which is the common assumption that a response that worked in a particular situation can be applied universally.[43]Principles, as he states, are "methods of inquiry and forecast which require verification by the event … Principles exist as hypotheses with which to experiment."[44]

Rigid adherence to theories and approaches may also become a kind of moral narcissism which shields our attention from the needs of the very others in the interests of which we claim to be acting. Judith Butler refers to this as a recoiling "from the other, from impressionability, susceptibility, and vulnerability."[45]And yet, generality and objectivity are taken to be strengths within prevalent moral approaches, which problematically distance the particularity and singularity of individuals.[46] Following these considerations, conventional perceptions and beliefs about the place of wild animals, as distinct from humans, must be challenged, as should approaches toward widespread un-nuanced interventions into the lives of wild animals. The cases of the seabird colony and the urban rock hyraxes break from the perceived pattern in terms of the dependencies that wild animals often have on humans. The problem I wish to highlight is not that approaches creating distinct divisions between humans and wild animals are prevalent, but rather that they are taken as universally applicable to all wild animals.

In accounting for a moral dilemma, we need to consider the full circumstances as far as is practically possible. It is crucial to recognize that this is an ongoing project of revision and improvement, and will realistically never come to a definitive end. However, to recognize that there is not such an end is not a reason to throw up our arms and not attempt to do our best. Rather, as Dewey maintains, this is one key reason why moral deliberation must be focused on the particularity of the present moment. As he states, "In morals, the infinitive and the imperative develop from the particle, present tense. Perfection means perfecting, fulfillment means fulfilling, and the good is now or never."[47] To focus too heavily on the future is to assume a present which never realizes the presumed ideal. To overemphasize the past is to assume that past approaches to solving moral dilemmas are necessarily also applicable objectively in the present: as a result, the present is not recognized in its particularity.

Consider the crucial point about neglecting the particularity of the present in the context of abolitionist animal rights approaches (such as are advocated by Gary Francione) who outright reject any and all use of animals by humans.[48] Francione opposes measures to improve the wellbeing of animals living within the contemporary animal industrial complex because, by his account, such approaches support the idea that there are better and worse ways to exploit animals, and they also fail to question the underlying moral wrong which is that animals are understood to be property, not individual bearers of rights. Yet, such an approach can be said to view the animals living and suffering in the here and now as merely a means to an end toward a vision of an idealized future in which animals might one day live completely independent of humans.[49] Francione maintains that domestication, for example, is a necessarily exploitative process, and holds that, although we have obligations to care for those domesticated animals who are already living, we should do everything in our power to end the process and live our lives largely apart from other animals.[50]

Although domestication is a complex topic, we might indeed say that many anthrodependent species are in the early stages of what Melinda A. Zeder refers to as the commensal pathway to domestication, which is a largely unintentional process which has manifested in the past through humans and other animals interacting and living in close proximity to one another.[51] With this in mind, positions such as Francione's, and indeed other idealistic, hands-off conservation approaches, which assume animals are necessarily better off being separate from humans, fail to recognize the widespread permeation of humans into the lives of other animals in the present context.

In the mitigation of COVID-19, much has been neglected in terms of the consideration of beings living in the here and now. The sheer scale of loss of human life as a result of COVID-19 is overwhelming, yet the loss of life transcends our species. Consider that as many as 17 million mink were killed in 2020 in just one country, Denmark, due to a mutated form of COVID-19 that spread quickly through these animals who were being raised for fur.[52] The mink are not alone in terms of animals infected with COVID-19.[53] Yet no animals who contracted COVID-19 were killed at the rates at which mink were. It is important to note that the deaths of the mink were not from the actual disease, but rather because of attempts to mitigate the spread in the crowded facilities they were being kept. Farmed mink are known to be particularly vulnerable to pathogens, and the spread of COVID-19 through mink farms was compounded by the suppressed mink immune systems that may have been caused by psychological distress from forcing these otherwise solitary animals into such crowded and confined environments.[54] Once again, this case illustrates the numerous power dynamics at play in compounding the spread of zoonoses, outside of the diseases themselves.

It is also crucial to account for the externalized impacts on animals from research and development for pharmaceuticals and vaccines in response to COVID-19. As I have highlighted in previous work, rhesus macaques, an anthrodependent species in some parts of the world, who were impacted by the absence of humans as a result of lockdowns, are also considered a crucial tool in animal testing trials for vaccines.[55] As Stacy Alaimo writes, "The pandemic witnessed not only a shortage of toilet paper, but a shortage of a strategic 'resource'—the rhesus macaques."[56]

Consider the case of the seabird colony at Stora Karlsö. Clearly, an approach which assumes that tourist presence was expected but not necessarily beneficial to the murres at this colony is mistaken. Yet, in considering the full circumstances of the situation, it is clear that not all residents of the seabird colony were harmed by the absence of tourists. The gulls and crows who fed on the murre eggs no doubt benefited from a new and easily accessible food source provided by the murres leaving their nests in fear.

Dewey maintains that we must attempt to account for the diverse environing conditions that make each situation unique, so as to consider it in its particularity. We must then also ask how humans are enmeshed in the circumstances at a site such as Stora Karlsö, beyond simply the presence or absence of tourists. For example, the large number of white-tailed sea eagles that returned are related to past harms and attempts to assist this once-threatened species. The pesticide dichlorodiphenyltrichloroethane (DDT), as well as other industrial chemicals, such as polychlorinated biphenyls (PCBs), once caused significant drops in white-tailed sea eagle populations. Following the banning of such chemicals and a successful recovery project, the numbers are of the white-tailed sea eagles are now higher than they were prior to being impacted by pesticides and industrial pollutants.[57] As a result, white-tailed sea eagles have negatively impacted numerous seabird populations in areas similar to Stora Karlsö.[58]

Even decisions that seem to be based on straightforward observations—for example, that the banning of DDT and PCBs is beneficial for all forms of life, or that successful recovery of a once-threatened species is an objective good—should be considered in their particular contexts. This is not to claim that either of these decisions was a mistake, but rather that we ought to explore potential consequences of our actions in ways that are more mindful to multiple perspectives. Chelsea Batavia and colleagues use the term "moral residue," to illustrate that most moral dilemmas—especially within a field such as conservation biology—will not give way to responses which are completely free from moral failings toward some beings.[59] Indeed, moral residue refers to an emotional recognition of the harms that come about from even the *best* possible response to such situations which do not exist as a simple binary of right or wrong.[60] As I show with the two case studies discussed, this is also often the case with situations which we do not originally see as moral dilemmas, or even explicitly moral in nature. It therefore behooves us to attempt to be more receptive to the ways in which the foundations for dependencies of other animals on humans are being laid. Such an approach aligns Dewey's emphasis on moral deliberation as an "imaginative dramatic rehearsal" whereby we imagine different paths of response to situations in their context, prior to acting to hopefully bring about an equilibrium.[61]

In the case of the seabird colony at Stora Karlsö, a profound impact came about with the removal of the regular tourist presence to curb the spread of the deadly SARS CoV-2 virus. We know that there have been numerous social reverberations in the human context from measures meant to curb the spread of COVID-19, such as widespread mental health challenges and financial instability. This highlights that the fallout from the pandemic has impacted some of the most vulnerable humans apart from the consequences of the disease itself. Yet, these impacts should not be seen as arguments against measures to curb the spread of COVID-19, but rather as indicative of the necessity for more sympathetic exploration of how to implement measures in a way which would more widely accommodate intersecting needs.

Given that numerous seabird colonies are impacted by the presence of increased numbers of white-tailed sea eagles, we might also ask—as an example of the philosophical fallacy that Dewey draws our attention to—if the application of what is taking place at Stora Karlsö would work outside of this particular context. Would an approach reframing tourists as seabird guardians be possible and/or beneficial in every situation across the range of the common murre? And to whom would it be beneficial? Given each particular situation, it is probable that, under some circumstances, it would make sense to use past approaches as guidance. However, given the wide range of the common murre, approaches would have to consider each particular population and the circumstances of their colony.

Conclusion

This chapter has drawn attention to the complicated nature within which we assume to be performing straightforward moral decision-making, and the need for an adaptive and sympathetic approach. The cases of anthrodependency that I considered complicated the perception of what counts as moral. Consider the approach to urban development in Jerusalem, which was largely responsible for inviting in the populations of rock hyraxes who have now adapted to a life in the city. Whereas one would typically not consider such city engineering practices through a moral lens, they clearly carry moral weight.

The presence of the rock hyraxes within Jerusalem comes at a cost to both animal and human lives. Although the hyraxes have benefited from plentiful food sources and a habitat well suited to them, the reality that they are reservoir hosts for the zoonotic disease leishmaniasis means they are also being killed as a consequence of the implications of this zoonosis for human populations. Importantly, none of the actions—which, in combination, have resulted in the current situation—were intentionally directed toward its current end. This emphasizes a key takeaway message: the pervasiveness of anthrodependency—and the relational spillover we become aware of in moments of zoonotic crises such as the COVID-19 pandemic—call upon us to consider the particular circumstances of each situation. In taking an approach sympathetic to the impact of decisions not previously thought to involve other animals, we also come to question the divide between moral and non-moral.

Peter Singer uses the metaphor of "the expanding circle" to illustrate a broadening circle of moral inclusion by which humans have widened the moral community to include each other at more complex levels: from family to nation to all within our species.[62]This extends beyond the arbitrary boundary point of the human species toward the inclusion of other animals. I want to propose that the relational spillover exemplified through anthrodependency also pushes us beyond focusing solely on *who* is included, but rather *how they already are*. A crucial part of our ongoing moral project must then involve assessing and reevaluating everyday human actions—perhaps especially those which initially seem ethically benign, neutral, or of no moral weight— in light of their broader implications for the more-than-human world. Crucially, if we recognize that we cannot help but be implicated in the lives of many wild animals, then it may be rightfully said that a truly negative ethical approach rooted in non-interference is impossible.

In the context of zoonotic disease mitigation, this must mean abandoning a unidirectional approach which merely prioritizes explicitly known threats to human health. Even in a purely ethically anthropocentric approach, this could have pitfalls. Although the path of zoonotic disease follows trajectories that we may be able to predict, when considerations are always framed in terms of the known risk to humans, the interests of animals themselves are often weighed secondarily. One example we can look to in this regard are rabies vaccination campaigns for wild animals.

Consider the Ethiopian Wolf Conservation Programme, which frames its approach to vaccination through One Health due to the close proximity of humans and livestock to the wolves' habitat, and due to the threat of domestic dogs to the endangered wolves.[63] The vaccination of Ethiopian wolves clearly benefits individual wolves as it prevents suffering through an excruciatingly painful rabies infection.[64] However, as is the case with many other vaccination campaigns for wild animals, their interests are incidental given that these animals "were vaccinated not for their own good, but to protect human interests by preventing the transmission of rabies to domesticated animals and humans, or to conserve populations of endangered species."[65]

Approaches toward the prevention of rabies transmission to humans are often framed explicitly toward maintaining a strict division/dualism between wild animals and humans. Consider the United States National Park Service's statement on One Health and rabies: "By protecting natural environments and their ecological properties and processes, and by appreciating wildlife from a distance, we can help protect ourselves from rabies—this is One Health in action."[66] However, even in considerations where humans and/or livestock are not currently in direct contact with such animals, it would be proactive to emphasize vaccinating other wild animals. Even if interests remained centered on humans, we cannot perfectly predict how human and animal lives will intersect in the future. Indeed—as shown by the cases of anthrodependency I have considered—humans are often implicated in the lives of other animals to degrees we are significantly unaware of. Furthermore, in the context of a disease such as rabies—which causes immense suffering—even if there is no immediate benefit to humans, practicing a sympathetic approach toward other animals would behoove us to intervene.

Writing in the early twentieth century, Dewey uses an apt example of the discovery of the relationship between bacteria and disease to illustrate the misconception of a perceived distinction between moral and non-moral. He states:

> At any moment, conceptions, which once seemed to belong exclusively to the biological or physical realm, may assume moral import. This will happen whenever they are discovered to have a bearing on the common good. When knowledge of bacteria and germs and their relation to the spread of disease was achieved, sanitation, public and private, took on moral significance it did not have before. For they were seen to affect the health and wellbeing of the community.[67]

This statement aligns with what Stacy Alaimo refers to as "trans-corporeality," an approach recognizing that "all creatures, as embodied beings, are intermeshed with the dynamic, material world, which crosses through them, transforms them, and is transformed by them."[68] Human presence and the environments we modify penetrate into the more-than-human world physically, socially, and morally. A recognition that human embodiment, health, sociality, and morality flow multidirectionally among our own species and the more-than-human world should lead us to approaches that are mindful to the necessary nature of our existence as always exceeding our perceptions.

One key realization tied to the COVID-19 pandemic has been widescale recognition of the connections between human health and the more-than-human world. And yet, most responses taking this seriously—such as One Health measures put forward by the United Nations, the World Health Organization, the United States Centers for Disease Control and Prevention, and others—take a primarily ethically anthropocentric approach to animals.[69]Although animals are considered in such measures, their inclusion is primarily instrumental toward human ends. One important result of One Health approaches is that they do recognize animals and the environment as necessary conditions for human health and wellbeing, yet they largely fail to consider humans themselves as the background conditions for many other species' wellbeing. However, as I have shown in this chapter, zoonotic diseases and measures meant to curb them can be seen as revelatory of the degree to which many wild animals are deeply dependent on humans, often previously outside of our awareness. Again, this is where breaking down divisions between the moral and the non-moral is so crucial to emphasize.

Anthrodependent animals are clearly already implicated in the lives of humans in a multitude of ways, yet often outside of our perception. Zoonoses often act as crisis points through which relational spillovers are revealed. I have argued that our task must be to sympathetically engage with possible scenarios of response which better account for the presence of animal others and the degree to which we are implicated in their lives. Furthermore, this task calls upon us to question the pervasive dualism between moral and non-moral, which serves to close off our perception to the implications of seemingly mundane and everyday decisions that oftenhave a profound moral impact on both other species and our own. To be human is to be a moral being who is necessarily implicated in the lives of others. As Dewey states, "if one is going to live, one must live a life of which these things form the substance."[70] We do not have a choice whether we are implicated in the lives of others or not. Thus, a crucial part of our moral project must be to sympathetically reflect on and broaden our understanding of the ways in which we are.

Notes

1 The term "Anthropause" was coined by: Christian Rutz et al., "COVID-19 Lockdown Allows Researchers to Quantify the Effects of Human Activity on Wildlife," *Nature Ecology and Evolution* 4 (2020): 1156–1159.
2 Ardern Hulme-Beaman et al., "An Ecological and Evolutionary Framework for Commensalism in Anthropogenic Environments," *Trends in Ecology and Evolution* 31, no. 8 (2016): 633–645.
3 Walzer and Braverman, this volume.
4 Deepa Lakshmin, "Amid the World's Strictest Lockdown, People Who Feed Stray Dogs Are Now Deemed Essential," *National Geographic*, May 8, 2020.
5 Ralf Buckley, "Conservation Implications of COVID19: Effects via Tourism and Extractive Industries," *Biological Conservation* 247 (2020): 108640.
6 National Park Service, "Complete Text of the Wilderness Act of 1964," accessed April 29, 2022, https://www.nps.gov/orgs/1981/upload/W-Act_508.pdf.
7 Val Plumwood, *Feminism and the Mastery of Nature* (New York: Routledge, 1993).
8 Sue Donaldson and Will Kymlicka, "Rights," in *Critical Terms for Animal Studies*, ed. Lori Gruen (Chicago: University of Chicago Press, 2018), 320–336.
9 Tom Regan, *The Case for Animal Rights* (Berkeley: University of California Press, 2004), xxxviii.
10 Lori Gruen, *Ethics and Animals: An Introduction* (Cambridge: Cambridge University Press, 2011), 182.
11 Martha Nussbaum, *Frontiers of Justice: Disability, Nationality, Species Membership* (Cambridge, MA: Harvard University Press, 2006), 379.
12 Jeff McMahan, "The Moral Problem of Predation," in *Philosophy Comes to Dinner: Arguments About the Ethics of Eating*, eds. Andrew Chignell and Terence Cuneo (New York: Routledge,2015), 273.
13 Clare Palmer, "What (If Anything) Do We Owe Wild Animals?," *Between the Species* 16, no. 1 (2013): 15–38.
14 Cara Giaimo, "Covid-19 Kept Tourists Away: Why Did These Seabirds Miss Them?," *New York Times*, January 21, 2021.
15 Jonas Hentati-Sundberg et al., "COVID-19 Lockdown Reveals Tourists as Seabird Guardians," *Biological Conservation* 254 (2021): 108950.
16 Ibid.
17 Ibid.
18 Ibid.
19 Ibid.
20 Noam Ben-Moshe and Takuya Iwamura, "Shelter Availability and Human Attitudes as Drivers of Rock Hyrax (*Procavia capensis*) Expansion Along a Rural-Urban Gradient," *Ecology and Evolution* 10, no. 9 (2020): 4044–4065.
21 Ibid.
22 Ibid.
23 Ibid.
24 Noam Ben-Moshe and Takuya Iwamura, "Rock Hyraxes in the City; The Bi-Directional Effects of Cultural Norms on Urban Wildlife," *Ecology and Evolution* (blog), accessed April 29, 2022, https://ecologyandevolution.blog/2020/04/17/rock-hyraxes-in-the-city-the-bi-directional-effects-of-cultural-norms-on-urban-wildlife/.
25 Ibid.
26 Ibid.
27 Ibid.

28 "Leishmaniasis," World Health Organization, 2022, https://www.who.int/health-topics/leishmaniasis.
29 Ben-Moshe and Iwamura, "Shelter Availability."
30 Zafrir Rinat, "Israeli Minister Permits Culling Rock Hyraxes to Prevent Leishmania Spreading," *Haaretz*, October 5, 2018.
31 "Zoonoses," World Health Organization, 2022, https://www.who.int/news-room/fact-sheets/detail/zoonoses.
32 Bjørn Ralf Kristensen, "Welcome to the Viralocene: Transcorporeality and Peripheral Justice in an Age of Pandemics," *Medium* (blog), May 19, 2020, https://medium.com/@bjornkristensen/viralocene-66a954260487.
33 Julia Kuznetski and Stacy Alaimo, "Transcorporeality: An Interview with Stacy Alaimo," *Ecozon@* 11, no. 2 (2020): 137–146.
34 Plumwood, *Environmental Culture* (New York: Routledge, 2002).
35 "Draft Convention on Animal Protection for Public Health, Animal Welfare, and the Environment," Convention on Animal Protection, October 20, 2021, https://www.conventiononanimalprotection.org/the-cap-treaty.
36 Jeff Sebo et al., "Open Letter: Sustainable Development Matters for Animals Too. World Leaders Have a Responsibility to Recognize That,"*CABI One Health* (2022), 1–2: 2.
37 Plumwood, *Feminism and the Mastery of Nature.*
38 John Dewey, Jo Ann Boydstom, and Patricia Baysinger, eds., *The Middle Works of John Dewey, Volume 14, Human Nature and Conduct 1922* (Carbondale: Southern Illinois University Press, 1988).
39 Ibid., 48.
40 John Dewey, Jo Ann Boydstom, and Barbara Levine, eds.,*The Later Works of John Dewey, Volume 7, 1925–1953*, eds. Jo Ann Boydstom and Barbara Levine (Carbondale: Southern Illinois University Press, 2008), 282.
41 Peter Singer, *The Expanding Circle: Ethics, Evolution, and Moral Progress* (Princeton: Princeton University Press, 2011).
42 Dewey, *The Later Works*, 270.
43 Dewey, *The Middle Works*, 123.
44 Dewey, *The Middle Works*, 164–165.
45 Judith Butler,*Giving an Account of Oneself* (New York: Fordham University Press, 2005), 99–100.
46 Jacques Derrida,*The Gift of Death* (Chicago: University of Chicago Press, 1996), 61.
47 Dewey, *The Middle Works*, 200.
48 Gary L. Francione,*Animals as Persons: Essays on the Abolition of Animal Exploitation* (New York: Columbia University Press, 2008).
49 Credit to Marcia Condoy Truyenque of Derecho Animal en Perú (Animal Law in Peru) for making this profound point in a personal conversation with me.
50 Francione, *Animals as Persons*, 13.
51 Melinda A. Zeder, "Pathways to Animal Domestication," in *Biodiversity in Agriculture: Domestication, Evolution, and Sustainability*, eds. Paul Gepts et al. (Cambridge: Cambridge University Press, 2012), 227–259.
52 Helen Briggs, "Denmark to Cull Up to 17 Million Mink Amid Coronavirus Fears," *BBC News*, 5 November, 2020.
53 "Animals and COVID-19," Centers for Disease Control and Prevention, 2022, https://www.cdc.gov/coronavirus/2019-ncov/daily-life-coping/animals.html.
54 Sonia Shah, "Animals that Infect Humans Are Scary. It's Worse When We Infect Them Back," *New York Times Magazine*, January 19, 2022.
55 Kristensen, "Welcome to the Viralocene."
56 Stacy Alaimo, "The Portal Was Already Here: Epistemological Rupture, Speculation, and Design in the Long 2020," in *For the Long 2020*, eds. Richard Grusin and Maureen E. Ryan (Minneapolis: University of Minnesota Press, 2023).

57 Mark J. Hipfneret al.,"Unintended Consequences: How the Recovery of Sea Eagle *Haliaeetus*spp. Populations in the Northern Hemisphere Is Affecting Seabirds," *Marine Ornithology* 40 (2012): 39–52.
58 Ibid.
59 Chelsea Batavia, Michael Paul Nelson, and Arian D. Wallach, "The Moral Residue of Conservation," *Conservation Biology* 34, no.5 (2020): 1114–1121.
60 Ibid.
61 Mark Johnson, *Morality for Humans: Ethical Understanding from the Perspective of Cognitive Science* (Chicago: University of Chicago Press, 2014), 90.
62 Singer, *The Expanding Circle*, 120.
63 "Projects – One Health," Ethiopian Wolf Conservation Programme, 2022, https://www.ethiopianwolf.org/disease-control.
64 For more discussion on rabies in the context of One Health see McHugh, Nadal, and Hinchliffe, this volume.
65 "Vaccinating and Healing Sick Animals," Animal Ethics, 2021, https://www.animal-ethics.org/vaccinating-healing-sick-injured-animals/.
66 "One Health and Disease: Rabies," National Park Service, March 27, 2018, https://www.nps.gov/articles/one-health-disease-rabies.htm.
67 Dewey, *The Later Works*, 282.
68 Stacy Alaimo, "Trans-Corporeality," in *Posthuman Glossary,* eds. Rosi Braidotti and Maria Hlavajova (London: Bloomsbury Academic, 2018), 435.
69 "One Health," Food and Agriculture Organization of the United Nations, 2022, https://www.fao.org/one-health/en/.
70 Dewey, *The Middle Works*, 58.

PART IV

Decolonizing One Health: Toward Postcolonial and Indigenous Knowledges

11

BIRDS AS SENTINELS OF THE ENVIRONMENT IN HONG KONG AND TAIWAN

Frédéric Keck

Avian influenza has been an exemplar in the implementation of a "One Health" approach in the past 15 years. Since the identification of the H5N1 virus in Hong Kong in 1997, the monitoring of domestic poultry and wild birds has become a model by which to understand the conditions for zoonotic transmission of pathogens from animals to humans and back to animals, such as SARS-Cov in 2003, H1N1 in 2009, and SARS-Cov2 in 2019. When the H5N1 virus spread from Asia to Europe in 2005, it raised concerns for a pandemic but also for food security, as it affected humans and domestic poultry, and for the health of wild birds who were suspected to carry the virus over long distances. In coordination with the World Health Organization, the World Organisation for Animal Health, and the Food and Agriculture Organization, the Wildlife Conservation Society launched the One World One Health initiative to promote the monitoring of emerging pathogens at the interface between humans, animals, and their environment. Their argument was that the surveillance of zoonoses should aim at improving not only the health of humans but also the wellbeing and diversity of animals.[1]

Indeed, the loss of diversity among wild birds is considered to be one of the main drivers in the emergence of highly pathogenic avian influenza, because viral mutations are attenuated or "diluted" when viruses infect close species, whereas the dramatic increase of the size of the poultry industry, with its genetic standardization and its high exposure to vaccines and antibiotics, is a factor for the amplification of new viruses.[2] However, little attention has been paid to environmental conservation in the historical accounts on One Health. The territory of Hong Kong, where measures to control avian influenza were precocious and

FIGURE 11.1 Birdwatchers release two black-faced spoonbills in the National Park of Tainan. Photo by Frédéric Keck, April 29, 2013.

DOI: 10.4324/9781003294085-16

innovative, tried at the same time to manage intense relations between humans and animals at the risk of zoonotic transmission and to preserve cultural practices such as live poultry markets. In the words of its experts in veterinary medicine and microbiology, a One Health strategy required "sensitive implementation to optimize food safety and food security, while safeguarding the economics of animal husbandry and the environment and remaining sensitive to cultural practices."[3] Although the Hong Kong government initially prioritized culling live poultry and closing poultry farms, experts recommended the surveillance of wild birds as a less authoritative and more respectful measure.[4] Surveillance meant regularly sampling wild and domestic birds to check if they carried viruses similar to H5N1.

This chapter shows how birdwatchers were enrolled in the monitoring of avian influenza in Hong Kong and Taiwan, and asks how this alliance changed the perspective on birds as reservoirs of pandemic influenza. I use the concept of "sentinel" to describe this enrollment, playing on the two senses of the term. A sentinel is a soldier bringing warning signals from the frontline with an enemy, in this case a potentially pandemic pathogen. But it is also a sensor of environmental threats that concern humans as well as birds themselves, such as loss of habitat, nuclear radiation, or climate change. A sentinel is more than an indicator of a statistical tendency: it allows biologists to imagine the transmission of a pathogen based on shared vulnerabilities between humans and animals.[5] Although the injunction to imagine the effects of a pandemic on vital infrastructures is part of the strategy of preparedness, the capacity for humans to imagine what it feels like for a bird to be affected by influenza is central to the One Health initiative.

The concept of sentinel connects two concerns: on the one hand, the concern for biosecurity, defined as the control of infectious pathogens threatening human health, and on the other hand, the concern for biodiversity, defined as the well-being of environments where humans and nonhumans coexist in multispecies relations. Although birds as reservoirs of pathogens were often considered by public health authorities to be threats to eradicate, wild birds in nature reserves were counted as lives that should be conserved.[6] If Hong Kong and Taiwan were configured as sentinel posts in a global politics of pandemic preparedness, the term "sentinel" was also used to describe local practices involving humans and nonhumans, such as keeping non-vaccinated chickens in poultry farms or collecting samples from wild birds in nature reserves.[7]

This chapter asks what One Health means for birdwatchers when they are enrolled in pandemic preparedness. How does the colonial and military genealogy of sentinel posts still inform contemporary mobilizations of birdwatchers, while they promote the democratic ideal of citizen science? When birdwatchers become experts for the surveillance of avian influenza, how do they integrate local knowledge to be attentive to birds' ways of life? How do they transform warning signals of extinction into sustainable forms of conservation, and how do they manage the risk of a false alarm? What are the conditions of success for a sentinel to build a sustainable collective for human and nonhuman lives?

Based on ethnographic research I conducted in Hong Kong and Taiwan between 2007 and 2013, I compare how the military model of birdwatching has been converted into a democratic form of environmental mobilization in these two territories, and I study how they exchange information about threats on birds by mixing military surveillance with democratic participation. I use the method of the sociology of public problems to understand how an unhealthy wild bird can become the starting point of an environmental mobilization.[8] By displaying a genealogy of One Health in military practices of environmental conservation, I also argue for a multiple and postcolonial history of One Health from the sentinel territories where it has been implemented.[9] The transformation of military practices into democratic mobilizations is also an appropriation of colonial models of surveillance by local forms of attention to birds, which I argue is also a condition for the democratization of One Health.

Birdwatching in Hong Kong as a Postcolonial Practice

When I arrived in Hong Kong in 2007 to conduct research on avian influenza, I wanted to describe how virologists collected samples in poultry farms and wild bird reserves to transform them into genetic sequences in the laboratory where they could read early-warning signals of pandemics. I sent an e-mail to the Hong Kong Birdwatching Society and asked how they collaborated with virologists in the management of wild bird reserves. Mike Kilburn, who was in charge of public relations for the Society, replied immediately. He was angry about the Hong Kong government's decision to close the Mai Po reserve for 21 days when a wild bird was found infected with highly pathogenic influenza in a three-kilometer radius around the reserve. The Mai Po marshes, a territory at the end of the Pearl River Delta and a site of passage for Chinese refugees as well as for migratory birds, had become a nature reserve when the government granted its management to the World Wildlife Fund in 1984. With 10,000 migratory birds feeding in the marshes every year, in 1995 it became a "Wetland of International Importance" under the Ramsar Convention. The first manager of the reserve, David Melville, held the position of Government Ornithologist between 1974 and 1980. The current measures taken by the Hong Kong government were perceived by Mike Kilburn as retaliations against what remained a colonial heritage.

The Hong Kong Birdwatching Society (HKBWS) was founded in 1957 by British officers who wanted to establish a list of the bird species on the territory for conservation purposes. In 2007, they argued that the wild birds found with H5N1 on the territory were resident species and not migratory, and that it was therefore irrational to close Mai Po and not bird parks in Kowloon. As urban visitors were suspected to be in contact with bird feathers or feces, birdwatchers said, "it is not possible to catch avian influenza through binoculars!" Birdwatchers used ornithology, alongside commonsense, to criticize the blaming of wild birds out of fear of avian influenza. They argued that it was easier to close a former colonial institution where a few thousand ecotourists come every year than the urban spaces in which millions of citizens are in contact with wild birds or live poultry. This economic rationality denied the knowledge they had built over two centuries as a group of passionate birdwatchers.

In May 2007, Mike Kilburn organized a joint conference between the HKBWS and the Department of Microbiology of Hong Kong University. They showed the journalists a map where the cases of influenza on wild birds were reported. It was clear that most of the cases occurred around the bird market of Kowloon. Mike Kilburn told me that wild birds were sold to Buddhist practitioners who released them for spiritual purposes, and that these birds died after being released from the stress of transportation into cages, some of them being infected with influenza. Live poultry markets and wild bird release were considered by the government to be politically dangerous issues, whereas nature reserves were not, even though they were regarded by birdwatchers as sentinel territories to do regular surveillance. Kilburn said:

> Mai Po is probably the most tested place in the world for wild birds, and no wild birds have been found with the disease. I know that one or two dead birds have been found in the area. But in terms of the claim that it's migratory birds that are spreading death to humans, it's nonsense. My frustration is that so little research has been done on the wild bird trade. But the fact is that birds are an easy way for the government to lay the blame elsewhere. If you say, "shoot the birds," you have to contend with a few green groups. If you say, "close down the poultry farms," you have to contend with the global agricultural industry.[10]

The Agriculture Fisheries and Conservation Department commissioned the HKBWS for a water bird survey of the Mai Po Marshes and for a study of egretries in the entire territory of the reserve, because birdwatchers had developed competences in identifying and counting bird species, through regular meetings, courses and competitions (such as an annual "bird race"), using field guides as standards of reference.[11] These competences allowed them to transform a bird encounter into a "record"—namely, a species identification which could be integrated into numbers of birds seen on the territory. Kilburn said:

> We've been collecting records of birds for fifty years in Hong Kong. This gives us an authority that nobody can question on birds, because the HKBWS was started by English birdwatchers who had this amateur birdwatching model: you write down the birds that you see and you submit your records to the society at the end of the year, and those records are available for anybody who wants to use them. The Conservation Department [admitted] to me that, when they try to catch up [on] an area of biodiversity, they don't see the point to try to compete with the birdwatching society.[12]

What the HKBWS did collectively in Mai Po, one of its members, Geoff Welsh, also did as an individual sentinel. Trained in birdwatching during his teenage years in England, he worked as a businessman in Hong Kong. After his retirement, he returned to what he perceived as a hobby, and invested in birdwatching the rigor he had learned during his career. He spent three days a week on Po Toi, an island south of Hong Kong, inhabited by only a few fishermen, where he counted seabirds. His data and pictures were then posted on the website of the HKBWS and greatly appreciated by other members of the association, because they gave the best demonstration of the extinction of some resident species and the arrival of new ones. Whereas other birdwatchers practiced monitoring as a hobby, randomly returning to the same places, Geoff Welsh did it systematically, turning a deserted island into a sentinel post. However, he was insistent about not presenting himself as an environmental militant, warning about climate change; he contented himself with producing valuable records, leaving the work of interpretation to others. "I love numbers," he told me, "I've done that all my life."[13] He presented himself as a "serious birdwatcher," in contrast with "amateur birdwatchers" who disrupted bird habitats when they came too close with their camera to post a nice picture on the HKBWS website.

This notion of a good distance between humans and birds that made possible a healthy and productive encounter was reframed in a local idiom by the head of the HKBWS. Chiu Ying Lam was the first Chinese member of the colonial society in 1976 and became its director in 1997 at the time of the handover of the British colony to Chinese sovereignty. He told me that he had first observed birds in the Hong Kong cemeteries, and that this link between birds and the spirits of ancestors led him to facilitate the discussion with Buddhist practitioners, whom he convinced to stop releasing birds. He also said that when he moved from observing stars to watching birds, he felt he was coming closer to the movements of his own heart, because he could communicate with birds when he looked into their eyes, which he could not do with stars. "The English say: can we take pictures? which species is it? The Chinese say: can we eat it? how does it taste? I'm interested to know what happens when I look at a bird and it looks at me."[14] Chiu Ying Lam was the first birdwatcher to talk to me about the effects of birdwatching on his health: he didn't speak of numbers of birds as indicating a biodiverse environment, but of the colors and sounds of birds which transformed localities, such as cemeteries, into meaningful, semiotic spaces.[15]

HKBWS' greatest success, according to Chiu Ying Lam, was the protection of the area of Long Valley, at the North of the New Territories, which attracted more members to the association. This agricultural wetland was threatened by a railway line planned by the Kowloon-Canton Railway (KCR). The HKBWS showed that Long Valley was home to more than 210 bird species, and argued that many of them had disappeared from similar habitats on the other side of the border with mainland China. They launched a public campaign between 1999 and 2001, and secured the diversion of the railway into an underground tunnel that did not affect the habitat. Chiu Ying Lam recalled:

> At that time, we had virtually no chance of winning since we were fighting the railway company, which was rich and powerful. We were also labeled as "a tiny group of bird-watchers." But [the] birdwatchers did whatever they could: some wrote letters, some offered ideas, some helped linking up with fellow NGOs and the media, some took legal actions, etc.[16]

The defense of Long Valley was successful. Experts from the network BirdLife International came to Hong Kong to examine the Environmental Impact Assessment of the KCR project, and showed that the conservation value of the wetland had been downplayed. The construction project was canceled and the railway was built underground. But the area was not protected, and Hong Kong birdwatchers are encouraged to take pictures of bird species and post them on the website. Unlike Mai Po, where records of migratory birds are sent to the Conservation Department, Long Valley has only resident species. Sharing a picture, in this case, doesn't mean producing public knowledge on nature, but participating with a collective engagement in the defense of the local environment. Even if they are less informative than a record, pictures are signs that a birdwatcher was there and that a meaningful encounter occurred.

Birdwatching in Taiwan as a Democratic Form of Mobilization

The defense of Long Valley was an opportunity for Hong Kong birdwatchers to exchange competencies in political mobilization with their counterparts in Taiwan, who were involved in similar movements to protect birds in their local habitats. Taiwanese birdwatchers had also been transformed from Western observers under military office to local practitioners promoting democracy and citizen science. Hong Kong and Taiwanese birdwatchers were trained in the Migratory Animal Pathological Survey (MAPS), led by ornithologist Elliott McClure and Colonel Charles Barnes from 1963 to 1971 for the US Army. This project required capturing and banding birds throughout the Far East so as to assess the spread of Japanese encephalitis. It involved 13 teams in 9 countries, and banded more than 1 million birds of 1,218 species.[17] Although the headquarters of this project were based in Tokyo and Bangkok, Hong Kong was the reference center where tags found on birds all along the migratory flyway were to be mailed. Because birds follow migratory flyways that cross borders, considering them as carriers of pathogens for humans was a way for US scientists to extend military control over the borders of allied territories in the context of the Cold War.

If this project was the first massive production of statistical data on East Asian birds, it was also the first attempt to convert a military concern for biosecurity into an environmental care for biodiversity. A former student of Cornell University who taught French and English at Tunghai University in Taichung, Sheldon Severinghaus, was in charge of the Migratory Animal Pathological Survey for Taiwan. Through this program, he trained "Chinese workers" to catch, band, and ring birds, such as Peter Chen, who studied ecology in the United States and later taught it at Tunghai University. Peter told me in our interview:

> We set up nests to put rings, with two numbers on the ring: an identification number and a mailing box in Hong Kong. We measured the weight, the body length, [the] wing length. We collected parasites for pathological survey, inside and outside. We put dry powder on the plumage and lay the bird on a sheet of paper, and the parasites would fall on the sheet. We cut the central nail of the leg and collected the grease. Then we sent samples to headquarters in Bangkok and Tokyo.[18]

The development of birdwatching in Taiwan thus ran parallel to the democratization of civil society. Since it was forbidden to declare an association under martial law in Taiwan, a "bird club" was set up in Taipei in 1973 for Westerners who wanted to practice birdwatching, under the tutorship of the Animal Protection Association sponsored by Chiang Kai-shek. Similar clubs were created in Taichung in 1975 and Kaohsiung in 1979. During the military regime, wearing binoculars in public was associated with spying activities, and reserved for authorized specialists.[19] The first non-Western member was Lucia Liu, who studied ornithology at Cornell before being hired at the Academia Sinica, and who told me:

> The first meetings were in the homes of Westerner[s], and usually there were a dozen Western people, two or three Chinese from what I can remember—[so] mostly Westerners. By the end of the 1970s, there were almost no Westerners; it was entirely Chinese. And it was difficult in those days: private cars were rare, binoculars were rare, people gathered and they took buses to wherever they were going, and they shared the few pairs of binoculars. In terms of birdwatching, those were really hard days. But people were really interested, they went every week, and the group just grew.[20]

The lifting of martial law in 1987 opened up not only the possibility of organizing birdwatching activities publicly, but also a growing desire of people to enjoy leisure outside of working hours.[21] The number of societies increased to 19 and formed a federation, which joined the network BirdLife International in 1996 and currently claims to have 5,000 members. Under the politics of indigenization led by the Democratic Progressive Party President Chen Shui-bian, the Chinese Wild Bird Society became the Wild Bird Federation of Taiwan around 2000; it returned to its previous name after the fall of Chen Shui-bian in 2008. In 1996, the creation of the nature reserve of Guandu, after a ten-year campaign against construction projects, was the federation's first major success.[22] This park, in the suburbs of Taipei, is home to shorebirds and attracts around 10,000 people in October for the federation's annual bird fair. It can be compared to Hong Kong's Mai Po for its role in popularizing birdwatching. The number of birdwatching societies and sites was an indicator of the "democratic health" of Taiwan, in the sense that spaces were opened for a meaningful encounter with nature.

In 1999, the Chinese Wild Bird Society supported the fight that villagers in Huben, Yunlin County, took up against a gravel extraction project. They argued that the Huben site was home to the Fairy Pitta (*Pitta nympha*), a colorful, migratory bird considered endemic to Taiwan. They launched an international petition, "Save the pitta's home. Stop the gravel extraction," and collected more than 10,000 signatures, including 95 perent of the villagers from Huben. The Council of Agriculture carried out a survey of the Fairy Pitta breeding population and estimated a population of 40 in Yunlin County. On June 14, 2000, newly elected President Chen Shui-bian declared, "If Taiwan lost the Fairy Pitta, we would not only lose the most beautiful thing in Taiwan, but the whole world would be a poorer place." Consequently, Huben Village was designated as an Important Bird Area (IBA) by BirdLife International, and as an ecological village by the ROC government. This international organization for bird conservation encouraged Taiwanese birdwatchers to come to Hong Kong to support their own movement against KCR in Long Valley. The global support of conservation thus led to local forms of attachment of these two territories at the borders of China.

However, local attachments also led to a decline in mobilization in Taiwan. Further studies led by the Institute of Endemic Species showed that the Fairy Pitta was spread all over the island, and attributed the overall decline of its population to deforestation in Indonesia, where it spends the winter. After a long process of Environmental Impact Assessment, the construction of a dam in the village of Huben was authorized in 2008. The failure of the birdwatchers' mobilization in Huben, which coincided with the ending of the first Democratic presidency in Taiwan, may come from the fact that it was organized around a "flagship species," that is, a species with a high iconic or symbolic value that attracts different actors to get "on board" with an environmental movement.[23] By restricting themselves to the meaningful encounter between inhabitants and one bird species in an insular territory, Taiwanese birdwatchers had lost the capacity of colonial, military practices of surveillance to connect human and nonhuman populations across borders. While the protection of a flagship species is weakened if a study shows that this species is not threatened in the way it was believed to be, the conservation of a sentinel territory, on a border where the effects of environmental destruction are visible in the long term, appears more sustainable.

Whereas the failure of the Huben movement resulted from an excessive alert, other birdwatcher movements in Taiwan led to a relative lack of an alert. Kinmen is an island on the shores of the China Sea, very close to Xiamen, which remained a property of Taiwan and became the site of several fights with the People's Republic of China.[24] The Wild Bird Society of Kinmen has started a petition against the planned construction of a bridge between Kinmen and Xiamen, which, they argued, would destroy the rich biodiversity of the island, preserved by a half-century of military presence. They obtained the support of the Wild Bird Society of Xiamen, who showed that, with the same ecosystem and a higher level of development, many species found in Kinmen had disappeared from Xiamen. But local inhabitants of the island support the construction project, arguing that after 50 years of military isolation, the bridge would bring tourists to enjoy the rich cultural heritage (Hakka houses and museums of war) as well as products of local vineyards. In contrast with Hong Kong, a military sentinel failed to be turned into an environmental sentinel because the threats were not perceived as high enough to trigger a local mobilization.

The success of Hong Kong birdwatchers in the protection of birds can thus be explained as a middle path between the two extremes that were observed in Taiwan: an excessive alert (the mobilization to protect the Fairy Pitta in Huben as an endemic species) and a low level of alert (the lack of engagement of Kinmen's inhabitants in conserving their territory). This shows that birds must be perceived as warning signals of environmental threats beyond the borders of a territory. But, even if Hong Kong and Taiwan share the same bird species, they don't perceive them as sentinels in the same way because of their historical and geographical differences. Since Taiwan has many more forest birds living in the mountains, Taiwanese birdwatchers emphasize the importance of "endemic species" to valorize their ecosystem and promote "eco-tourism," whereas Hong Kong birdwatchers tend to emphasize migratory birds—the perception by a trained birder of a "first-seen" in a familiar habitat—by contrast with resident birds. In Hong Kong and Taiwan, birds become meaningful sentinels when they cross borders between species and territories and send warning signals of environmental threats on a larger scale. In the same way as a sentinel cell can overreact or underreact in the presence of a pathogen at the frontline of the immune system,[25] birdwatchers build a sentinel when they react adequately to the perception of a bird that crosses a border. The capacity of conservation to cause mobilization relies not only on the capacity of local experts to count birds but also on their capacity to perceive birds as signs of threats.

Releasing Wild Birds as a Healthy Connection Between Territories

How does the perception of a bird as a sentinel of environmental threats develop a healthy relationship with the territory? I will now use the concept of therapeutic landscape to describe how birdwatchers perceive birds not only as signs in a disrupted environment but also as living beings in need of care.[26] Birdwatching, I argue, contributes to a One Health approach because it is attentive to the way birds live, get sick, and die in an environment, the vulnerabilities of which they reveal by their diseases. I will take the example of the collaboration between birdwatchers in Hong Kong and Taiwan to protect a bird species traveling between the two territories.

Black-faced spoonbills, present only in East Asia, are classified by the International Union for the Conservation of Nature as an endangered species, due to the destruction of their feeding sites on the migratory pathways. In 1990, their number was estimated to be 2,000, but, thanks to the conservation efforts between Japan, South Korea (where they breed during the summer), Taiwan, and Hong Kong (where they migrate in the winter), their number had increased to 3,000 by 2010. The Mai Po marshes in Hong Kong and the Taijiang National Park in Taiwan (north of Tainan) are two major feeding sites on the migratory pathway of black-faced spoonbills, as well as the Fujian coast in mainland China. In winter 2002–2003, 73 black-faced spoonbills were found dead with botulism in Taijiang National Park. Botulism is a major epizootic for migratory sea birds, particularly because of the high concentration of the bacterium responsible present in wetlands, which are considered to be shelters on their flyways.[27] A vaccination campaign for botulism was consequently organized by the Wild Bird Society of Tainan. To prevent vaccination from causing harm to the birds, decoys were built on which the staff could train. Since then, drills are regularly organized with these decoys for bird protectors to learn how to manipulate spoonbills with care.

On the morning of April 29, 2013, I saw the three black-faced spoonbills that had been captured the night before. Ten wooden decoys were packed on the grounds; they had been posted in the marshes to attract the real birds, who were then caught in cages. The birdwatchers told me I was lucky: they had not been able to catch any black-faced spoonbills these last months, and now they caught three on the one day. Hong Kong birdwatchers often told me about their attempts to trap black-faced spoonbills, but they were often unlucky. The goal of the traps was to equip these endangered birds with satellite trackers in order to follow their migratory trajectory. Decoys are used by hunters to trap birds by luring them; but here the lure allowed birdwatchers to take the birds' perspectives and transform them into sources of information regarding the species migratory trajectories on a satellite map.[28]

As it was very sunny that morning, birdwatchers had found shelter in a Taoist temple on the banks of the marshes. They had switched on some Buddhist music to soothe the birds. With very quiet gestures, a man equipped with gloves and masks was attaching the satellite tracker around the waist of the spoonbill, while a woman was holding the bird gently but firmly. Five other birdwatchers were watching attentively, taking pictures and making comments on the reactions of the bird. The man explained to me that he had to be very cautious because it was a young spoonbill, so the weight of the satellite tracker had to be imposed on its body in such a way that it would not hamper its growth or unbalance its flight. After the three spoonbills were equipped in the same way, and banded with colored tags, they were released in the closest pond. The birds walked slowly to the middle of the pond, opened their wings and flew away.

When I observed the release of a black-faced spoonbill in Taiwan, I thought about the discussions I had had in Hong Kong with Mike Kilburn and Chiu Ying Lam about the Buddhist practice of bird release. Birdwatchers reproached Buddhist practitioners, who invoked birds' souls, of not being attentive enough to the signs of sickness as a result of the birds not being released in the proper environment. Indeed, Chinese birdwatchers published books about scientific release to explain in which environments illegally traded wild animals could be released without damage. The satellite antenna carefully attached to the body of the bird transformed it into a sentinel of the fate of this endangered species, and informed humans about the effect of the reconstruction of their habitat. Releasing a bird with an antenna from a Taoist temple was a way to inscribe it in a therapeutic space, where a careful relation between ornithologists and birds was built and extended through the production of signs of extinction.

The language of sacrifice is often used to describe the ethical dilemmas raised by environmental tracking. Ornithologists know that wild birds might die because of satellite tracking, and resist the idea that these birds are sacrificed for the sake of the whole species they work to conserve.[29] The attention of ornithologists was not turned toward a symbolic species for the sake of which individual birds are killed but toward sentinel birds who survive as witnesses of the fate of their species. In the same way, birds in Long Valley were considered as sentinels of a fragile environment which Hong Kong birdwatchers wanted to protect, whereas, in Taiwan, the Fairy Pitta was considered to be a symbol of democratic mobilizations. Flagship species are above the lives of individuals and sometimes justify their deaths, and sentinel species capture the signs sent across species borders and political borders of a common environmental threat.

This leads me to conclude by asking what kind of agency can be conceived for animals in a "more-than One Health." In the examples I have described, birds are quite passive: they live their lives, and their movements are perceived by ornithologists as signs of the fate of their species. To borrow terms from Michel Foucault, while the politics of zoonoses is often a power to "make animals die," when it is allied with conservation under the banner of One Health it becomes a power to "let them live." Yet this apparent passivity of animals may be such only through the framework of what Foucault called pastoral power, which manages animals as populations. What I have described through the military genealogy of conservation in Hong Kong and Taiwan is closer to the techniques of hunters who imitate animals and defer their death—that is, do as if death came from outside in a process of identification.[30] The anthropology of more-than-human worlds has thus showed that the shape of an insect or the sound of a monkey are interpreted by hunters as signs sent by animals which can be interpreted to anticipate the animals' future movements.[31] Birdwatchers certainly refrain from being compared to hunters and yet they share some of their techniques to identify signs from the animals.

When they use decoys to lure birds or when they take pictures of birds to signal a common habitat, birdwatchers impose some risks to individual birds for the sake of their species. Yet they place themselves at the same level as birds in reciprocal relations, and not above them in a form of protective control. A more-than-human approach to One Health thus does not raise the health of a certain species above others but is attentive to signs exchanged between species in practices of communication. A One Health approach often considers animals as reservoirs from which humans can anticipate future pandemics. But the nature reserves where ornithologists observe birds appear as therapeutic landscapes where birds are in relation with humans.

The logic of conservation pushes One Health to the limit by integrating animals into public health strategies through the signs of their vulnerabilities. The comparison I have made here between two societies faced with the challenges of conservation has shown that this relation between humans and birds as sentinels is always mixed with more symbolic and political forms of integration. If more-than One Health is an ideal when taking care of humans and animals, it is realized in more than one territory: hence the importance of analyzing how conservation strategies are implemented in local contexts, combining military and democratic genealogies.

Notes

1 See the website of this initiative: "One World—One Health," Wildlife Conservation Society, 2021, https://oneworldonehealth.wcs.org/. For an update on the One Health program since the 2005 declaration, see Irus Braverman's interview with Chris Walzer in this volume.
2 Rodrick Wallace, Deborah Wallace, and Robert G. Wallace, *Farming Human Pathogens: Ecological Resilience and Evolutionary Process* (New York: Springer, 2009).
3 Leslie Sims and Malik Peiris, "One Health: The Hong Kong Experience with Avian Influenza," *Current Topics in Microbiological Immunology* 365 (2012): 281–298.
4 On authoritative measures taken in South Korea to control zoonoses, see Kiheung Kim and Myung-Sun Chun, this volume.
5 Frédéric Keck and Andrew Lakoff, "Sentinel Devices," *Limn* no. 3 (2013), https://limn.it/articles/preface-sentinel-devices-2/; Andrew Lakoff, *Unprepared: Global Health in a Time of Emergency* (Oakland: University of California Press, 2017); Frédéric Keck, *Avian Reservoirs: Virus Hunters and Birdwatchers in Chinese Sentinel Posts* (Durham: Duke University Press, 2020).
6 On the extension of the biopolitics of surveillance and counting in nature reserves, see Rafi Youatt, *Counting Species: Biodiversity in Global Environmental Politics* (Minneapolis: University of Minnesota Press, 2015), and Irus Braverman, "Governing the Wild: Databases, Algorithms, and Population Models as Biopolitics," *Surveillance and Society* 12, no. 1 (2014): 15–37.
7 Peter Doherty, *Sentinel Chickens: What Birds Tell Us About Our Health and the World* (Melbourne: Melbourne University Press, 2012).
8 John Bowen et al., eds., *Pragmatic Inquiry: Critical Concepts for Social Sciences* (London: Routledge, 2020).
9 See Hinchliffe, this volume.
10 Mike Kilburn, interview by author, Hong Kong Central, September 25, 2007.
11 John Law and Michael Lynch, "Lists, Field-Guides and the Organization of Seeing: Birdwatching as an Exemplary Observational Activity," *Human Studies* 11, nos. 2–3 (1988): 271–303.
12 Ibid.

13 Geoff Welsh, interview by author, Aberdeen, Hong Kong, July 15, 2012.
14 Chiu Ying Lam, interview by author, Hong Kong, December 8, 2008.
15 Andrew Whitehouse, "Listening to Birds in the Anthropocene: The Anxious Semiotics of Sound in a Human-Dominated World," *Environmental Humanities* 6, no. 1 (2015): 53–71.
16 Timothy Choy, *Ecologies of Comparison: An Ethnography of Endangerment in Hong Kong* (Durham: Duke University Press, 2011).
17 Elliott McClure, *Migration and Survival of the Birds of Asia* (Bangkok: White Lotus Press, 1974).
18 Peter Chen, interview by author, Taichung, April 27, 2013.
19 In South Korea, environmental protection was also promoted by military surveillance of the border. See Eleana J. Kim, *Making Peace with Nature: Ecological Encounters Along the Korean DMZ* (Durham: Duke University Press, 2022).
20 Lucia Liu Severinghaus, interview by author, Taipei, April 30, 2013.
21 Robert Weller, *Discovering Nature: Globalization and Environmental Culture in China and Taiwan* (Cambridge: Cambridge University Press, 2006).
22 Shui-Yan Tang and Ching-Ping Tang, "Local Governance and Environmental Conservation: Gravel Politics and the Preservation of an Endangered Bird Species in Taiwan," *Environment and Planning A* 36, no. 1 (2004): 173–189.
23 Diogo Veríssimo et al., "Birds as Tourism Flagship Species: A Case Study of Tropical Islands," *Animal Conservation* 12 (2009): 549–558.
24 Michael Szonyi, *Cold War Island: Quemoy on the Front Line* (New York: Cambridge University Press, 2008).
25 Jacques Banchereau and Ralf Steinman, "Dendritic Cells and the Control of Immunity," *Nature* 392 (1998): 245–252.
26 Richard Gorman, "Therapeutic Landscapes and Non-Human Animals: The Roles and Contested Positions of Animals Within Care Farming Assemblages," *Social and Cultural Geography* 18, no. 3 (2017): 315–335.
27 Robert M. Wilson, *Seeking Refuge: Birds and Landscapes of the Pacific Flyway* (Seattle: University of Washington Press, 2010).
28 On the notion of lure, see Hinchliffe, this volume.
29 Etienne Benson, *Wired Wilderness: Technologies of Tracking and the Making of Modern Wildlife* (Baltimore: Johns Hopkins University Press, 2011).
30 Rane Willerslev, *Soul Hunters: Hunting, Animism, and Personhood Among the Siberian Yukaghirs* (Berkeley: University of California Press, 2007).
31 Eduardo Kohn, *How Forests Think: Toward an Anthropology Beyond the Human* (Berkeley: University of California Press, 2013).

아프리카돼지열병(ASF) 확산 방지를 위해 최선을 다하겠습니다
ASF 차단 울타리 「출입문 닫기」 캠페인
"문 닫고, 소독하고, 신고하고"
– 환경부 야생동물질병관리원 국립생태원 –
1
차단 울타리 출입 후 출입문을 반드시 닫아 주세요
2
출입 후에는 반드시 소독해주세요
3
훼손(찢어지거나 구멍난 곳 등)된 울타리는 반드시 신고해주세요!
※ ASF 차단 울타리
야생멧돼지의 이동을 차단하여 아프리카열병(ASF)의
전파·확산을 예방하기 위해 국가 및 지방자치단체가 설치한 중요 시설

12

THE SPATIALIZATION OF DISEASES

Transferring Risk onto Vulnerable Beings

Kiheung Kim and Myung-Sun Chun

Introduction

In South Korea, more than 7.4 million livestock animals have been culled to prevent foot and mouth disease (FMD), avian influenza (AI), and African swine fever (ASF) in the past decade.[1] The elimination of risks by killing potentially infected animals has been justified to secure health and safety for humans and livestock animals. Furthermore, neoliberal policy has supported the proliferation of intensive farms in the livestock industry and globally standardized preventive interventions against livestock diseases, including mass culling, to protect the trade and consumption of livestock products. Despite unceasing social and ethical debates on preventive culling and inhumane culling methods, the authorities have maintained such practices as a form of preventive intervention, appealing to its effectiveness.

Emerging infectious diseases are caused by changes in pathogen–host relationships in a specific space and time.[2] In controlling these emerging diseases, especially zoonotic infections, conventional disease interventions and responses revealed their limits due to the anthropocentric approach and lack of consideration of the complex environment. Therefore, global health organizations have put forward the concept of "One Health," a holistic, interdisciplinary approach to solve these new kinds of problems in order to promote the health of humans, animals, and the ecosystem.[3] In the One Health approach, animals and ecosystems that have been instrumentalized and ignored in conventional

FIGURE 12.1 An African swine fever education card distributed by the South Korean government from 2021. The text reads: "We'll do our best to prevent ASF. ASF Blocking Fence Door Closing Campaign: 'Close the door, disinfect and report.' Ministry of Environment, National Institute of Wildlife Disease Control and Prevention, National Institute of Ecology." Used with permission.

DOI: 10.4324/9781003294085-17

health policy play a more central role.[4] However, the One Health strategy is not utterly free from an anthropocentric and pathologic view, emphasizing contamination and controlled forms of practice over interspecies health.[5] The One Health strategy can even exacerbate the vulnerabilities of humans and animals because disease interventions at the interface between humans and animals may create vulnerable others, such as wild boar hunted to eliminate possible AFS-infected individuals or frontline workers culling AI-infected poultry.

We suggest that considering shared risks and vulnerabilities between humans, animals, and ecosystems is the first step for a practical solution to overcome the anthropocentric weakness of the One Health strategy and move toward interspecies health. This chapter traces the origin of authoritative and centralized disease control policies, focusing on the period of colonization and the military government period in South Korea. The chapter draws on historical material produced by related authorities, national statistics, white papers, governmental reports, news articles, and academic research on the responses to the infectious diseases of humans and animals during the Korean Empire (1897–1910), the Japanese colonial period (1910–1945), and the Republic of Korea (1945–present). In addition, we reference historical data to secondary materials like medical and veterinary history books on infectious diseases in Korea. Based on the historical materials, we will show how the space-based containment has been adopted as a public strategy for infectious diseases, in particular in pre-colonial, colonial, and post-colonial contexts, and how the vulnerabilities of humans, animals, and wildlife in ecosystems were intensified in the process of disease prevention even in the One Health framework.

Shaping the Modern Policy of Disease Prevention in South Korea

Repeated and cyclic outbreaks of novel infectious diseases suggest that these events are unique. In other words, such unprecedented events have not occurred previously, thus requiring the envisioning of an unprecedented event, including responses to outbreaks.[6] However, the use of imagination is always based on previous "disease experience."[7] For instance, COVID-19, which originated in the Wuhan region in China during the winter of 2019, spread quickly and became a pandemic. The mysterious, unknown disease triggered extreme responses from different governments. The Chinese government implemented one of the harshest lockdowns worldwide in the early stages of the pandemic, which was considered an authoritarian measure by most Western countries.[8] Ironically, when the Italian government failed to contain the devastating spread of the infection, most European countries and the US followed aggressive measures. By contrast, some East Asian countries, including Korea, Japan, and Taiwan, decided to implement measures to avoid the challenging lockdown policy.[9] Every government must implement different containment measures based on their public health capacity, economic recovery, sociocultural conditions and their past disease experience, such as outbreaks of SARS (Severe Acute Respiratory Syndrome), avian influenza, and MERS (Middle East Respiratory Syndrome).

Whereas the European experience of a high mortality rate and complete cessation of social and economic activities caused fear and confusion, the Korean experience of no mandatory lockdown has been regarded as a relative success. For example, the mortality rate of Korea in November 2020 was 0.94 out of 100,000 people, whereas the British case was 7.5, and 1.93 for the United States.[10] Many academics have found that centralized and aggressive test-and-trace systems played key roles in producing relative success in containing COVID-19.[11] The main assumption of the containment strategy in European countries has been closely associated with the behavior of individuals. On the contrary, South Korea's approach is quite different: the containment strategy focuses not on individual behavior, but on a specific spatial cluster, which can be called a space-based containment strategy.[12] The early achievement may be attributed to experience with previous coronavirus outbreaks and cohesive communities with varying levels of social control.[13] However, this containment strategy was not equally effective against the highly contagious Omicron variant, with its weak pathogenicity. Therefore, South Korea had to change its strategy to strengthen responsible sanitary behavior of individuals without space containment. China's persistent zero COVID-19 strategy seemed to fail to contain the more contagious Omicron variant.[14] Before discussing them further, it is necessary to find the root of the assumptions embedded in the entire framework of the containment strategy in South Korea.

The Origin of Space-Based Containment Policy at Ports and Railways in Modernized Korea

In the Middle Ages in East Asia, quarantine and restriction of the movement were considered insignificant for public health owing to the small scale and low frequency of the population movements, except in the cases of war or migration. However, after Korea shifted its international policy from seclusion to openness in 1876, the public health responses to infectious diseases arriving from outside the Korean Peninsula needed to be established. When the Korean Empire[15] was established in 1897, Korea experienced a major outbreak of cholera, which precipitated the building of a modern public health system. Because it was considered exogenous to Korea, the quarantine of border and harbor cities became a pivotal measure to prevent the spread of the disease. When the Korean Empire was annexed by the Japanese Empire in 1910, there was a very serious outbreak of the pneumonic Manchurian plague. The Japanese colonial government implemented more sophisticated quarantine measures of building quarantine centers in the border city of *Shineujoo* (新 義 州). The modern hygiene regime began with the principle that the infectious agent should be managed by closing the border and isolating potentially infected individuals who could spread the disease.[16]

The modern hygiene policy in Korea began with opening ports and managing the population. Modern hygiene policy did not dominate the management of port areas because the concept of hygiene and its institutionalization were not yet well established. In 1887, the first guideline for containing infectious diseases, called *Onyokjangjung* (瘟 疫 章 程, "quarantine guideline for plague")—which prohibited people from leaving or offloading from a ship—originated from the region where the plague had spread. Korea's quarantine measures provoked resistance from other countries. It was difficult to convince those countries that the real entity of the plague was based on traditional *Wenyilun* (瘟 疫 論or plague theory) in which "Plague-Qi" invaded a body and caused "Fever."[17]

When the Korean Empire opened its ports to the imperial powers, including Japan and Russia, the ports were set up as leased territories where imperial laws and legal regimes were applied. In 1876, Korea had its first diplomatic relationship with Japan. One of the diplomatic terms was the establishment of leased territories, mainly ports. Japan set up three leased territories where the Japanese administrative rules were applied. In addition, the Russian naval fleet occupied a small island called *Jeolyoung-do* (絶 影 島) outside Busan port and demanded a leased territory, a demand which was rejected by the Korean government. As cholera broke out in the port city of Busan in 1886 and spread to the capital city of Seoul, the quarantine measures were tightened at the western port city of Incheon, and consulates in each country strengthened the sanitary administration of their settlements. For instance, in Busan port, disinfection measures were taken for Koreans entering and leaving settlement territories. Nevertheless, there was a new trend of modernized hygiene policies that elite leaders tried. One member of the reform party, Kim Ok-kyun (1851–1894), published an ambitious document, *Chidoyaklon* (治道略論, or "strategies for governing logistics"), establishing a hygiene police system in urban settings.[18] Another reformist and politician, Yu Gil-joon, proposed to establish a hygiene department to manage infectious diseases after the Gabo Reformation (甲 午 改 革, 1894 Reformation). The purpose of the department was to "govern all public health issues, including preventing infectious diseases, vaccinations, and quarantine."[19]

At the time, the general method of containment was spreading disinfectants (coals were thought to be the most effective disinfectant at the time) and isolating patients in quarantine hospitals. Animals, including cattle and chickens, were also allowed to enter Japanese settlements after staying in quarantine stations for one week.[20] The Japanese jurisdiction over leased territory in the Korean Empire led to the establishment of a protected space in a contaminated nation. Koreans were regarded by the Imperial countries, such as Japan, Russia and Britain, as potential spreaders of infectious diseases that should be controlled. The defense method of creating a safe zone was reproduced in the quarantine strategy of the Japanese Government-General of Korea following annexation in 1910.

Quarantine measures were established under the assumption that the plague (*on-yok* or *wenyi* in Chinese, 瘟 疫) had entered from outside Korea.

Simultaneously, there was a special uncontaminated territory that formed a strict boundary between contaminated and uncontaminated locations based on Japanese-ruled spaces in Korea during the end of the Joseon dynasty and Korean Empire. Following the annexation, the Japanese Government-General implemented various measures of containment, quarantine, and hygiene by shaping a safe zone in colonized Korea. This safe zone was established by creating quarantine stations in ten port and border cities to prevent the spread of the cholera epidemic, starting from southern China in 1919.[21] The Japanese Government-General of Korea established a local vigilance union to recruit a new workforce to contain cholera.

Another detailed example of controlling a specific space to respond to infectious diseases from outside involves the two large epidemics of the Manchurian plague, in 1910 and 1920. The Japanese Government-General of Korea implemented two critical measures to prevent the influx of the plague soon after the annexation of the Korean Empire. Initially, the movement of Chinese seasonal migrant workers ("coolies"), who were blamed for the spread of the plague, was blocked and controlled by the public health authority. Quarantine stations were established, and quarantine measures were implemented at railway stations in border cities and trade ports coming from China. Furthermore, academics misidentified the plague spreading in Manchuria as a bubonic plague. As a result, many suspected that rats and fleas were vectors of spreading the plague. Thus, the Japanese Government-General of Korea believed that it was logical to launch a rat-catching and -exterminating campaign in newly colonized Korea.[22]

However, scientists soon realized that the first Manchurian plague was not a bubonic plague transmitted by rats through skin-to-skin contact; rather, it was a pneumonic plague that spread through the respiratory tract.[23] Nevertheless, the Japanese Government-General of Korea maintained its position that the plague could be controlled through the rat-catching campaign. This stance changed with the spread of the second Manchurian plague in 1920. After the outbreak of cholera, which caused great damage to colonial Korea in 1919, the quarantine policy was tightened in Shineuju, the border city between Manchuria and Korea, to block the influx of plague patients from Manchuria. A strict quarantine policy for Chinese workers was imposed on the border town. As the rat-catching policy was a biopolitical strategy for the discipline of colonial Koreans, a new quarantine system, called the "household quarantine (戶 口 防 疫)" system, organized every household as a unit for containing the disease.[24]

The Establishment of the Immunized Zone in the Colonial Territory

The opening of the ports ushered in a new awareness of infectious diseases coming from outside Korea. As the era of the discovery of microbes progressed, the modern hygiene policy gained critical weapons in the form of vaccines and the isolation of pathogens. In the 20 years after Robert Koch discovered the causative agent of tuberculosis, a series of causative agents for major human and animal diseases were identified.[25] Modern hygiene policies were applied similarly to infectious diseases in livestock. Even before the modern sanitation system for livestock was established in the Joseon dynasty, trading cattle practices existed with neighboring countries, such as Japan and China. They were able to quarantine livestock shipped into Korea and to ban export animals from Korea for the purpose of preventing infectious diseases in their countries. Therefore, Korean farmers learned the concept and importance of quarantine on a practical level.

With the onset of colonial rule in Korea, sanitation policy turned Korea into a border zone of the Japanese Empire, setting up a space of quarantine to contain infectious diseases. Both human and nonhuman animals on this border thereby became marginalized and vulnerable beings, because they were strictly controlled, contained, and objectified during the quarantine process. Without establishing modern medical personnel and facilities, improving nutritional status, and establishing infrastructure such as water and sewage facilities, it was difficult to obtain effective results for preventing infectious diseases by deploying colonial hygiene policies consisting of personal hygiene or discipline of the population.[26] However, countermeasures against livestock infectious diseases achieved relative success, resulting in increased livestock productivity.

Traditional infectious diseases on the Korean Peninsula often followed a geographical vector. For instance, the cattle plague "was spread from west to south" before it spread to the entire nation.[27] As Japan won the Russo-Japan War, the Japanese government conducted a large survey of livestock diseases on the Korean Peninsula.[28] The survey showed that the livestock disease from China was introduced to Japan through the Korean Peninsula as anticipated, which would not only be a threat to domestic livestock in Japan, but also the biggest obstacle for using Korean livestock. Therefore, the Japanese Government-General of Korea established a "dual quarantine system" that implemented quarantine in Korea and re-quarantine in Japanese import ports.[29] In other words, colonized Korea became a buffer zone to block the influx of infectious diseases coming from China, especially Manchuria. In Busan, a quarantine station for exporting cattle and a serum manufacturing facility for rinderpest were established. The quarantine station was a symbolic institution that controlled and monitored infectious diseases of livestock within the Japanese Empire.

The development of modern veterinary medicine in colonial Korea allowed trading of healthy, safe cattle to the Japanese Empire. In 1908, when rinderpest again erupted in Korea, the Japanese Resident-General in Korea deployed quarantine officers to control trading ports. A year later, the Export Cow Quarantine Act was enacted to check for diseases of cattle exported from Korea, stipulating the establishment of quarantine stations and examinations before quarantine. Along with the hygiene police force, Japanese veterinarians oversaw livestock quarantine in Korea. In Korea under Japanese rule, the "cattle of Joseon" became the "cattle of the Empire." However, as young cows were loved by Japanese consumers, mature cows became scarce in Korea, resulting in a decrease in the productivity of the "cow of the Empire" on the Korean Peninsula.[30]

Moreover, the Korean Peninsula itself became a huge, immunity barrier installed outside the borders of mainland Japan. Japan created an immunity zone to prevent the spread of infectious diseases in the border areas between China (Manchuria) and Korea. The campaign for injecting an immune serum started mainly in areas severely damaged by rinderpest in the Tuman River and Yalu River basins. When the serum manufacturing facility produced a vaccine, the Japanese Government-General of Korea vaccinated approximately 8,000 cattle in North Hamgyeong Province and North Pyongan Province in 1924. From December 1924 to April 1925, the cattle plague was prevalent, but the vaccinated cattle were protected. From 1925 to 1930, approximately 190,000 cows in the region received vaccinations.

As the rinderpest immunity zone was established, an animal disease immunity zone was created for other diseases, such as blackleg. The sanitary department of the Japanese Government-General of Korea encouraged the creation of a local organization to control and monitor animal diseases. The space-based containment strategy achieved its own success, and the number of outbreaks of rinderpest gradually decreased after large-scale outbreaks in 1908 and 1920.[31] Strong space-based quarantine through mass vaccination based on administrative districts was implemented only in livestock epidemics. As large-scale production of livestock vaccines was possible, the spatial quarantine strategy was feasible in colonial Korea. The Korean Peninsula became a buffer zone to protect mainland Japan from infectious animal diseases.

Infectious Disease-Free Policy in Korea: Routinization of Mass Culling

Colonial legislation for livestock disease prevention remained in effect until 1961, when the new Korean legislation was issued. However, the legislation structure and system have not changed significantly. Thanks to the sanitary policy, outbreaks of old-generation epizootics, such as rinderpest, FMD, and anthrax, decreased. However, industrialized livestock farming of pigs and chickens has resulted in other epidemics. After the Korean War in 1953, local authorities and research institutions were responsible for screening, diagnosing, and culling infected livestock. National authorities established a long-term plan to eradicate classical swine fever, Newcastle disease, pullorum disease, fowl typhoid, bovine tuberculosis, and brucellosis. The goal was to declare Korea free from the abovementioned infectious diseases.[32] Additionally, global free-trade pressure led the national authority to protect the domestic livestock industry. The primary strategy for promoting the domestic livestock industry was to prevent livestock diseases.[33] One of the most critical diseases, classical swine fever, infected thousands of pigs annually. Following the "successful cases of advanced countries" through the broad vaccination and culling program, by 2000 the disease should have been eradicated in

South Korea.[34] Eradication of a livestock disease through a diagnosis–culling–vaccination strategy was similarly adopted for other diseases. Local governments do not have the authority to decide or change the disease prevention process.

FMD outbreaks in 2010 and 2011 raised questions regarding the basic rapid diagnosis–culling–vaccination strategy against epizootics. Since the last outbreak in 1934, Korea did not experience any FMD outbreaks until 2000. Thus, the agricultural authority in Korea did not have any institutional memory or experience of preparing for the sudden outbreak of this nearly forgotten animal disease, but the main methods of containment, aggressive culling, and vaccination, were implemented. In particular, ring vaccination, which is delivered to surround the epicenter of the outbreak, prevented the spread of the disease and kept the number of culled animals to a minimum. Despite the successful control in 2000, when the vaccination policy was implemented, the South Korean agricultural industry had to wait for the recommencement of the meat trade. What the authority learned from the disease experience was disillusionment with the vaccination-centered policy. Therefore, preventive culling was prioritized to quickly achieve a disease-free status for the meat trade.

The preventive-culling-based strategy for containing the disease had a devastating effect during the 2010 FMD outbreak. Nearly 150 thousand cattle and 3.3 million pigs were killed and buried. The number of animals culled was beyond expectation, causing catastrophic disasters in rural communities. The public was upset about the inhumane and brutal processes of killing animals in the name of prevention and the rural residents suffered from the leachate from the burial sites. The public, observing aggressive and indiscriminate mass culling, were aware that animal diseases could impact "lifescapes," articulating spatial, emotional, and ethical dimensions of the relationship between landscape, livestock, farming, and rural communities.[35] After the 2011 FMD outbreak, criticism of the centralized and aggressive culling of animals grew. However, preventive culling of all animals in potentially contaminated spaces is still prominent, and more frequent than other alternatives.[36]

Marginalized Wild Animals and Livestock Animals

Although the authorities needed to shift their main controlling policy to mass vaccination during the FMD outbreaks in 2010 and 2011, the culling policy has been maintained in the situation of outbreaks of AI and ASF. As the preventive culling made domesticated animals vulnerable to being killed in the FMD outbreaks, wild animals became subject to culling due to the potential risk of infection. In these cases, mass culling is applied as a preventive measure to identified or domesticated animals at risk and to wild animals at unknown risk. However, unclear demarcation of boundaries between the uninfected and infected makes both types of animal vulnerable beings, even though they are not confirmed to be infected.

Every year, AI pathogens are detected in wild birds at migratory bird sites throughout the country. In the early 2000s, a few local communities tried to burn the sites to prevent AI infection in chickens and ducks. The ecological value of wild birds has become a pivotal reason to protect them from mass culling.[37] However, poultry farms near the sites could not escape marginalization. Preventively culling potential host chickens and ducks to control AI is the only alternative, as vaccination against the highly pathogenic AI is not feasible due to the fear of virus mutation to strains pathogenic toward humans. These fowl, on the border between farms and the wild, were considered to potentially have had contact with the wild birds carrying the virus. If birds on a farm were detected as being positive for the AI virus during routine monitoring surveillance or if farmers reported suspected symptoms, the entire flock of birds was killed by the local government. From 2014 to 2018, approximately 70 million domestic birds were culled. In the winter of 2020 alone, 30 million fowl were culled and buried.[38] Mass culling and burials have been routinized for the winter season, creating local companies dedicated to culling and disinfecting contaminated farms. Such firms take culling orders from local governments and hire cheap and unskilled people, including foreign workers. Some local governments have attempted nonaggressive elimination of animals on the border. They compensated the owners for the economic loss if the owners voluntarily closed the duck farms during the winter because domesticated ducks were more sensitive to AI infection.[39]

In the case of the African swine fever (ASF) outbreak in 2019, no vaccines were available. After the first ASF outbreak in 2019, the ASF prevention strategy was to block the spread of the disease in the south of Korea. To decrease the wild boar population, the mass hunting of wild boars (not identified as being infected) between the contaminated and safe spaces is allowed and rewarded by local governments. As well the mass hunting, defensive physical walls are constructed to block the spread of the disease. A total of 326 kilometers of fences were built by the central authority, crossing the Korean Peninsula close to the Military Demarcation Line between South and North Korea to prevent the movement of wild boar. Preventive culling and fences seem to effectively block the spread of ASF.[40] However, such mass hunting has been criticized for disturbing the natural movement of wild boars and for spreading ASF more rapidly among the wild boar population. This aggressive method stopped the possible interactions between other animals, such as rodents, birds, and humans, who cross the border between contaminated and safe spaces.

The Spatialization of Diseases

The aggressive spatial strategy of the central government relies on three important assumptions. First, it requires a centralized method that can implement containment measures seamlessly while minimizing or ignoring the intervention of the local government or civil sectors. Regarding the outbreak of FMD in 2010, local governments demanded the central government loosen its monopoly on authority of inspection and examination, but the central government's position remained unchanged;[41] it still monopolizes the right to manage tests. The second assumption is that quarantine capabilities must be secured by the mobilization of civil servants and the military, which is reminiscent of a wartime mobilization system. This mobilization method is centered on the military, police, and civil servants. Its origins are in the colonial era's "sanitary police" and also influence the role of public health doctors and officials in the current COVID-19 situation.[42]

The third assumption and perhaps the most important, is that the centralized strategy is dependent on spatial containment. As discussed, the spatialization of disease is a principal characteristic of the modern health system in East Asia. Since outbreaks of modern, infectious diseases in humans and animals like cholera, Manchurian plague, and rinderpest, containment policy has not been focused on movements or behavior of individuals. Rather, health authorities focus on space (urban space, ports, and railway stations), territories (immunized zones), and collective movement (banning or controlling movement of workers). This approach has been effective in the case of outbreaks of a series of animal infectious diseases. The centrality of space and containment were well illustrated by the case of COVID-19.

Space is not only physical. It is also a network of social, cultural, and economic institutions. Since the first COVID-19 case was reported in January 2020, Korea has regarded the infection not as a result of individual behavior, but as a result of collective infection in a specific space. In the case of collective infection, when a space or group is specified, swift and easy epidemiological tracking is possible. In addition, tracking and isolating possible patients and their contacts can be easily accomplished. The space-based containment system is also the core of the prevention and control of COVID-19. As described, the prototype of the current COVID-19 prevention is found in the control strategy for human and animal infectious diseases during the colonial period.

In the context of East Asia, public hygiene and prevention of infectious diseases led to the spatialization of disease, body, and medicine.[43] The public health authority attempted to control and block the influx of infectious diseases by reshaping the space and creating boundaries. At the time, the primary method used by the colonial authorities to prevent infectious diseases such as cholera, the Manchurian plague, and rinderpest was controlling specific spaces. The prototype was maintained in the current quarantine strategies. The space-based system creates boundaries that separate normal space from pathological space through the spatialization of disease.

This boundary-making produces vulnerable beings. Under the name of "swift and coordinated centralized quarantine," humans and animals, who are vulnerable to infection, are easily evicted from the normal space, marginalized, and excluded. The most representative case is the "swift and intensive" preventive culling implemented by the central government to prevent highly pathogenic avian influenza (HPAI) since 2003. Between 2010 and 2011, 6.5 million poultry were culled due to HPAI, and 61.2 billion won (US$50 million) was required to compensate for livestock damage.[44] From the coordinated mobilization of public health workers to the diagnostic process managed by the central government and the flexible responses, the Korean government's specific responsive strategies are commonly applied to both animal and human infectious diseases. In this process, animals in the suspected infection space becoming the target of preventive culling are displayed as pathological entities.

Not only humans and animals are located in the suspected space; those involved in the control and preventive measures are also excluded from normal life. Civil servants and military personnel are required to mobilize quickly to achieve the effective, preventive culling used for FMD or HPIA. In the case of the FMD outbreak in 2010, nearly two million people were mobilized to prevent the disease.[45] The role of public health doctors, nurses, and public health workers has been crucial in coping with the COVID-19 pandemic. For instance, instead of compulsory military service, public doctors, who are under the control of the state, are tasked with on-site mobile examinations and collecting samples from symptomatic patients. In particular, the public health doctors and nurses played a decisive role in controlling the dangerous situation in Daegu, where the number of COVID-19 patients surged in March 2020. The government ordered the mobilization of 742 new public health doctors, exempting them from the 4-week compulsory military training.[46] The mobilization system seen during COVID-19 has similarities in its strategic principles to the mobilization system already used in the prevention strategies for FMD and AI. However, because personnel are in the same space as the infected people or animals, their status becomes pathological too. In a situation reminiscent of the wartime mobilization system, they are required to perform low-waged, highly intensive labor and are being subjected to unilateral sacrifice.

The centralized "containment–isolation strategy" for infectious human diseases, including COVID-19, showed high efficiency through the intensive control of a certain space. This successful strategy becomes the "Korean-quarantine (K-quarantine) model." However, space-based strategies of containment isolation are not always successful. Preventive strategies shaped over a long period of time and past disease experience cannot adequately respond to the ever-changing variations in the outbreak of infectious diseases. For instance, the central authority announced that the so-called "zero COVID-19" policy led by the central government, which aimed to eradicate the infectious disease based on "containment–isolation" and "social distance," would be abandoned in November 2021. Instead, the main strategy would be shifted to "living with COVID-19."[47]

The basic assumption of the zero COVID-19 policy was based on the eradication of pathogens through strict and aggressive containment–isolation. However, the newer policy seeks coexistence with the disease, and the coexistence policy cannot be maintained if the principle of eradication still plays a role as a background assumption. Unlike the strong and aggressive state-led containment–isolation policy, coexistence with the pathogen is not feasible unless the fundamental assumptions are reconsidered. Rather, without having policies to induce citizens' voluntary participation and cooperation, the "gradual recovery of normal life" would likely fail. However, when the virulence of the Omicron variant was obviously weaker than the previous dominant variants like the alpha and delta variants, the health authority announced a long-term plan for the recovery of normal life in February 2022.

The effectiveness of state-led, aggressive, and centralized policies is undeniable. It was because of the state-led policy that South Korea was able to establish the K-quarantine model while avoiding large-scale prevalence of COVID-19 in the early stages of the pandemic. However, the centralized and uniform policy ultimately failed to avoid major negative consequences, revealing the vulnerabilities of humans and animals facing risk. To solve the problem of excessive state-centered and human-centered quarantine policy, the One Health initiative should be considered to encourage collaboration/participation between various relevant actors.

Exacerbated Vulnerabilities Under the One Health Strategy

In the One Health initiative, the interfaces between humans and animals (livestock), humans and nature (wild animals), and animals and nature are recognized to be critical risk points for the evolutionary emergence of new pathogens. However, the distinction between nature and society, and between livestock and the wild, is dynamic.[48] Hygiene and public health policies shape and reshape artificial boundaries of safe zones immunologically and socio-culturally, and push vulnerable individuals out of the boundaries and eliminate them. There is little concern about whether humans, domestic animals, and wildlife fairly share the risks under the One Health strategy. Indeed, it is difficult to avoid criticism for instrumentalizing animals and ecosystems to promote human health. In a human-made health system, the hierarchy of humans, animals, and other species in the ecosystem is not treated as equal. Particularly, in the case of zoonosis, in which humans and animals are susceptible or animals are suspected to transmit diseases to humans, irreversible damage that can occur in the process of controlling the disease is solely targeted at animals.[49]

Diseases among animals are also categorized as based on human interest, and harm to animals is hardly considered in the process. Specifically, wildlife roaming in natural ecosystems has received the lowest status. The One Health strategy attempts to correct this hierarchy, but ends up reinforcing it. In the process of implementing the control and management of infectious diseases, some species or individuals are included in the system, while other species are strategically driven out and remain on the boundaries of the health system. This banishment is due to the influence of social, ecological, economic, cultural, and human–animal relations.

Modern hygiene policies apply similar approaches to infectious diseases in humans and animals. In the early twentieth century, advances in modern microbiology, immunology, and antibiotics played decisive roles in reducing the incidence of traditional infectious diseases. Large-scale epidemics in the twentieth century occurred mainly in livestock because of the rapid expansion of the industrialization of farming sectors after the Second World War. Therefore, the social experience of epidemics has been shaped by livestock epidemics. The prevention of a series of diseases driven by coronaviruses has similarities with the prevention method of animal infectious diseases. In the case of zoonotic diseases, animals are regarded as carriers of the infectious agent and are thus treated as pathogens themselves. In modern public health, responses to infectious diseases are systemized, specialized, and standardized. In this sense, various professionals participate in the control and prevention process. In the case of livestock, large-scale mass culling is often carried out as a control method and workforces are mobilized for this culling.

Table 12.1 shows how traditional and modern hygiene and current public health regimes have responded to infectious diseases in which the vulnerabilities of humans and animals are reinforced by the expansion of the One Health strategy.

Heterogeneous actors are mobilized between the normal and the pathological boundaries in the course of outbreaks of new infectious diseases. They are not only veterinarians, public health workers, and outsourced workers, but also nonhuman animals (livestock or wildlife) that become victims of the diseases. According to historical experience, sudden outbreaks and death fears have exacerbated social discrimination. In the pre-modern era, when the plague spread in an area, people fled to avoid the epidemic. Some people evacuated to a safe and isolated area (refugia) provided by individuals and the state.[50] When the servants of the noble family became ill, all family members fled to the refugia. If the patient belonged to the lower class, they had to move to a remote and isolated shelter. It was thought that diseases could be transmitted from the lower class to the upper class.[51] Thus, isolation of the lower class was believed to be a more effective way of controlling the plague. In this process, the lower class became one of the vulnerable groups.

The problem of severe trauma among health workers and veterinarians is another important social concern.[52] Most health and essential workers share the same problem with workers who participate in the prevention and control of animal diseases.[53] A large-scale system of mobilization of temporary and flexible labor forces increases the problem of outsourcing risks.[54] The outsourcing of risks can be extended to nonhuman actors like livestock and wild animals. Low-pathogenicity avian influenza virus is not a radical threat to chickens and ducks; however, farmers' economic burden from the decline in productivity and lowered quality of goods due to the disease is prioritized over animals' lives. The risk of the disease falls on the bird, and the burden of the buried carcasses falls on the ecosystem. The number of burial sites created by the FMD epidemic in 2010–2011 reached 4,799, and most were built around available farmland, causing serious environmental problems in rural areas and scarring communities.[55] Although the system of control and prevention of infectious diseases has been maintained by the centralized and strong space-based system of mobilization, the risk was outsourced to the region and affected farming communities.

Table 12.1 The response to infectious diseases and emergence of vulnerable beings. Compiled by the authors.

Regimes	*Traditional Regime*	*Modern Hygiene Regime*	*Current Public Health Regime*
Main disease	Plague (wenyi/onyok)	Plague/rabies/rinderpest	FMD/MERS/avian flu/ African swine fever
Socioeconomical background	Class system	Colonial sanitary rules Modernization of livestock	Neoliberal industrialization
One Health perspectives	None	Animals as a vector Perception of zoonoses	Expansion to wildlife
Boundary	Safe zone (refugia)	Immune zone	Disease-free zone
Control of space	Distancing individuals	Boundary-making	Space-based prevention
Mechanism of marginalization	Exposure to disease Hierarchical opportunities of treatment	Protection of the Imperial productivity Creation of buffer zone	Outsource of risks Eradication policy
Vulnerable beings	People in the lower classes	Colonized people and animals	Health workers/ livestock/wildlife

Conclusion: Interspecies Health and Outsourced Vulnerabilities

Emerging infectious diseases that have periodically hit Korean society revealed the limits of the One Health strategy with its anthropocentric view. Including animals and ecosystems as actors according to the One Health strategy may intensify hierarchizing human and nonhuman animals, marginalizing vulnerable beings, and outsourcing health risks to them. One Health's purpose is to overcome the anthropocentric medical approach to solve current and future context-dependent health issues. However, nonhuman animals and ecosystems remain invisible to health policies. Additionally, the critical aspects of the socio-cultural context and human–animal relations are underestimated. Still, the spatialization of diseases works in the field of One Health.

As previously discussed, the disease experience of Korean society plays a critical role in building prevention strategies in human and animal diseases. Disease control strategies for both humans and animals are interrelated and inter-productive. Therefore, One Health must weigh more on the interaction and shared vulnerability-based approach than on the interface-based (space-based) approach. Moreover, this change should not remain fixed on ethical debates, but should shift with respect to political and practical implementation. The One Health strategy can reveal and, at the same time, exacerbate the shared vulnerability of humans and nonhumans. It can maintain their biological, geographical, cultural, and socially linked vulnerabilities.

In mass culling cases in South Korea, the vulnerability of livestock farmers and animals that could not escape from the risk area was linked to the vulnerability of frontline workers for mass culling and extended to the vulnerability of community members who lived or worked close to the culling and burying sites. Centralized and aggressive authorities attempted to make these vulnerabilities into personal problems. Mental health support, such as trauma counseling and frontline workers' education, were suggested as solutions. However, focusing on interspecies vulnerability, One Health can respect local experience and knowledge and build a more detailed disease control strategy for humans and nonhumans in the same health community.

In 2017, a chicken farm in the avian influenza outbreak area filed a lawsuit against a preventive killing order, and the application for dismissal of the execution, which was rejected in the first trial, was accepted by the appeals court. The animals tested negative, and culling was not carried out.[56] Because the farm was a certified animal welfare farm, animal protection organizations and the public who were against aggressive culling supported the farm's resistance. Therefore, this case was perceived as an animal protection movement. However, the case directly revealed the overlapping vulnerability of local governments who do not have the authority to make decisions regarding culling and other preventive methods on farms located near the migratory birds' sites and of nonhumans facing mass culling. Additionally, the public—with psychological vulnerability, after witnessing repeated mass culling scenes for decades—showed their political will to change the situation.[57] Overlapping vulnerability is not likely to be alleviated by disease prevention measures to minimize economic loss and achieve an early disease-free status.

Current preventive measures, which are centralized and ignore vulnerability, may achieve short-term outcomes. However, the outcomes can show a distorted reality and disturb creative preventive measures against human and nonhuman diseases. Considering the new pandemics affecting multiple species, the outcomes should be reevaluated from the perspective of interspecies health[58] in the shared ecosystem. This is because multiple One Health values entangle interaction and vulnerability for humans and nonhumans in the shared ecosystem.

The One Health strategy extended the scope of disease prevention and the actors involved in it to humans, animals, and ecosystems, as against a conventional human health approach. But still, there is a risk, and actual instances, of instrumentalizing other-than-human actors and transferring the risk to them. The centralized space-based containment model has secured the health of relatively advantaged ones (such as humans and those of higher socioeconomic status) by discriminating and marginalizing vulnerable ones (such as animals and those of lower socioeconomic status). The reverse idea, aimed at reducing interspecies vulnerabilities, will be a way of moving toward more-than-human health or interspecies health.

Acknowledgements

This chapter was presented by the authors at the symposium of the Korean Association of Science and Technology Studies (November 13, 2020): Kiheung Kim, "Spread of Infectious Diseases and Diagnostic Technology: COVID-19 and FMD"; Myung-Sun Chun, "One Health Perspectives and Response to Epizootics." They were supported by the National Research Foundation of Korea, funded by the Ministry of Education (Grant No. 2019S1A5A2A03047987).

Notes

1 Jongku Kang and Minseong Kim, "Fact Check, 100 Million Livestock Animals Were Culled for 20 Years [Korean]," *The Korea Economy Daily*, October 22, 2019.

2 Annemarie Mol, *The Body Multiple* (Durham: Duke University Press, 2003), 1–28; Steve Hinchliffe, "More than One World, More than One Health: Re-Configuring Interspecies Health," *Social Science & Medicine* 129 (2015): 28–35.

3 Louise H. Taylor, Sophia M. Latham, and Mark E. J. Woolhouse, "Factors for Human Disease Emergence," *Philosophical Transactions: Biological Sciences* 356, no. 1411 (2001): 983–989.

4 World Organisation for Animal Health, "Tripartite and UNEP Support OHHLEP's Definition of One Health," December 1, 2021, https://www.oie.int/en/tripartite-and-unep-support-ohhleps-definition-of-one-health. The tripartite of One Health (World Health Organization, World Animal Health Organisation, and Food and Agriculture Organization) each has a different "One Health" strategy in priority and

purpose. WHO presents One Health as an efficient tool against zoonotic diseases and antimicrobial resistance, and for food hygiene. Meanwhile, the OIE stresses the importance of the human–animal–environment interface in global health issues. The FAO promotes the One Health strategy in a local capacity.

5 Hinchliffe, "More than One World."

6 Andrew Lakoff, *Unprepared: Global Health in a Time of Emergency* (Oakland: University of California Press, 2017), 135–161.

7 Kiheung Kim, "Construction of the Covid-19 Disease-Scapes: Human-Animal Disease Experience and Space-Centred Containment," *ECO* 25, no. 1 (2021): 83–130.

8 Amy Qin, Steven Lee Myers, and Elaine Yu, "China Tightens Wuhan Lockdown in 'Wartime' Battle with Coronavirus," *New York Times*, February 6, 2020.

9 Dylan Scott and Jun Michael Park, "South Korea's Covid-19 Success Story Started with Failure," *Vox*, April 19, 2021.

10 "Mortality Analyses," Johns Hopkins University & Medicine Coronavirus Resource Center, 2022, https://coronavirus.jhu.edu/data/mortality; UK Health Security Agency, "Covid-19 Confirmed Deaths in England (to March 31, 2022): Report, UK Health Security Agency, 2022, https://www.gov.uk/government/publications/covid-19-reported-sars-cov-2-deaths-in-england/covid-19-confirmed-deaths-in-england-to-31-march-2022-report.

11 Mark Ryan, "In Defence of Digital Contact-Tracing: Human Rights, South Korea and Covid-19," *International Journal of Pervasive Computing and Communications* 16, no. 4 (2020): 383–407; Jongeun You, "Lessons from South Korea's Covid-19 Policy Response," *The American Review of Public Administration* 50, nos. 6–7 (2020): 801–808.

12 Kim, "Construction of the Covid-19 Disease-Scapes"; Choon Key Checkar and Hyomin Kim, "Covid-19 Exceptionalism: Explaining South Korean Responses," *East Asian Science, Technology and Society: An International Journal* 16, no.1 (2022): 7–29.

13 Guangyu Lu et al., "COVID-19 in Germany and China: Mitigation Versus Elimination Strategy," *Global Health Action* 14, no. 1 (2021): 1875601.

14 Luke Taylor, "Covid-19: Hong Kong Reports World's Highest Death Rate as Zero Covid Strategy Fails," *British Medical Journal* 376, no. 420 (2022): 35177535; Vincent Ni, "China's Zero-Covid Policy Is Not Sustainable, WHO Director General Says," *The Guardian*, May 10, 2022.

15 Although the Joseon dynasty was managed independently between 1392 and 1897, it was a perfunctory client kingdom of the Qing dynasty of China, which should be understood in the international diplomatic context. However, as Western and Japanese pressures were growing, King Gojong proclaimed independent from the Chinese influence and declared the "Great Korean Empire" (大 韓 帝 國) in 1897.

16 Kyu-Hwan Shin, "Unexpected Success: The Spread of Manchurian Plague and the Response of Japanese Colonial Rule in Korea, 1910–1911 [Korean]," *Korea Journal* 49, no. 2 (2009): 165–182.

17 Dong-Won Shin, *History of Modern Korean Medicine* [Korean] (Seoul: Hanwool, 1997), 116–117.

18 In-sok Yeo et al., *History of Medicine in Korea* [Korean] (Seoul: Yeoksagongkan, 2018), 258.

19 Korean Society of Infectious Diseases, *History of Infectious Diseases in Korea* [Korean] (Seoul: Kunja Publishing, 2009), 327–330.

20 Hanmin Park, "Cholera Epidemic and Quarantine of Open Ports in Joseon in 1886 [Korean]," *Korean Journal of Medical History* 29, no. 1 (2020): 43–80.

21 Kyu-Hwan Shin, "Maritime Quarantine and East Asia: The Prevention of Cholera in Colonial Taiwan and Korea, 1919–20 [Korean]," *The Journal of Chinese Historical Researches* 124 (2020): 187–217.

22 Kyu-Hwan Shin, "The First and the Second Pneumonic Plague in Manchuria and the Preventive Measure of Japanese Colonial Authorities (1910–1921) [Korean]," *Korean*

Journal of Medical History 21, no. 3 (2012): 449–476; Kyoungsun Jeon, "The Plague Pandemic and Prevention in Manchukuo (1933) [Korean]," *The Journal of Chinese Historical Researches* (2018): 275–307.

23 Se-Kwon Jeong, "How Was a Science on Epidemic Disease Discussed? The Pneumonic Plague in Manchuria and the International Plague Conference, 1910–1911 [Korean]," *Critical Review of History* 133 (2020): 350–379; Jeong-Ran Kim, "Malaria and Japan's Frontier: Manchuria, 1910s–1940s," *Social Science Diliman* 14, no. 2 (2018): 26–48; Mark Gamsa, "The Epidemic of Pneumonic Plague in Manchuria, 1910–1911," *Past & Present* 190, no. 1 (2006): 147–183; Christos Lynteris, ed., *Framing Animals as Epidemic Villains* (London: Palgrave MacMillan, 2019).

24 Shin, "The First and the Second Pneumonic Plague."

25 Stanley Maloy and Moselio Schaechter, "The Era of Microbiology: A Golden Phoenix," *International Microbiology* 9, no. 1 (2006): 1–7.

26 Eun-Kyung Choi, "Anti-Tuberculosis Policy of the Government General of Korea During Japanese-Colonial Period (1910–1945): From Simple Restriction to Active Enlightenment [Korean]," *Korean Journal of Medical History* 22, no. 3 (2013): 713–757.

27 Myung-Sun Chun, "Geographic Distribution of Livestock Infectious Diseases in Korea During the Japanese Colonial Period [Korean]," *Journal of Cultural and Historical Geography* 31, no. 1 (2019): 57–70.

28 Hatsuo Tokishige, *Investigation Trip Report on Rinderpest and Other Livestock Infectious Diseases in Korea* [Japanese] (Tokyo: Agriculture Bureau, Department of Agriculture, 1907).

29 Yoon-Geol Jang, "Control of Korean Livestock Resources by the Japanese Empire—The Establishment of the 'Dual Quarantine System' and Its Significance [Korean]," *The Korea-Japan Historical Review* 69 (2020): 63–94.

30 Chaisung Lim, "Korean Cattle and Colonial Modernization in the Japanese Empire: From 'Cattle of the Peninsula' to 'Cattle of the Empire,'" *Korea Journal* 55, no. 2 (2015): 11–38.

31 Siyoung Lee, *History of Veterinary Medicine in Korea* [Korean] (Anyang: National Veterinary and Quarantine Service, 2010), 385.

32 Korean Veterinary Medical Association, *The 50 Year History of Veterinary Medicine in Korea* [Korean] (Seongnam: Korean Veterinary Medical Association, 1998), 221–227.

33 Ok-Gyeong Kim, "1991 Livestock Sanitary Policy [Korean]," *The Korea Swine Journal* 13, no. 2 (1991): 62–66.

34 Korean Veterinary Medical Association, *The 50 Year*, 232–234.

35 Ian Convery et al., "Death in the Wrong Place? Emotional Geographies of the UK 2001 Foot and Mouth Disease Epidemic," *Journal of Rural Studies* 21, no. 1 (2005): 99–109; Eun-Jung Choi and Myung-Sun Chun, "Response to Risk of Foot and Mouth Disease [Korean]," *The Journal of Rural Society* 25, no. 1 (2015): 271–315; Kiheung Kim, "Multiplicity of Pathogen: Construction of Foot-and-Mouth Disease in White Papers in Korea [Korean]," *ECO* 19, no. 1 (2015): 133–171; Kiheung Kim, "The FMD Diagnostic Kit as a Boundary Object and Boundary Making: Conflicts and Negotiations Between the State-Centered and Decentralized Sociotechnical Orders [Korean]," *Journal of Science and Technology Studies* 18, no. 2 (2018): 307–342.

36 Young-Soo Kim and Jongwong Yoon, *A Culling–Vaccine Dilemma: Anthropocentric Preventive Measures* [Korean] (Seoul: Mobl Books, 2021).

37 Taehoon Lee, "Birds? They Are Fortunes from the Sky [Korean]," *Chosonilbo*, November 4, 2005, https://www.chosun.com/sit0e/data/html_dir/2005/11/04/2005110470415.html.

38 Ministry of Agriculture Food and Rural Affairs, "Daily Report of High-Pathogenic Avian Influenza Outbreak (2021.5.26)," accessed June 1, 2021, https://www.mafra.go.kr/mafra/2625/subview.do.

39 Jiin Bae et al., "Improvement of HPAI Prevention Policy [Korean]," *KERI Agriculture Policy Focus* 143 (2017): 1–31.

40 Junsoo Kim, "Pig War – Toward the More-than-Human Territoriality through African Swine Fever [Korean]," *Journal of Cultural and Historical Geography* 31, no. 3 (2019): 41–60.
41 Kiheung Kim, "Multiplicity of Pathogen: Construction of Foot-and-Mouth Disease in White Papers in Korea," 133–171.
42 Jin-Gyu Kim, "The Formation Process of Sanitary Association System in the Japanese Colonial Period – Focused on the Police-Government Organization [Korean]," *Journal of Korean History Hankuk Saron* 66 (2020): 79–139; Keunsik Jung, "Formation, Change, and Legacy of the Colonial Sanitary Police in Modern Korea [Korean]," *Social History* 90 (2011): 221–270; Jae-Seong Choe, "Propaganda of Sanitation by Japanese Police Organizations [Korean]," *The Journal of Korean-Japanese National Studies* 40 (2021): 5–53.
43 Kyu-Hwan Shin, *Red Doctors in Peking* (Seoul: Yeoksagongkan, 2021), 481.
44 Ministry of Agriculture Food and Rural Affairs, *High-Pathogenic Avian Influenza White Paper* (Sejong: Ministry of Agriculture Food and Rural Affairs, 2018), http://lib.mafra.go.kr/Search/Detail/29638.
45 Dong-Kwang Kim, "What Is the Korean 2010~2011 Foot and Mouth Disease Epidemic? Focusing the Nationalistic Slaughter Policy and Its Implication [Korean]," *The Journal of Academy of Democratic Society and Policy* 20 (2011): 13–40.
46 Kyeondae Ha, "Without Breaking, Public Health Doctors Are Working [Korean]," *Medi:Gate News*, March 7, 2020; Kyeonmuk Oh, "World Surprising Covid-19 Mass Diagnosis Thanks to Public Doctors' Mobile Check-Up [Korean]," *The Korea Economic Daily*, March 23, 2020; Shinjeong Ko, "Urgent Input of 742 Public Health Doctors at Covid-19 Sites [Korean]," *Doctors News*, March 5, 2020.
47 Kim, "Construction of the Covid-19 Disease-Scapes"; Myung-hee Yang and Hyungang Kim, "The Critical Discourse Studies of "with Corona" Editorials – Focused on National Newspaper and Economic Newspaper [Korean]," *URIMALGEUL: The Korean Language and Literature* 91 (2021): 97-124. The shifting policy from zero-COVID to living with COVID was initially not effective in 2021. Due to the outbreak of the Omicron variant, the authorities ceased the transitional plan temporarily until February 2022.
48 Frédéric Keck, "Bird Flu: Learning Lessons from Traditional Human-Animal Relations," *The Conversation,* December 15, 2016.
49 Joost van Herten, Bernice Bovenkerk, and Marcel Verweij, "One Health as a Moral Dilemma: Towards a Socially Responsible Zoonotic Disease Control," *Zoonoses & Public Health* 66, no. 1 (2019): 26–34.
50 Korean Society of Infectious Diseases, *History of Infectious Diseases in Korea,* 327.
51 Bokgyu Kwon, "On the Epidemic Diseases of the Early Choson Period [Korean]," *Korean Journal of Medical History* 8, no. 1 (1999): 21–32.
52 Hyomin Park, Myung-Sun Chun, and Yunjeong Joo, "Traumatic Stress of Frontline Workers in Culling Livestock Animals in South Korea," *Animals* 10, no. 10 (2020): 1920.
53 Cho-Rom Lee, "At Corona-19 Corona Management Hospital, A Study on the Legal Issues of Labor Relations Laws in Hospital Workers [Korean]," *Labor Law Forum* 30 (2020): 209–252.
54 Park, Chun, and Joo, "Traumatic Stress of Frontline Workers," 1920.
55 Seonhee Moon, *Burial: A Record for Burial Sites of Culled Animals Because of Infectious Diseases* [Korean] (Seoul: Chaekgongjang Dubulu, 2019); Kim and Yoon, *A Culling-Vaccine Dilemma.*
56 Junhee Kim, "Court Said, 'Ban Culling Chickens Before the End of the Trial' [Korean]," *The Joongang*, May 18, 2017.
57 Euy-Ryung Jun, "Pity and Compassion: On the Discourse of Animal Welfare and Multispecies Vulnerability [Korean]," *Korean Cultural Anthropology* 52, no. 3 (2019): 3–43.
58 Hinchliffe, "More than One World."

13

RABIES ON ICE

Learning from Interspecies Suffering in Arctic Canada

Susan McHugh

My interests in the ways in which One Health might better attend to the affective dimensions of successful rabies-eradication efforts start from a sense of personal obligation to tell the story of a particular dog. In 1973, I was severely bitten by Duke, the neighbors' yard dog. I was four years old. This incident brought the dog and me to the front lines of modern rabies control in the USA, an experience that reflects complex and changing histories of human–canine relationships, including the transformation of dogs into primarily pets in the urban–industrial imaginary. I now understand that Duke was immediately impounded by animal control and killed because he was not just considered vicious for leaving a chunk of my face hanging by the skin. Left unvaccinated and permanently tied outside the neighbors' house, the dog was potentially exposed to infected raccoons, bats, skunks, or foxes, among the other wild animal vectors seen in the woods and fields that then still surrounded our housing development. I dimly recall wandering into the range of Duke's tether with absolutely no sense of these or any other dangers to come.

Rapid rabies testing was acutely warranted in my case because I was bitten several times in the head, and the virus is neurotropic, replicating first in muscle and then in nerve cells to move steadily toward the central nervous system, where its victims eventually die of brain inflammation. The general rule is that the closer the proximity of the site of exposure is to your brain, the sooner you will die, and it is an unequivocally horrible death. Today, it is well documented that children's small statures make them more likely to be bitten in the head, and that the incubation period for the disease is much shorter for them than for adults. If not the municipal dog officer, then surely the visiting South American

FIGURE 13.1 Traditional qamutik (sled), Cape Dorset. Photo by Ansgar Walk, 1999 (cropped). CC BY-SA 2.5.

DOI: 10.4324/9781003294085-18

plastic surgeon who skillfully stitched my face back together would well have recognized the immediate and grave dangers that I faced.

My memories of the fateful incident remain unsure. When growing up, all I knew for certain was that the scars on my face proved what the neighbors' children constantly reminded me: that I killed their dog. I did not do so literally, of course, but effectively, as the older kids held forth in grim detail. Time and again, I was told that "the Dog Officer had to cut off Duke's head to prove that our dog didn't give you rabies." Such was our folk knowledge of what the World Health Organization (WHO) website verifies: "[f]or post mortem diagnosis, the gold-standard diagnostic technique is to detect rabies virus antigen in infected tissues, preferably brain smears or touch impressions collected from a biopsy, by fluorescent antibody test."[1] I cannot imagine that anyone in my social circle had learned that in school or anywhere else. Decades later, as a career academic researcher in canine cultural studies, I learned *why* Duke died the way that he did; still, somehow, everyone I grew up with knew that the dog *was* decapitated. Furthermore, we knew that it was all done in order to spare me, the neighbor kids said, from being "stabbed over and over again with giant needles" in my belly, which to my young ears sounded like a fate worse than death.[2]

Such was our fast-and-dirty understanding of the painful, tedious protocol of the nerve tissue vaccine (NTV) that was the post-exposure prophylactic (PEP) available at that time. Developed in the nineteenth century by Louis Pasteur and Émile Roux, NTV is an unequivocal game-changer—the first live, attenuated vaccine ever developed for humans and the first PEP regimen too[3]—but its limitations include that it requires multiple subcutaneous booster injections to produce an effective immune response. After I was bitten, it would be another decade or so before the four-shots-to-the-arm-over-four-weeks cell culture vaccine (CCV) would have become available, which is much safer as well as more easily tolerated than NTV. Once the CCV became widely available, the WHO discouraged administration of NTV,

> because, in addition to requiring one shot a day for fourteen consecutive days (plus three booster doses in the following weeks) in the abdomen, it is not as effective as CCV and has a high rate of adverse side effects (from abdominal swelling to neuroparalytic complications that can sometimes be life threatening).[4]

Again, it seems unlikely that, back when this mattered most to my little world, anyone was going around spreading news of these debilitating, even deadly, side effects to rabies treatment on our street, especially within earshot of little ones like me. No one remembers me having a phobia before or since, yet one strangely distinct memory of a hospital scene persists, featuring me repeatedly screaming, "NO NEEDLES!" Nor does anyone recall whether I received any PEP shots in those first hours, but, even if I did, then I certainly was not administered a full regimen. Afterward, all of us children seemed to understand without explicitly being told that my successful medical and Duke's lethal veterinary treatments were administered primarily in order to spare me added suffering and possible death, and all in the name of public health. Still, the neighbors' children made it clear that they never forgave me.

I do not remember a time in which remorse for Duke's death was not part of who I am. Nevertheless, I was deep into decades of researching and writing about dog stories before it occurred to me that the fatal convergence of his story with mine might have set me on an academic course-of-no-return in canine research. All of this is a roundabout way of explaining how a deeply meaningful, personal revelation emerged through encountering one of Deborah Nadal's central findings in *Rabies in the Streets: Interspecies Camaraderie in Urban India*. Quite unusually for researchers in this area, Nadal includes informants in the demographic most likely to die of rabies: poor children from the Global South. Their perspectives are astonishingly fresh and critical of problematic dimensions all too common in rabies stories, namely 1) fear for the cure, 2) fear for the vector, and 3) both as fueled by conceptualizing health as separable from affective bonds that link people to dogs as more than just vehicles of contagion. Like me, the destitute street children colloquially known as "ragpickers"—who, it must be underscored, are at highest risk of everyone to contract rabies—more often than not wish no harm to dogs. Together, we resist the anthropocentric logic that makes dogs disposable, instead expressing and cultivating sympathy for them in exposure, suffering, and death.

What is more, the at-risk child caring about canine victims runs directly counter to thinking about rabies as justifying zoophobia. Actively resisting the assumption that managing zoonoses is incompatible with zooambivalence, let alone zooieya—the significances of which Susan Merrill Squier elaborates on in this volume—is more than just a personal perspective, but coming to this understanding has required putting my own affluent, white settler privilege in context. In Nadal's explanation of the halting efforts of One Health initiatives to stem the tide of rabies in a challenging socioecological context, I began to grasp how, in order for a One Health approach to work, there needs to be a shared baseline understanding: that it is managing what happens not just to bodies but more importantly to affective bonds, including cross-species ones, that make or break community efforts to control rabies.

My own story of rabies bleeds into larger narratives of how treating this disease has come to concern human and canine care practices, and more specifically the many different ways a dog suspected of being rabid can be cared for, not just feared and killed, through culturally sensitive extensions of the One Health paradigm. Informed by research in literary and interdisciplinary animal studies, this chapter adapts methods of personal criticism to advance an understanding of One Health as being better served by attending to the deep entanglements of human and animal lives in the interest of multispecies justice. By seeking to "understand the types of relationships that humans ought to cultivate with more-than-human beings so as to produce just outcomes," the project of multispecies justice more generally aims to deconstruct and decolonize legal, representational, and deliberative processes.[5] In the context of rabies control, I argue that deconstructing and decolonizing treatments of un/healthy assemblages of dogs, people, and disease are instructive for making a more diverse, equitable, and inclusive One Health work.[6]

To explore how ignoring this complexity aids and abets exterminationist logics that extend far beyond immediate public health risks to the human, the bulk of this chapter focuses on a case study in which rabies-control efforts have become caught up in profound transformations of local human–dog relations. In earlier projects, I researched how a grassroots multispecies justice project, which was focused on recovering stories of the mid-twentieth-century disappearance of Inuit sled dogs in Arctic Canada, also documented cultural genocide of Inuit, inspiring massive-scale truth and reconciliation work of and for Indigenous peoples across Canada.[7]

Here, I focus on the historic links of these developments to rabies-control efforts. I conclude by considering how these inform contemporary, localized adaptations of One Health methods to provide what is needed for rabies prevention in Inuit communities today, not least of which is cultural sensitivity to the high value placed on children and dogs learning how to live together through supervised play. Adding to sociologist Frédéric Keck's consideration of the quantitative impacts of "subjective effects of culling on those exposed to animal diseases,"[8] I delve further into how effective One Health mitigations of the risks of zoonoses is informed by cultural sensitivity to the effects of qualitative losses as well. The next section takes a brief detour to elaborate on why and how rabies management remains an unevenly shared concern.

Human–Canine Affective Compromises: Beyond a Personal Viewpoint

Considering affective dimensions of rabies control opens up another dimension of what Steve Hinchliffe characterizes as the enticement/ trap "lure" of One Health:[9] human dominance of and affection for dogs are part of the history of rabies prevention and cure, as are neglect of and harm to humans, dogs, and other animals. It is hard to imagine how it could be otherwise. Persisting as "the most lethal infectious disease known to humans," rabies is also one of the oldest known diseases. Although rabies is conventionally referred to in the singular, multiple zoonotic viruses actually cause it.[10] When rabies is left untreated, however, the outcomes are uniformly the stuff of nightmares.

Modern medicine has made the disease preventable, but treatment is tricky because it only works if rabies vaccine (and possibly human rabies immunoglobulin) can be administered before any symptoms appear. Moreover, the window between being exposed and becoming symptomatic can vary from as little as several days to several years, during which time it remains clinically undiagnosable. It is so difficult to verify by *in vitam* techniques that such testing is discouraged by the WHO.[11] Seeking treatment immediately after any suspected exposure is therefore vital, ideally within a 48-hour window. Virtually all will die after symptoms appear. Worldwide, deaths are at an estimated rate of 59,000 humans annually, nearly half of whom are children less than 15 years old.[12] Shadowing these numbers are far less easily estimable losses of members of other species, including different domestic, feral, and wild mammals, who are also susceptible to lethal infection.

Folk remedies like "the hair of the dog that bit you" attest to ancient knowledge of the most familiar vector of the disease. And they prove just as ineffective at stopping the spread of rabies as the wholesale slaughters of man's best friend that were instituted long ago in ancient Greece and China.[13] Nonetheless, dog-eradication programs persist in modern worlds, guided by hygienic visions of purification from disease, sometimes even remaining as direct hangovers of colonialist-era measures to control peoples along with their dogs, and all regardless of their scientific invalidity.[14] As ecologists have abundantly demonstrated, the presence of community dogs effectively keeps "interlopers"—or non-local canine stranger-dangers—from entering the neighborhood. Even so-called passive killing strategies, involving the dogs' removal from densely populated urban environments to other locations, create an ecological vacuum almost immediately filled by new dogs, priming a more volatile environment for risky interactions among humans and dogs who are strangers to each other, including bites through which saliva and blood readily transfer the virus between the species.

Ignoring dogs also is not a viable option wherever rabies has become endemic because we so consistently insinuate ourselves into each other's lives, and apparently have done so since the beginning of human settlement. That said, some relationships prove riskier than others. Unvaccinated dogs over whom people claim ownership but who are not kept from roaming at will are particularly concerning when it comes to spreading rabies because they come into close proximity with humans along with potentially infected feral dogs and wild animals, effectively looping unwitting people and other animals into the cycle of infection.[15] Focusing on management rather than eradication of dog populations, One Health approaches to rabies control today are taking great strides in promoting the benefits for humans and dogs alike. Combining canine spay and neuter and vaccination programs, success comes at considerable costs, if also yielding considerable returns on investments. That dog bites account for 99 percent of disease transmission to humans in regions where such programs have yet to be funded is a pressing argument for expanding these programs. But they have not made considerable inroads to date in countries like India with longstanding traditions of respecting off-leash canines as community members, where infection rates remain the highest, and where the very concept of free-ranging dog ownership has become contentious.

More precisely, in the developed world, dog vaccination and stray animal control programs instituted during the 1940s are considered the benchmark of success, but the uneven replication of this model worldwide warrants further scrutiny of the factors involved, for instance, in making dogs no longer the primary vector of the disease in the United States.[16] Rabies eradication was drastically ramped up in response to the conclusions of ecologist Alan Beck's pioneering studies of free-ranging dogs in the city of Baltimore in the 1960s and 1970s. Beck documented the significant ecologically adaptive advantages for unowned dogs who behaved like owned pets, and who therefore considerably complicated the question of what kind of dog was the source of most bites and consequently rabies exposure in an average modern city. Among Beck's many recommendations were that rabies vaccinations for dogs be not only mandated but also state-subsidized, which prompted public health officials across the United States to move quickly to implement sweeping measures to control canine-borne zoonoses that continue to be in effect in most municipalities today.[17]

The Centers for Disease Control confirm that "the number of human rabies deaths in the United States has been steadily declining since the 1970s, thanks to animal control and vaccination programs, successful outreach programs, [and] public health capacity and laboratory diagnostics," all of which have helped to reduce dependence on the also important availability of modern "rabies biologics"—human rabies vaccines and immunoglobulins—for treatment.[18] I linger in this list over the pivotal contributions of successful outreach, for Nadal's study confirms my experience that the importance of learning how to live with dogs all too often gets downplayed as a factor in mitigating disease, even dismissed as driven by misplaced human affection for man's best friend.

Liking dogs or not is beside the point, for, as Cary Wolfe elaborates, "when the animal is taken seriously, not just as another topic of study among others but as one with unique demands,"[19] it marks a powerful biopolitical challenge to the primacy of human interests, not least of which is to exclude many peoples from consideration as human. As Irus Braverman abundantly demonstrates in the Palestinian context, settler states often "control" particular populations of animals in the name of ecology to protect "native" organisms but their selfsame eradication strategies simultaneously hurt Indigenous human populations, as is particularly evident in situations in which people are made to lose their livelihood along with the animals targeted for destruction.[20] An ongoing concern for the developing world, One Health approaches immediately exacerbate these problems by "valu[ing] nonhuman animals primarily for the sake of humans, which can lead to policies that harm and neglect nonhumans unnecessarily."[21]

But there are more insidious effects of devaluing Indigenous peoples' traditional ways of living with animals as well. By failing to attend to the multiplicity and integrity of human–animal relations—along with their uses and abuses in maintaining hierarchical dualisms among humans and other animals—One Health interventions risk empowering the forces of multispecies injustice.[22] Building respect into the story of effective rabies prevention—respect regarding dogs, as well as for what living with dogs demands of people in specific socio-ecological circumstances —can safeguard against the risk that One Health does more harm than good.

Discounting Community Valuations of Canines and Cultural Genocide

My research on the uneven legacy of Inuit sled dogs—scientifically designated a subspecies, *Canis lupus familiaris borealis*, and known long before in Inuktitut as "qimmit"—in Arctic Canada inspires me to dig deeper into questions of how One Health approaches appear to be finding success through local adaptations. Why are innovative, site-specific strategies only recently being developed in order to address the particular threats that rabies presents to Indigenous human, canine, and wild animal populations in the region? An obvious answer is that the threat is configured differently there: although globally the most common reservoir animal for the rabies virus is *Canis lupus familiaris*, the species commonly known as the domestic dog, in the far north its main reservoir has become the Arctic fox (*Vulpes lagopus*).

A unique strain referred to as the Arctic rabies virus variant (ARVV) remains predominant in the circumpolar north. Although its distribution is not limited to the Arctic region, epidemiological knowledge of the variant remains nascent. One concerning factor is its apparent persistence amid relatively low population densities of reservoir species in the Arctic, particularly foxes.[23] Inuit cultural traditions of keeping free-ranging dogs and enlisting them in the work of trapping foxes for the fur trade nonetheless make qimmit primary points of rabies exposure, so managing their health and safety is a critical go-between for environmental and human health. Yet local rabies management today faces further challenges through the twentieth-century history of how the Canadian settler state catastrophically failed Inuit and their dogs, all too often in the name of rabies prevention.

"The ways in which scientific research is implicated in the worst excesses of colonialism remains a powerful remembered history for many of the world's colonized peoples," Indigenous studies scholar Linda Tuhiwai Smith explains.[24] A retrospective view of the scientific goal of pursuing rabies control highlights the challenges of decolonizing planetary-scale initiatives like One Health, particularly in the case of Inuit, for whom rabies and attempts at its eradication both become implicated in the decimation of traditional human–animal relationships. The predominant narrative of successfully combating rabies is premised on a biopolitical regime of population management that involves not just vaccination but also sterilization or removal of "strays." The term "strays" already implies ownership as a dog's normative status, with the absence of ownership being a license to kill.[25] But, for the few *Homo sapiens* working with *Canis lupus familiaris borealis*, a different set of premises defies such terminology, and instead underpins a mutualistic relationship ideally suited for living off the land and sea ice. Inuit maintained lifeways with sled dogs well into the twentieth century when, along with Eurowhite settlers and their dogs, rabies arrived. Inuit persistence with qimmit presents profound challenges to the story that disease wiped out the dogs who were then replaced by snowmobiles, challenges that were first systematically mounted by the Qikiqtani Truth Commission (QTC)—an Indigenous-led social justice project that began in 2005.

The QTC gathered testimonies that overwhelmingly linked the disappearance of Inuit sled dogs—whose numbers fell from tens of thousands in the 1950s to a few hundred in the 1970s— to evidence of cultural genocide. Long before the advent of snowmobiles, qimmit were viewed as essential to Inuit lifeways, and this remains the case today, in principle, if no longer practice. For semi-nomadic people who traditionally did not use snowshoes or skis while living off the land and sea ice, qimmit provided a crucial means of mobility, livelihood, protection, and identity. Most people misrecognize them today through the canine actors (both qimmit) used to portray tamed wolves—more specifically, the now-extinct dire wolves (*C. dirus*)—in the TV series *Game of Thrones*, and with good reason.

Temperamentally, qimmit are characterized as extremely active as well as highly reactive—qualities guaranteed to make them either neurotic pets in cities or excellent company in bear country—and as highly social with their own pack and people, although aloof with outsiders. Conflicting views of these dogs as desirable or dangerous map readily onto insider/outsider cultural positionings, but all who have experience living with them agree that such dogs are not stereotypically dependent companion animals, nor are they akin to free-ranging or feral street dogs. In traditional practice, Inuit sled dogs are semi-feral working animals, who thrive in a special sort of working partnership of people and dogs, enabled by highly specialized adaptations to subsistence living in the Arctic. For these reasons and others, their disappearance within a generation was devastating.

The QTC's inquiry into why these dogs disappeared from their homeland provides a poignant window into how cultural and historical factors complicate not only the Eurowestern conventional separate-and-exterminate approach to zoonotic disease control but also in this case the Inuit Traditional Ecological Knowledges (TEK), a term that at once encompasses subsistence hunting knowledge and competencies necessary for successful adaptation to changing Arctic conditions, along with the need for culturally sensitive approaches to understandings of how Inuit and canine health are interrelated. Rabies threatened not only lives but also the unique interdependence of people and dogs foundational to Inuit culture, and in ways that are instructive for expanding appreciation for how people who build lives together with dogs understand zoonotic contagion.

The myriad of ways that Inuit people have depended on their dogs makes all the more poignant the story of qimmiijaqtauniq—literally "many dogs (or dog teams) being taken away or killed" and frequently translated now as the Mountie Sled Dog Massacre, or more simply "the dog slaughter."[26] The stories of these events that emerge through the records and reports of the QTC are exceptional for many reasons. A rare, large-scale project in which Indigenous people take charge together of decolonizing their own story, the QTC resulted in the public reconstruction of an oral history of the extermination practiced by white officials charged with moving Inuit into permanent settlement. The project includes testimony from hundreds of Inuit about dog teams intimately identified with particular people and families, groupings that glimpse ways of seeing animal and human lives together as sources of biopolitical power.

The QTC's collective story of death and other losses detail an understanding of life with animals before and beyond colonialism that grows from a spiritually and physically sustaining relationship. Testators' understandings of the significance of disruptions to human–dog relations are not easily grasped by outsiders like me, who are, after all, not their initial or primary audience. Recovering the details of why and how great quantities of these dogs died within a couple of decades therefore entails piecing together a complex story of how killing dogs, even for the well-intended purpose of rabies eradication, accelerated Canadian nationalism, capitalism, and cultural genocide. So how did this all come about?

In the year 2000, five years after the first formal requests for a national-level explanation were submitted by Inuit groups, elders testified to a Canadian House of Commons committee that, "to diminish our numbers as Inuit, our dogs were being killed" by members of the Royal Canadian Mounted Police (RCMP).[27] The Canadian Parliament subsequently commissioned a self-study in which the RCMP, colloquially known as Mounties, exonerated themselves from the accusation that actions of their own officers caused a precipitous decline of the dog population during the period under which they were charged with ruling Inuit territory. The disappearance is not disputed by anyone, but the cause has been hotly debated. Responsibility for qimmit deaths, not their consequences for people, was the narrow focus of the Mounties' self-study, and their *Final Report* concedes that only eight Inuit are represented in the approximately 150 eyewitness statements that they reviewed in order to draw their own conclusions. In short, Inuit versions of the events were conspicuous by their absence from the first official account of how qimmit disappeared.

Starting in 2007, the QTC collected accounts from more than 350 Inuit through written testimonies and many more interviews and statements recorded at public hearings, together voicing a strikingly different story. Reviewing these and other documents, the *QTC Final Report* concluded that "Government records, police patrol reports, scholarly research, newspaper and magazine articles from the 1950s, 1960s, and 1970s [corroborate testimonies] that dogs were killed in the Baffin Region often without due regard for the safety of and consequences on Inuit families and because Qallunaat (white people) were scared of dogs."[28] Never simply countering the Mounties' review of the facts, the QTC paints a richer picture of the regional introduction of a disproportionately powerful minority charged with settling a historically semi-nomadic people within spaces that once provided them freedom. The stories of dogs being killed bleed into those of children taken from their families and prevented from learning their language and cultural heritage in residential schools where they were starved, beaten, and sexually abused, and where in as-yet-untold numbers they disappeared into unmarked graves. Thus emerged the first wide-scale documentation of some of the human costs amassed ostensibly to assure Inuit political freedom to participate in federal democratic processes, freedom that largely failed to materialize.

A common thread across these stories is that eliminating Inuit sled dog teams effectively condemned their people to imprisonment in the new settlements, but the stories of what exactly happened are complicated. When Mounties shot dogs according to policies that were created with no Inuit input—often "at random, without warning, and without consideration of the consequences"[29]—independent hunters were initially shocked and immobilized, fearful of being killed themselves. As their stories continue, it emerges that immobility means that they were also thenceforth reduced to lives of dependency and menial service in settlements in which they were effectively silenced by a combination of fear of reprisals and grief for their dogs.

Settler-colonial ideals of hygiene haunt these accounts. Some Inuit shot their own dogs in advance of relocation when informed that the teams would be prohibited in the settlements, and others were forced to abandon their teams in the open, often on short notice when evacuated for medical treatment.[30] For, just as qimmit demonstrated little resistance to common European dog diseases, Inuit people proved particularly vulnerable to the spread of new-to-them pathogens long endemic in Eurowhite populations, like tuberculosis. Some even had guns put in their hands and were commanded to shoot their own dogs for reasons unknown. But more testators recount how subtle strategies of coercion sealed their fates. Lured by promises of jobs, housing with modern conveniences, healthcare, and education that would lead children to a better life, altogether too late they realized that they had lost far more than they could ever gain.[31]

How exactly the dogs came to be slaughtered within settlements involves an equally complex convergence of factors, only more explicitly charged with the fear of rabies. Resisting pressures to assimilate, some Inuit attempted compromises. But supplementing life on the land with wage income during off-seasons precariously hinged on both flexible employment as well as the adaptation of qimmit to settlement life. All too often, hunters who became wage-laborers stopped feeding their dogs within the settlements, although not seasonally, as in pre-contact conditions, but when they inevitably ran out of time and meat to keep them properly. Amid the steadily growing crowds, a new situation emerged.

Some dogs must have starved, but because opportunistic scavenging outdoors year-round is part of the qimmit way of life, leaving such animals to fend for themselves in settlements with open dumps was not an automatic death sentence.[32] Selected to succeed as foragers, they cleaned up meat scraps and other garbage that would otherwise attract more fearsome wildlife like polar bears and wolverines. For Mounties and others charged with maintaining public order, however, these advantages were outweighed by the significant risks of daily living around large, semi-feral dogs regularly accustomed to killing and eating wild animals susceptible to rabies infection.

Citing evidence that dogs were shot "by the hundreds [and] perhaps thousands," the QTC's *Final Report* identifies a major factor to be the perceptions of non-Inuit or "Qallunaat [who] considered the dogs to be a danger to inhabitants" of their communities.[33] The RCMP's *Final Report* lends credence to the fear factor with descriptions like the following: "The Inuit sled dog is a large and aggressive animal that can pose a danger to public safety, particularly when diseased or starving."[34] The specter of the rabid beast thus shaped white settlers' self-perceptions of extermination practices as necessary to their success in running the new settlements.

The feared plague of rabies never materialized, but Qallunaat responses in several documented outbreaks of canine distemper and other diseases that humans could also contract indicate disturbing changes over time. During the early years, RCMP officers working in the region, who owned and cared for sled dogs, clearly recognized the value of qimmit, in some cases providing their own husky dogs' pups to Inuit families otherwise facing destitution when dogs died. They inoculated thousands of qimmit to fight their abundantly clear decimation from the newly introduced diseases, simultaneously keeping rabies infections among humans well at bay. Yet the RCMP *Final Report* notes too that the period in question was one in which the government was phasing out sled dogs from official duties,[35] making it an open question as to whether the presence of Inuit sled dogs became equally devalued from settler perspectives as these changes took effect.

The history of dog regulation in the region is telling. Although dog attacks were extraordinarily rare, considering how people lived in constant close proximity with qimmit, increasingly severe governance provisions followed swiftly on the heels of incidents in which people were mauled or killed, particularly when the victims were white women or children.[36] A series of regulations instituted largely as amendments to the misleadingly named Ordinance Respecting Dogs ranged from coercive to confusing from Inuit perspectives.[37] The requirements to tie or muzzle loose dogs on pain of ruinously high fines to be paid within very short allowances of time seem lifted straight from more overtly colonial contexts, in which the effect similarly is to privilege settlers' dogs and to make it virtually impossible for Indigenous people to move into settlements while maintaining dogs in the old ways. In addition to being scientifically disproven as measures to stem spread of diseases by free-ranging dogs, these policies moreover reflect at best ignorance of and at worst an active menace to local lifeways, such as the dogs' need to have their muzzles free to eat snow in order to hydrate, feed, clean, and protect themselves from predators. Inuit testimony that dogs entitled to a period of impounding instead were often shot point-blank, sometimes in harness, contributes to lingering senses of confusion and resentment.

Even for Inuit trying to follow the letter of the law, tying up their dogs was not a viable option. The QTC clarifies, "Inuit were particularly critical of Qallunaat who had no knowledge of the impact of chaining dogs on the behavior of working animals,"[38] but there is evidence that some knew well what the outcome would be. In a letter dated 1960, Northern Service officer W. G. Kerr avers, "I personally do not think that 'wandering' dogs create any greater hazard than does the normal automobile traffic of southern Canada. … It is also my experience that a tied-up dog, if approached by children, is more dangerous than a 'wandering' one."[39]

Such was my own experience with Duke. Far less able to fend for themselves, and far more vulnerable to attacks by dogs and other animals, tied dogs also tire more easily in harness, as many testators note. It does not take much familiarity with dogs to guess how many formerly free dogs put in chains became neurotic, self-destructive, vicious, and otherwise unworkable. What does remain unfathomable, particularly to eyewitnesses, is why many dogs kept in compliance with the Ordinance Respecting Dogs were also shot.[40] For Inuit, the incidents additionally mark the moments at which poverty became destitution, if also eventually the motive for a beleaguered community to rally together.[41]

Rabies persists as a minor but real threat throughout the settlement history of eastern Arctic Canada, not least because the Arctic fox populations that had become the disease reservoir were also central to the fur trade that outsiders eagerly moved in to control in the early twentieth century.[42] But the disease's history as a mitigating factor in the events known as the Mountie Sled Dog Massacre remains an important reminder that not all the dogs were killed by humans. Some qimmit surely were infected by settler dogs and, along with them, transmitted disease among themselves and to wildlife before inoculation campaigns were instituted by the RCMP, not so much ironically as coincidentally. Vaccination was a practice that likewise benefited settlers' dogs, not to mention prevented horrible deaths from rabies for all but one person to date in the region. Like the canine-restraint policies, however, the colonial history of the enactment of these and other veterinary protocols reflects no engagement with Inuit practices.

Caught up with the acts of cultural genocide documented by the QTC, the official reliance on culling and immunization during disease outbreaks remains controversial because it contrasts sharply with Inuit TEK of canine and human sickness. An unnamed QTC testator reflects on how Inuit used to treat diseased dogs: "Some were recognized as dangerous or certain to die, and quickly dispatched. In others, the disease was allowed to run its course, in the expectation that most would die but enough would survive for rebuilding teams."[43] Although Inuit were prepared to work with the forces of natural selection to attempt to produce disease-resistant dogs—and their having done so across millennia surely contributed to the persistence of *Canis lupus familiaris borealis* as one of the rare kinds of dog to have survived another plague that wiped out almost all New World dogs upon European contact[44]—the settler ideology of wholesale culling and replacement with immunized dogs prevailed. Inuit understandings of sick dogs as capable of bearing illness away from human individuals and communities were utterly disregarded in favor of eradication practices that deprived people of their family dogs along with effectively creating ecological niches filled by feral dogs, compounding rather than resolving the dangers of rabies spread in settlement communities.

Less easily understood by outsiders like me are the ways in which the social significances of dogs in Inuit life contribute to the imbrication of cultural genocide and the dog slaughter in these stories. Inuit TEK of human–dog relations weaves together practical and spiritual understandings. A person is a dog's *inua*, a word that is "formed using the radical *Inu(k)* (person) and the grammatical affix *a* (possessive), and [that] literally translates to 'my own self.'"[45] Though often interpreted as a dog's "owner" or "master," *inua* is the same word for the animal god or protective spirit specific to every other animal species and so grounds "strong affectionate bonds" in a unique sense, both physical and psychic, of "symbiosis between dogs" and their people.[46] Inuit sled dogs thereby are understood as contributing to the mental and physical health of individuals, but how they maintain community ties proves even more challenging to western-rationalist perspectives.

Traditional Inuit affirm that, in addition to immortal spirits and mortal bodies, humans have name-souls that influence character traits and can live on when the name is transferred to another.[47] Inuit culture's ritualized, gender-neutral naming practices make it easier for each newborn to be bequeathed the name of a valued community member, ideally a person who is elderly or dying to ensure that their name-soul continues to live long past their mortal self. That person does not exactly live on through the baby, but rather shares something special, for the baby gains "the identity of the spirit associated with it," and eventually "become[s] that person in adulthood."[48] Name-souls are an essential mechanism for sharing and instilling character, and, what is more, maintaining a sense of kinship beyond blood ties that proves crucial for gluing together a semi-nomadic society traditionally structured as small, far-flung families who might gather in bigger groups at most seasonally, for a few months of the year.[49]

Qimmit were given human names for many reasons—to maintain intimacy with human relatives, to bear human illnesses or even grudges, or to indulge a child sharing their own name as a sign of "infatuation" with their very first dog.[50] But in the old days, under conditions conducive to low birthrates as well as to high risks of untimely demises, dogs became essential as short-term keepers of the name-souls of those who suffered untimely deaths, name-souls that needed to wait to be passed on later, when infants were forthcoming. Understood yet not elaborated by the QTC testators, the cross-species name-soul sharing is one of many culturally specific traditions that help explain why dogs hold such a special place in Inuit society, why the names of dogs were not readily shared with outsiders, and why the random indiscriminate killings of dogs by outsiders continue to hurt. How many ancestors' name-souls were lost forever because there were no namesakes to take them when dogs bearing them were abruptly killed, and with no warning? How painful, even impossible, is it to name the name-souls that no longer have any human bearers, that were unable to be transferred from the dying to the living? Although I don't pretend to know what those feelings are, researching these cultural aspects has helped me to understand why the

QTC testimonies often include versions of the statement, "I remember the day my dogs were shot," or "I remember when my father's dogs were shot," as well as why these lines are frequently spoken through tears.

How do these histories inform twenty-first-century conditions in the region, where distrust remains high among Inuit toward veterinary professionals?[51] How might the persistence of Inuit TEK emphasizing human–canine connections inform the low incidences of rabies infections in humans and dogs, in glaring contrast to other regions of the world where close living conditions have otherwise proven conducive to spreading the disease? How can compromises be negotiated, and for the benefit of humans, dogs, and wild animals at risk of rabies infection? Some answers are emerging through adaptations of One Health principles to the findings of the QTC, suggesting how multispecies justice initiatives can engage communities in framing as well as addressing rabies as a shared threat that can be mitigated by sharing responsibilities for fostering the conditions under which human–canine relationships flourish.

Adapting One Health to the QTC's Findings

I acknowledge that my particular alignments of childhood half-knowledges of rabies with Inuit TEK, cultural traditions, and spiritual understandings of human–canine relationships are troublingly random, and, in many other ways, vexed by reaching for comparison across disparate global settings that in turn are tenuously aligned by the prevalence of peoples recently uprooted from traditional communities. But I cannot think of another way to emphasize how important it remains to attend to cross-species affective bonding in effectively addressing the persistent threat of this deadly disease. Rabies connects peoples and dogs across millennia, providing a material link to the deeply shared lifeways through which we have come to be unable to know our own species apart from their species. Killing dogs in the name of human health has failed to address their need to live alongside and never completely controlled by humans, and in turn our need to let them be dogs. This makes it all the more compelling to contemplate how One Health is being adapted to local and Indigenous efforts to address rabies through outreach efforts focused on socialization, particularly caring for the education of children on living together with dogs in communities.

Contemporary rabies research in the Nunavik region of Québec offers hope in new developments of collaborative methods of both identifying and addressing rabies as a public health concern that Inuit communities can work together with outsiders in order to address. Just a quick survey of academic articles published after the QTC's 2010 *Final Report* identifies studies with authors representing membership of both scientific and Inuit communities, together seeking to address how rabies threats might be more effectively mitigated in Inuit communities. In the context of growing Indigenous resurgence locally and globally, it is particularly interesting that these studies often identify One Health as a useful starting point for growing community support for zoonotic disease management.

But their careful attention to salient details of the histories uncovered by the QTC also helps to clarify what Hinchliffe identifies as key factors in moving toward a more-than-one-health paradigm, namely, "the socio-economic conditions that 'configure' the disease."[52]

Reviewing recent data identifies patterns of suspected exposure in young children bitten by free-ranging dogs. Rather than pathologize child–qimmit contact, however, researchers emphasize responsibly managing it, including by "improv[ing] education on rabies and dog bites in the villages."[53] Teaching emerges as a complex concern, through survey and interview research findings that dog bites are frequently underreported, in no small part because "interventions targeting dogs are a sensitive subject" for Inuit communities due to the lingering traumas of the Mountie Sled Dog Massacre.[54] While a general consensus among Inuit surveyed points to free-ranging dogs as feared, informants' comments support other strong quantitative indicators that such dogs remain highly valued in the community. Moreover, informants clearly express concern—even, in one case, assume—that dogs identified as biters will be killed indiscriminately, as in the past. Small wonder that, despite risking possible exposures of children to rabies, dog bites appear likely to have been underreported in the community. While the situation makes it hard to gauge whether the frequency of dog bites is increasing, these academic studies chart pathways forward.

The most recent article that I found explicitly breaks down the One Health approach to focus on a description of not just the human but also the more-than-human community's risk of rabies exposure. Including extensive observations of daily canine–human interactions in a predominantly Inuit community, authors of the study identify highly context-specific behaviors and determinants in order to address a well-documented, global problem, namely that culturally insensitive education largely fails to prevent children from being bitten by dogs. What is more, the researchers' participatory approach directly addresses this problem by enlisting community members in identifying and prioritizing risks not just to serve human health needs but also to consider how the needs of wildlife and dogs are best served, for instance, by teaching children that randomly untying dogs may seem fun but ultimately proves problematic for everyone.

As one anonymous Inuit informant captures it, "negligence isn't better than aggression" in addressing the threats by and to qimmit.[55] Informed by Inuit TEK, teaching proper etiquette with free-ranging dogs as well as avoidance of risky behaviors, especially among children, is key in helping rabies prevention to gain momentum in the region.[56] What stands out to me most in these studies is the expressed need for increasing adult supervision of children at play with dogs. For it is a factor significantly lacking in my own childhood experience, as well as those of so many of Nadal's informants, but one that none of us had identified as such. Is this because Inuit TEK of how to care for children and each other with semi-feral working dogs has loomed so large in the stories of the QTC? There is hope in these findings for One Health and decolonial efforts becoming mutually empowered by multispecies justice projects, and allowing for the negotiation of more workable solutions than the all-or-nothing, top-down models that have been the cause of so much harm already.

Notes

1 "Diagnosis: Rabies," World Health Organization, 2022, https://www.who.int/teams/control-of-neglected-tropical-diseases/rabies/diagnosis.
2 I was startled to read nearly verbatim descriptions of what I remembered from my own childhood, recorded and translated by Deborah Nadal in her interviews with twenty-first-century Indian informants. See Nadal, *Rabies in the Streets* (University Park: Penn State University Press, 2020), 212–213.
3 For stunning archival accounts of the depth and breadth of rabies-related human and canine suffering alleviated by the advent of NTV, see especially Chapter 2 of Chris Pearson, *Dogopolis: How Dogs and Humans Made Modern New York, London, and Paris* (Chicago: University of Chicago Press, 2021). Pearson's attention to the mixed reception for the pioneering vaccine has eerie resonances, especially when citing dissenters of the time using the anti-science, xenophobic rhetoric that is proving so deadly today.
4 Nadal, *Rabies in the Streets*, 212.
5 Danielle Celermajer et al., "Multispecies Justice: Theories, Challenges, and a Research Agenda for Environmental Politics," *Environmental Politics* 30, nos. 1–2 (2020): 119–140, 119.
6 See Steve Hinchliffe, this volume.
7 See Chapter 5 in Susan McHugh, *Love in a Time of Slaughters: Human–Animal Stories Against Genocide and Extinction* (University Park: Penn State University Press, 2019), and "'A Flash Point in Inuit Memories': Endangered Knowledges in the Mountie Sled Dog Massacre," *ESC: English Studies in Canada* 39, no. 1 (2013): 149–175.
8 Frédéric Keck, *Avian Reservoirs: Virus Hunters and Birdwatchers in Chinese Sentinel Posts* (Durham: Duke University Press, 2020), 28. See also Keck, this volume.
9 See Hinchliffe, this volume.
10 Multiple lyssaviruses have been identified as the cause, according to Arnaud Tarantola, "Four Thousand Years of Concepts Relating to Rabies in Animals and Humans, Its Prevention and Its Cure," *Tropical Medicine and Infectious Disease* 2, no. 2 (2017): 5.
11 "Diagnosis: Rabies," World Health Organization, 2022, https://www.who.int/teams/control-of-neglected-tropical-diseases/rabies/diagnosis.
12 "Take a Bite Out of Rabies!" Centers for Disease Control and Prevention, May 4, 2022, https://www.cdc.gov/worldrabiesday/feature/index.html.
13 McHugh, *Dog* (London: Reaktion Books, 2004), 42–43.
14 See Nadal, *Rabies in the Streets*, 92–94, and also Robert Gordon, "Fido: Dog Tales of Colonialism in Namibia," in *Canis Africanis: A Dog History of Southern Africa*, eds. Lance von Sittert and Sandra Swart (Boston: Brill, 2008), 173–192.
15 Nadal, *Rabies in the Streets*, 80.
16 Larry J. Anderson et al., "Human Rabies in the United States, 1960 to 1979: Epidemiology, Diagnosis, and Prevention," *Annals of Internal Medicine* 100 (1984): 728–735, 728.
17 Alan Beck, *The Ecology of Stray Dogs: A Study of Free-Ranging Urban Animals* (West Lafayette: Purdue University Press, 1973), xi, 72.
18 "Human Rabies," Centers for Disease Control and Prevention, September 22, 2021, https://www.cdc.gov/rabies/location/usa/surveillance/human_rabies.html.
19 Cary Wolfe, "Human, All Too Human: 'Animal Studies' and the Humanities," *PMLA* 124, no. 2 (2009): 564–575, 566–567.

20 Irus Braverman, "Wild Legalities: Animals and Settler Colonialism in Palestine/Israel," *PoLAR: Political and Legal Anthropology Review* 44, no. 1 (2021): 7–27. See also McHugh, *Love in a Time of Slaughters: Human–Animal Stories Against Extinction and Genocide* (University Park: Penn State University Press, 2019).
21 Jeff Sebo et al., "Sustainable Development Matters for Animals Too: Governments Have a Responsibility to Recognize That," *CABI One Health*, 2022, https://www.cabi.org/wp-content/uploads/Sustainable-development-matters-for-animals-too_CABI-One-Health.pdf.
22 Hinchliffe, "More than One World, More than One Health: Re-Configuring Interspecies Health," *Social Science and Medicine* 129 (2015): 28–35, 30–31.
23 Toril Mørk and Pål Prestrud, "Arctic Rabies: A Review," *Acta Veternaria Scandinavica* 45, no. 1 (2004): 1–9, 1.
24 Linda Tuhiwai Smith, *Decolonizing Methodologies: Research and Indigenous Peoples* (London: Zed Books, 2012), 1.
25 See Nadal, this volume.
26 Qikiqtani Inuit Association, *QTC Final Report* (Inhabit Media Inc.: Iqaluit, 2013), 24.
27 Mr. George Koneak (Elder, Makivik Corporation), "38th Parliament of Canada, 1st session, Standing Committee on Aboriginal Affairs and Northern Development," March 8, 2005, http://www.parl.gc.ca/HousePublications/Publication.aspx?DocId=1682867&Language=E&Mode=1, 1115.
28 Qikiqtani Inuit Association, *QTC Final Report*, 39.
29 Qikiqtani Inuit Association, *Qimmiliriniq: Inuit Sled Dogs in Qikiqtaaluk. Thematic Reports and Special Studies, 1950–75* (Iqaluit: Inhabit Media Inc., 2013), 50.
30 Ibid., 21.
31 Padloping native Jacopie Nuqingaq succinctly explains of the family's relocation to Qikiqtarjuaq, "When we got here, our dogs were slaughtered, and we had no choice." Ibid., 13.
32 Ibid., 22.
33 Ibid.
34 "Final Report: RCMP Review of Allegations Concerning Inuit Sled Dogs," Royal Canadian Mounted Police, May 30, 2006, http://publications.gc.ca/collections/collection_2011/grc-rcmp/PS64-84-2006-eng.pdf, 14.
35 Ibid., 15.
36 Qikiqtani Inuit Association, *Qimmiliriniq*, 19.
37 For instance, the *QTC Final Report* notes that requiring dog handlers to have surpassed "the age of maturity, 16, was meaningless … [because] for Inuit, maturity was measured by abilities, not age." Qikiqtani Inuit Association, *QTC Final Report*, 23.
38 Ibid., 24.
39 Makivik Corporation, *Regarding the Slaughtering of Nunaviik "Qimmit" (Inuit Dogs) from the Late 1950s to the 1960s* (Kuujjuak: Makivik, 2005), http://pubs.aina.ucalgary.ca/makivik/CI232.pdf, 11.
40 Qikiqtani Inuit Association, *QTC Final Report,* 23.
41 Makivik Corporation, *Regarding the Slaughtering*, 15.
42 Richard C. Rosatte, "Rabies in Canada: History, Epidemiology, and Control," *Canadian Veterinary Journal* 29, no. 4 (1988): 362–365, 362.
43 Qikiqtani Inuit Association, *Qimmiliriniq,* 21.
44 Barbara van Asch et al., "Pre-Columbian Origins of Native American Dog Breeds, with only Limited Replacement by European Dogs, Confirmed by mtDNA Analysis," *Proceedings of the Royal Society Biological Sciences* 280, no. 1766 (2013): 20131142.
45 Francis Lévesque, "Sixty Years of Dog Management in Nunavik," *Medicine Anthropology Theory* 5, no. 3 (2018): 199–200.
46 Frédéric Laugrand and Jarich Oosten, "Canicide and Healing: The Position of the Dog in the Inuit Cultures of the Canadian Arctic," *Anthropos* 97, no. 1 (2002): 90.

47 Pauktuutit Inuit Women of Canada, *Inuit Way: A Guide to Inuit Culture* (Ottawa: Pauktuutit Inuit Women of Canada, 2006), 16.
48 Lee Guemple, "Gender in Inuit Society," in *Women and Power in Native North America*, eds. Laura Klein and Lillian Ackerman (Norman: University of Oklahoma Press, 1995), 27.
49 Qikiqtani Inuit Association, *Qimmilirniq,* 15.
50 Laugrand and Oosten, "Canicide and Healing," 92.
51 Lévesque, "Sixty Years of Dog Management," 207.
52 Hinchliffe, "More than One World," 29.
53 Cécile Aenishaenslin et al., "Characterizing Rabies Epidemiology in Remote Inuit Communities in Québec, Canada: A 'One Health' Approach," *EcoHealth* 11 (2014): 343–355, 354.
54 Cécile Aenishaenslin et al., "Understanding the Connections Between Dogs, Health and Inuit through a Mixed-Methods Study," *EcoHealth* 16 (2019): 151–160, 152.
55 Géraldine-G. Gouin et al., "Description and Determinants of At-Risk Interactions for Human Health Between Children and Dogs in an Inuit Village," *Anthrozoos* 34, no. 5 (2021): 723–738.
56 Sarah Mediouni et al. likewise conclude that bites are "provoked" by children's "misinterpretation of dogs' signaling behavior," warranting a "targeted prevention program" in "Epidemiology of Human Exposure to Rabies in Nunavik: Incidence, the Role of Dog Bites and Their Context, and Victim Profiles," *BMC Public Health* 20 (2020): 1–13, 10.

AFTERWORD

Among Animals, and More: One Health Otherwise

Warwick Anderson

No matter how many blood samples are taken, no matter how definitive the results of genetic testing, many people, including a few scientists, continue to deny that the novel coronavirus SARS-CoV-2 crossed over naturally as a zoonosis from one animal species to another, to humans, in Wuhan, China, around the end of 2019. Instead, many people have chosen to believe that a virus so deadly, constituting so grievous a biosecurity breach, must have been engineered in a Chinese laboratory, then released either accidentally or deliberately. It is inconceivable to them that the COVID-19 virus could become so well adapted to its human bodily niche without some ingenious scientific intervention.[1] Even Jeremy Farrar, the director of the Wellcome Trust, admitted to entertaining conspiracy theories about Chinese viral manipulation and laboratory leakage, despite his previous experiences studying natural outbreaks of dengue and avian influenza in Vietnam. Later, he explained away his gullibility and paranoia as the result of "extreme stress," which apparently is not conducive to appreciating ecological reasoning.[2]

Early genetic sequencing of SARS-CoV-2 indicated that the virus had been transmitted from crowded and stressed animals to humans, probably in Wuhan's Huanan market, findings since confirmed repeatedly.[3] Yet many people remain convinced that the virus must be a human thing. In contrast, ecological explanations that incorporate the agency of other organisms and disturbed environments are always "open to interpretation," which means amenable to obfuscation and repudiation. Accordingly, integrative framings of health and disease, such as One Health, remain contested, seemingly inconsistent with our ingrained sense of

FIGURE A.1 Cellular mimics using porous silica nanoparticles encapsulated by lipid bilayers. Nanotechnology Image Library collection. Credit: Mona Aragon, Carlee Ashley, and Jeffrey Brinker, National Cancer Institute, National Institutes of Health.

human dominance and unrivaled agency, incompatible with common assumptions of human separation from other animals and the rest of nature—they represent challenges to what makes us so very modern. As Irus Braverman and others point out here, the emergence of a novel coronavirus in Wuhan should rather have attuned us to disease ecology, to the importance of One Health, compelling us to address potentially pathogenic processes such as habitat destruction, overcrowding, mobility, amplification of contact between organisms—but it did not, or not enough at any rate, or perhaps just not yet.

A collaboration between those committed to connecting nonhuman and human animal health and disease, One Health emerged in the early twenty-first century as part of the entangled thorny bank of global health activities. Twenty years or so earlier, "global health" began to displace the older international health services, which had clustered around the beleaguered World Health Organization (WHO), still largely predicated on post-World War II nation-based regimes.[4] In the transformative neoliberal world order, global health offered a flexible repertoire of biosecurity and humanitarian interventions, new modes of preparedness and response against emerging infectious disease like AIDS and against residual threats to human health such as malaria and tuberculosis.[5] Critics of ramifying global health governance deplored its technocratic style, reliance on standardized metrics, simplistic modularity, dependence on philanthropic and nongovernmental organization (NGO) initiatives, resort to public-private partnerships, and common disregard of underlying structural and environmental causes of disease. Despite humanitarian claims, global health often seemed designed to protect some privileged humans from contracting the diseases of other humans, while preserving the health of global capitalism.[6] In the 1990s, however, scattered radical epidemiologists and public health experts started to warn of the dangers that anthropogenic climate change and planetary environmental degradation, the collapse of the earth's life-support systems, posed to human health—indeed, to the existence of many species. In seeking to scale up and systematize environmental health, they drew on post-war planetary thinking and disease ecology, until then an elite but marginal interest in infectious disease research.[7] As "Planetary Health," the study of the impact of human-induced global heating on population health took off around 2015, with the support of the Rockefeller Foundation, the Wellcome Trust, and *The Lancet,* offering a critical alternative to conventional global health.[8] Similarly, EcoHealth and One Health were gaining momentum during this period, examining environmental and multispecies influences on human health, though rarely on a planetary scale.[9] This problem of scale has become crucial in appraising the various proxies for global health, or rather the alternatives or supplements to it, including One Health.

One Health was finding its own special niche in global health's entangled bank.[10] As an "edge effect," it has come to occupy a boundary habitat, the borderlands between the more stable and settled communities of veterinary and human medicine. Such contact zones or conceptual range margins are especially revealing since they exemplify selection pressures at the utmost gradient of environmental and cultural stress.[11] Thus, we might profitably follow sociolinguist Mary Louise Pratt who suggests we consider critically what happens in colonial contact zones, by which she meant "social spaces where disparate cultures meet, clash, and grapple with each other, often in highly asymmetrical relations of domination and subordination."[12] Thinking through One Health as borderlands, or the conceptual range margins of multiple fields, should focus attention on what "boundary objects" proliferate there—including the concepts of "species" and "animal." As the Berlin Wall teetered in 1989, sociologists Susan Leigh Star and James R. Griesemer imagined boundaries as possible points of communication, as interfaces that assembled objects bearing significance in multiple social worlds. Such boundary objects are "plastic enough to adapt to local needs and constraints of the several parties employing them, yet robust enough to maintain a common identity across sites. They are weakly structured in common use and become strongly structured in individual-site use."[13] According to Star and Griesemer, boundary objects might be things, organizational forms, concepts, protocols, or procedures—whatever permitted communication across the boundaries of social worlds, or in the borderlands. Attention to the multiple boundary objects of One Health would give us a clearer picture of its contemporary practice. Efforts to expand the register to include other "boundary subjects" might open the field even wider, to more heterogeneous and agential entanglements.[14]

The chapters collected in this volume show us multiplicity and diversity under the One Health rubric, even though its different forms and modes stem from commonality of purpose. As Abigail Woods notes, One Health formally developed in the early years of the twenty-first century through the conjunction of veterinary medicine and epidemiology or human public health, focusing particularly on zoonoses or the transmission of infectious diseases from other animals to humans. In the first decade of the twenty-first century, epidemics of SARS and avian influenza, where the role of intermediate animal hosts was evident, prompted concern about risks to human populations of cross-species transmission of viruses and other microbes, which demanded integrated approaches to animal and human health—just as the advent of AIDS in the 1980s had once put the spotlight on viral emergence and the need for coordinated global health responses.

Of course, there were many precedents for One Health. The profusion of animal models of human disease over the past hundred years and more was surely a powerful conceptual antecedent.[15] Moreover, many straightforward contributions of arthropods and intermediate hosts to human disease patterns were obvious to anyone trained in tropical medicine and parasitology since the 1890s.[16] In the 1930s, disease ecologists, such as F. Macfarlane Burnet in Melbourne, Australia, and Karl Friedrich Meyer in San Francisco, California, USA, had investigated the spread of psittacosis from stressed birds to proximate humans, even as they looked for other bird reservoirs for common human diseases.[17] But most veterinary pathologists and epidemiologists along the North Atlantic littoral in the twentieth century seem to have given little heed to the interspecies paradigm. From the 1960s, however, a few research veterinarians in North America and Western Europe did begin to advocate, as we know, for One Medicine, attempting to refigure animal health as a human medical problem, thereby belatedly recognizing a truism of the colonized world.[18] No surprise, then, that the "new" One Health derived from, and initially flourished in, diverse imperial settings and institutions, whether under the aegis of Calvin W. Schwabe in California and Lebanon and Hawai'i or the Swiss Tropical and Public Health Institute or the Wildlife Conservation Society surveying the new American empire.[19] Several of its pioneer figures boasted extensive experience in imperial borderlands such as the western rangelands of the settler United States or neo-colonial African savannahs. In effect, they were bringing a vernacular understanding of interspecies relations and environmental stressors back to imperial centers.

Understandably, One Health has been criticized for human bias or anthropocentrism, just like Planetary Health and the other modalities of global health. Researchers and practitioners usually emphasize the risks that other animals present to human health and how to prepare against them and control them. As many authors demonstrate here, interest in the wellbeing of other animals often dwindles to how they might matter to human population health. But the seemingly inevitable recurrence and reconstitution of human sovereignty, or at least self-interest, is not the only concern. A special position is accorded to humans, to be sure, but there is also the problem of the broader ambiguity of the category of "animal" and how that is populated and enacted in One Health. In practice, One Health strives to confer greater agency and influence on other animals, but what, or who, counts as animal? The field remains oriented around a kind of vertebrate charisma, or even eukaryotic glamor, that limits its organismal reach and ecological range.

It puzzles me that microbiology, which developed in the late nineteenth century, is not regularly regarded as fundamental to configuring One Health—instead, microbes are generally regarded merely as transmissible items of no particular value.[20] We are still fixated on the dichotomy between single-cell prokaryotic organisms (such as bacteria), with no nucleus, limited to parasexual gene recombination, and multicellular eukaryotic organisms (plants and animals) capable of meiotic sex, a distinction popularized only in the 1960s by Roger Stanier and C. B. "Kees" van Neil.[21] This has allowed bacteria to be segregated into a separate kingdom, Monera, meaning single or solitary.[22] Using ribosomal RNA as a phylogenetic probe, Carl R. Woese in the late 1970s challenged such morphological classifications, showing they possessed no evolutionary justification and disguised the biological heterogeneity of each class.[23] As Woese later reflected: "This prokaryote-eukaryote dogma has closed our minds, retarded microbiology's development, and hindered progress in general. Biological thinking, teaching, experimentation, and funding have all been structured in a false and counterproductive and dichotomous way."[24] Or, as historian Jan Sapp puts it, bacteria had been "defined largely in negative terms: they lacked a nucleus, lacked mitosis, lacked [heteronormative] sex."[25] Yet One Health appears committed to perpetuating discrimination between races of animals and bacteria, keeping them separate and unequal.

Why should we continue to divide the living world according to antiquated and artificial European scientific classifications, thus making arbitrary and arguable inclusions and exclusions? One consequence of bacterial segregation is the failure to recognize the human body as multiply "animal," as a multi-organism assemblage, maintaining instead the illusion of its animal singularity. Human gut microbiota consist of hundreds of species of bacteria and archaea, more densely packed together than in any other known habitat, interacting with the body's digestive processes and immune system. Yet One Health tends to treat the human body as just one animal among others, a singular, bounded, static entity, engaging riskily with other similarly autonomous eukaryotic agents, which may or may not carry contaminants and the risk of disease.

Our perceptions of risk are still embedded in this dated animal ontology, rather than embracing a post-animal process ontology, which would expand the scope and quality of relationality. It may be a useful exercise to try to imagine the "animal" otherwise, as borderlands not boundary, as co-immune not immune, as interconnected and networked, hybrid and cyborg.[26] To invoke what Rosi Braidotti calls nomadic thought, it may be timely to argue for a "radically immanent intensive body [that] is an assemblage of forces, or flows, intensities and passions that solidify in space, and consolidate in time, within the singular configuration commonly known as an 'individual' self." This would represent "an ecological philosophy of non-unitary, embodied subjects of multiple belongings."[27] It would also possess the advantage of congruence with recent processual understandings of the vertebrate immune "self."[28]

Some may propose "multispecies" health as a simpler alternative to monist and animalistic One Health, but species typologies might still interdict or occlude recognition of the full variety of ecological interrelationships.[29] This persistence of typological thinking in One Health, "black-boxing" animals and species as boundary *objects*, distinct from evolutionary *processes*, surprises and disturbs me—though not, according to the contributors to this volume, very many of its practitioners.

Even as we consider more encompassing and heterogeneous conceptions of "animal" agency and connectedness, there remains the pressing problem of who get to be valued as authors of the One Health master narrative. Imperial and settler-colonial antecedents and inciters of contemporary One Health may have enabled greater sensitivity to human entanglements with a limited range of charismatic animals and celebrity species, but they also muted or suppressed other ways of appreciating the relatedness of organisms and environments. There are lingering assertions of white mastery, albeit undercut with anxiety and apprehension, in what John Law and Stephen Hinchliffe have called the "one-worldist" ontology of One Health.[30] As the chapters collected here attest, we need now to think otherwise about what constitutes animals, health, and relatedness, to attempt to decolonize One Health in the sense of gathering more knowledges together to appreciate better our connected ecologies, fashioning more eclectic matters of health concern.[31]

This means more than simply "permitting" Indigenous participation in One Health or looking for Indigenous analogies with existing veterinary and medical concepts or seeking to adapt conventional intellectual formations to out-of-the-way locations and marginalized communities—important as these advances would be. Rather, it requires genuine epistemic decolonization, different sensibilities and apprehensions, thoroughgoing questioning of concepts and enactments of identity, value, temporality, and space—a renewed collaborative effort to think otherwise about dynamic, heterogeneous, ecological configurations and relations. The chapters collected in this book should stimulate us to take further steps down this path. As Arundhati Roy observed as COVID-19 began to spread, "historically, pandemics have forced humans to break with the past and imagine their world anew. This one is no different. It is a portal, a gateway between one world and the next."[32] And maybe between One Health and the next?

Acknowledgements

For comments on earlier drafts, I'm grateful to Irus Braverman, Danielle Celermajer, Sophie Chao, and Eben Kirksey.

Notes

1 E.g., Nicholson Baker, "The Lab Leak Hypothesis," *New York Magazine*, January 4, 2021.

2 Jeremy Farrar and Ajana Ahuja, *Spike: The Virus vs. The People—The Inside Story* (London: Profile Books, 2021), 62.

3 Edward C. Holmes et al., "The Origins of SARS-CoV-2: A Critical Review," *Cell* 184, no. 19 (2021): 4848–4856; George Gao et al., "Surveillance of SARS-CoV-2 in the Environment and Animal Samples of the Huanan Seafood Market," *Research Square* (2022): https://doi.org/10.21203/rs.3.rs-1370392/v1; Michael Worobey et al., "The Huanan Market was the Epicenter of SARS-CoV-2 Emergence," *Zenodo* (2022): https://doi.org/10.5281/zenodo.6299600; and Jonathan E. Pekar et al., "SARS-CoV-2 Emergence Very Likely Resulted from at Least Two Zoonotic Events," *Zenodo* (2022): https://doi.org/10.5281/zenodo.6291628. On resistance to ecological explanation, see Lyle Fearnley, "Agnatology of Virology: The Origins of Covid-19 and the Next Zoonotic Pandemic," *International Review of Environmental History* 8 (2022): 121–130.

4 Theodore M. Brown, Marcos Cueto, and Elizabeth Fee, "The World Health Organization and the Transition from 'International' to 'Global' Health," *American Journal of Public Health* 96 (2006): 62–72.

5 Randall M. Packard, *A History of Global Health: Interventions into the Lives of Other Peoples* (Baltimore: Johns Hopkins University Press, 2016); and George Weisz and Noémi Tousignant, "International Health Research and the Emergence of Global Health in the Late Twentieth Century," *Bulletin of the History of Medicine* 93 (2019): 365–400.

6 For example, Andrew Lakoff, "Two Regimes of Global Health," *Humanity: An International Journal of Human Rights, Humanitarianism, and Development* 1 (2010):

59–79, and Andrew Lakoff, *Unprepared: Global Health in a Time of Emergency* (Oakland: University of California Press, 2017); Vincanne Adams, "Against Global Health? Arbitrating Science, Non-Science, and Nonsense Through Health," in *Against Health: How Health Became the New Morality*, eds. Jonathan M. Metzl and Anna Kirkland (New York: New York University Press, 2010), 40–58; and Anne-Emanuelle Birn, "Gates's Grandest Challenge: Transcending Technology as Public Health Ideology," *The Lancet* 366 (2005): 514–519. In contrast, Paul Farmer tried to inject reasoning from social medicine into global health, recognizing structural violence as a driver of health disparities: see *Infections and Inequalities: The Modern Plagues* (Berkeley: University of California Press, 1999).

7 Warwick Anderson, "Natural Histories of Infectious Disease: Ecological Vision in Twentieth-Century Biomedical Science," *Osiris* 19 (2004): 39–61, and Warwick Anderson, "Postcolonial Ecologies of Parasite and Host: Making Parasitism Cosmopolitan," *Journal of the History of Biology* 49 (2016): 241–259.

8 James Dunk and Warwick Anderson, "Assembling Planetary Health: Histories of the Future," in *Planetary Health: Protecting Nature to Protect Ourselves*, eds. Samuel S. Myers and Howard Frumkin (Washington, DC: Island Press, 2020), 17–35; and Warwick Anderson and James Dunk, "Planetary Health Histories: Toward New Ecologies of Epidemiology," *Isis* (forthcoming, December 2022).

9 B. A. Wilcox et al., "Introduction," *EcoHealth* 1 (2004): 1–2. See Henrik Lerner and Charlotte Berg, "A Comparison of Three Holistic Approaches to Health: One Health, EcoHealth, and Planetary Health," *Frontiers in Veterinary Science* 4 (2017): article 163.

10 Abigail Woods and Michael Bresalier, "One Health, Many Histories," *Veterinary Record* 174 (2014): 650–654; and Abigail Woods et al., *Animals and the Shaping of Modern Medicine: One Health and its Histories* (Cham: Palgrave Macmillan, 2018).

11 Warwick Anderson, "Edge Effects in Science and Medicine," *Western Humanities Review* 69 (2015): 373–384.

12 Mary Louise Pratt, *Imperial Eyes: Travel Writing and Transculturation* (London: Routledge, 1992), 4.

13 Susan Leigh Star and James R. Griesemer, "Institutional Ecology, 'Translations' and Boundary Objects: Amateurs and Professionals in Berkeley's Museum of Vertebrate Zoology, 1907–09," *Social Studies of Science* 19 (1989): 387–420, 393.

14 On the concept of boundary subjects, see Warwick Anderson, "Traveling White," in *Reorienting Whiteness*, eds. Leigh Boucher, Katherine Ellinghaus and Jane Carey (London: Palgrave Macmillan, 2009), 65–72.

15 Rachel A. Ankeny and Sabine Leonelli, *Model Organisms* (Cambridge: Cambridge University Press, 2021).

16 Warwick Anderson, *Colonial Pathologies: American Tropical Medicine, Race, and Hygiene in the Philippines* (Durham: Duke University Press, 2006).

17 Mark Honigsbaum, "'Tipping the Balance': Karl Friedrich Meyer, Latent Infection and the Birth of Modern Ideas About Disease Ecology," *Journal of the History of Biology* 49 (2016): 261–309.

18 Jakob Zinsstag et al., "From 'One Medicine' to 'One Health' and Systemic Approaches to Health and Well-Being," *Preventive Veterinary Medicine* 101 (2011): 148–156; and R. G. W. Kirk and Michael Worboys, "Medicine and Species: One Medicine, One History?" in *Oxford Handbook of the History of Medicine*, ed. Mark Jackson (Oxford: Oxford University Press, 2011), 561–577.

19 Angela Cassidy, "Humans, Other Animals and 'One Health' in the Early Twenty-First Century," in *Animals and the Shaping of Modern Medicine: One Health and its Histories*, ed. Abigail Woods (Cham: Palgrave Macmillan, 2018), 193–236.

20 Though if we follow Louis Pasteur craftily enrolling his microbes, they would be included as lively actors: see Bruno Latour, *The Pasteurization of France*, trans. Alan Sheridan (Cambridge, MA: Harvard University Press, 1993). See also Heather Paxson and Stefan Helmreich, "The Perils and Promises of Microbial Abundance:

Novel Natures and Model Ecosystems, from Artisanal Cheese to Alien Seas," *Social Studies of Science* 44 (2014): 165–193; and Salla Sariola and Scott F. Gilbert, "Toward a Symbiotic Perspective on Public Health: Recognizing the Ambivalence of Microbes in the Anthropocene," *Microorganisms* 8 (2020): 746. Recent examples of thinking with microbes—the microbial turn—can be found in Charlotte Brives, Matthäus Rest and Salla Sariola, eds., *With Microbes* (Manchester: Mattering Press, 2021).

21 Roger Stanier and C. B. van Neil, "The Concept of a Bacterium," *Archiv für Mikrobiologie* 42 (1962): 17–35. Stanier and van Niel attributed the distinction to Édouard Chatton, a French cytologist, in his *Titres et Travaux Scientifiques (1906–37)* (Sète, Italy: E. Scottano, 1938). Joshua Lederberg and Edward Tatum had shown that bacteria reproduce through genetic recombination in "Gene Recombination in *Escherichia coli*," *Nature* 158 (1946): 558. Since they are nucleated, single-celled protozoa like amoebae could be considered animal in this schema; while parasites, as causes of disease, may be plants or animals. Non-nucleated blue-green algae were initially classified as plants but became cyanobacteria by the 1970s.

22 R. H. Whittaker, "New Concepts of Kingdoms of Organisms," *Science* 163 (1969): 150–163.

23 Carl R. Woese and G. E. Fox, "Phylogenetic Structure of the Prokaryotic Domain: The Primary Kingdoms," *Proceedings of the National Academy of Sciences* 74 (1977): 5088–5090.

24 Carl W. Woese, "There Must be a Prokaryote Somewhere: Microbiology's Search for Itself," *Microbiological Reviews* 58 (1994): 1–9.

25 Jan Sapp, "The Prokaryote-Eukaryote Dichotomy: Meanings and Mythology," *Microbiology and Molecular Biology Reviews* 69 (2005): 292–305, 296. The status of viruses, as infectious nucleic acid, and prions, as infectious protein, raises even more tricky questions about what constitutes a proper life form, not just a proper animal. See André Lwoff, "The Concept of a Virus," *Journal of General Microbiology* 17 (1957): 239–253.

26 E.g., Jacques Derrida, *The Animal that Therefore I Am*, trans. David Wills (New York: Fordham University Press, 2008 [1997]) and Donna J. Haraway, *When Species Meet* (Minneapolis: University of Minnesota Press, 2007). See also Pierre-Olivier Méthot and Samuel Alizon, "What is a Pathogen? Toward a Process View of Host–Parasite Interactions," *Virulence* 5 (2014): 775–785; and John Dupré and Stephen Guttinger, "Viruses as Living Processes," *Studies in History and Philosophy of Biological and Biomedical Sciences* 59 (2016): 109–116.

27 Rosi Braidotti, "Posthuman, all too Human: Towards a New Process Ontology," *Theory, Culture and Society* 23 (2006): 197–208, 201, 203.

28 Warwick Anderson and Ian R. Mackay, *Intolerant Bodies: A Short History of Autoimmunity* (Baltimore: Johns Hopkins University Press, 2014).

29 See Ladelle McWhorter, "Enemy of the Species," in *Queer Ecologies: Sex, Nature, Politics, Desire*, eds. Catriona Mortimer-Sandilands and Bruce Erickson (Bloomington: Indiana University Press, 2010), 73–101; and Étienne Balibar, "Human Species as a Biopolitical Concept," *Radical Philosophy* 2 (2021): 3–12.

30 Stephen Hinchliffe, "More than One World, More than One Health: Re-Configuring Interspecies Health," *Social Science and Medicine* 129 (2015): 28–35.

31 Warwick Anderson, "Decolonizing Histories in Theory and Practice," *History and Theory* 59 (2020): 369–375, and Warwick Anderson, "Finding Decolonial Metaphors in Postcolonial Histories," *History and Theory* 59 (2020): 430–438. See also Bruno Latour, "Why Has Critique Run Out of Steam? From Matters of Fact to Matters of Concern," *Critical Inquiry* 30 (2004): 225–248.

32 Arundati Roy, "The Pandemic is a Portal," *Financial Times*, April 3, 2020.

INDEX

Note: Page numbers in **bold** and *italics* refer to tables and figures.